P9-APV-491

Salt:
The Brand Name Guide to Sodium Content

From the Center for Science
in the Public Interest

Bonnie F. Liebman,
Dr. Michael Jacobson,
and Greg Moyer

Recipes by Robin Rifkin,
Food Consultant, Pritikin Center

WARNER BOOKS

WARNER BOOKS EDITION

This Warner Books Edition is published by arrangement with Workman Publishing Co., Inc., 1 West 39th St., New York, N.Y. 10018.

Cover design by Gene Light
Cover photo by Bill Cadge
Book design by Nicola Mazzella

Warner Books, Inc.
666 Fifth Avenue
New York, N.Y. 10103

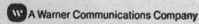 A Warner Communications Company

Printed in the United States of America
First Warner Books Printing: October, 1985
10 9 8 7 6 5 4 3 2 1

Acknowledgments

Producing an extensive reference work that is complete and current is no small task. We never would have been able to write and compile *The Brand Name Guide to Sodium Content* without the help of Karen Jeffrey and James Gollin, both staff members of the Center for Science in the Public Interest. They gallantly put up with recalcitrant companies and disobedient computers, among other obstacles, to produce the brand-name listings.

We also thank Samuel and Rose Liebman and Jan Zimmerman for meticulously checking each of the thousands of sodium levels appearing in the book. Few people know as much about sodium in foods as do these three.

Editing any book is a difficult task, but a book that attempts to include every food is in a class of its own. Suzanne Rafer of Workman Publishing, and Katherine Ness spent many hours deliberating both the big and tiny questions that arise in writing such a book. Their good humor and concern made the process more bearable for all of us.

We are also indebted to Miriam Diamond, proprietor of The Low-Sodium Pantry in Bethesda, Maryland, and to David Kimelman, president of The Montgomery County (Maryland) Stroke Club, Inc., for putting us in touch with stroke and heart attack victims. Many graciously answered our requests for information, telling us that they would be only too glad to help if it might prevent the suffering of others.

We would also like to thank the many hypertension researchers who have provided the Center for Science in the Public Interest with guidance over the years. Drs. Henry Blackburn, Herbert Langford, Edward Freis, William Castelli, Louis Tobian, Norman Kaplan, and many others have been generous with their time, patience and wisdom. Needless to say, these hypertension experts are not responsible for the opinions expressed by CSPI in this book.

Finally, we wish to thank Eric Flamm, Patricia Hausman, Bambi Young, and the other friends and co-workers who helped us keep this job in perspective.

Dr. Michael Jacobson
Bonnie F. Liebman
Greg Moyer

*Center for Science
in the Public Interest*

Contents

INTRODUCTION

Salt: A Front-Page Issue

The University of California at Santa Cruz once gave a course entitled "The Chicken." It was based on the concept that you can take almost any subject—even a chicken—and use it as a probe to understand a multitude of issues. In the case of the chicken, one can learn about animal husbandry, biology, marketing, cooking, cultural patterns, and so on.

Such a course might just as well have focused on salt, for once you put those little white crystals under the microscope of scrutiny, you can see the vast array of questions that need to be asked. Everything from taste perception to chemistry to cooking to high blood pressure to political pressure is part of the sodium story.

At the Center for Science in the Public Interest (CSPI), we examine the scientific literature to identify chemicals or foods that affect human health. Back in 1977, we discovered that the medical community had already reached a consensus regarding salt: it is a major contributor to high blood pressure in millions of Americans. But government, industry, and health professionals were doing nothing to help people reduce their intake of sodium. High blood pressure is an important issue because it greatly increases the risk of suffering a stroke or heart attack, two of the major killers in the U.S.

When scientific research is at odds with the policies and practices of government and industry, CSPI moves in. Knowing that food manufacturers are essentially impervious to polite letters suggesting they use less salt, we focused on the Food and Drug Administration (FDA). In 1978, we petitioned the FDA to require that all food labels list the amount of sodium (the troublesome part of salt) in each serving of a product. Moreover, we urged the FDA to limit to safe levels the amount of sodium in processed foods. The FDA denied both requests, ultimately forcing us to sue them. Our next step was to begin to educate consumers directly about the potentially fatal consequences of a steady diet of high-sodium foods. This book is part of that educational effort.

Though we believed we were presenting noncontroversial, agreed-upon health facts, we were not surprised that government officials and food industry executives ignored or disputed our recommendations. After all, many government officials would much rather sidestep an issue than subject themselves to the inevitable buzzsaw of lobbying from companies that produce salt and high-salt foods. And companies making millions of dollars from the sale of high-sodium foods would do anything to avoid changing their winning formulas. For the past five years, we have sought to make salt a front-page issue (it actually was a cover story of *Time* magazine in 1982) so that companies would see a competitive advantage in reducing the sodium content of their foods and listing sodium content on the labels.

Though the Reagan administration has a well-known antipathy for new regulations that place even the slightest burden on industry, Arthur H. Hayes, the FDA commissioner between 1981 and 1983, used his office as a bully pulpit to encourage food processors to reduce and label the salt content of their products. Simultaneously, the heavy publicity surrounding salt led some companies to produce and market low-sodium or no-salt-added products. Suddenly we've seen a happy proliferation of low-salt canned soups, cured meats, club soda, canned vegetables, and potato chips. And, some companies have begun listing sodium on the label, but the majority have not. The FDA has refused to require sodium labeling, let alone order reductions in high-sodium foods. Thus the need for this book.

The first several chapters explain in detail why you should be concerned about salt . . . and what government, researchers, and industry have been doing to provide or hide information about salt. About the time you're ready to stop reading and head for the kitchen, we provide a chapter filled with tips on how to reduce your own intake of sodium, whether you are shopping in a supermarket, eating at a restaurant, or cooking a meal at home. And then, to make low-sodium cooking delectable and accessible, the section of recipes by Pritikin nutritionist Robin Rifkin should make your taste buds happy.

We believe that Part 3 of this book contains the most exhaustive and up-to-date listing of the sodium contents of foods, both natural and processed, ever compiled. The listing covers everything from apples to zucchini, from Arby's to Wendy's, from Cool Whip to Kool-Aid. We hope you will find the categories into which we have divided the thousands of foods convenient for comparing the various types or brands of foods that you eat.

Dr. Michael Jacobson
Bonnie Liebman
Greg Moyer

ABOUT THE AUTHORS:

Each of this book's eminent authors has been on the staff of the Center for Science in the Public Interest (CSPI), a nonprofit, consumer advocacy group that focuses on food safety issues.

Bonnie F. Liebman, M.S. is Director of Nutrition at CSPI and a specialist in the field of excess salt, sugar and fat consumption. Her expert testimony has been solicited by Congressional committees and numerous journalists. She appears regularly on the major morning talk shows. Ms. Liebman received her masters degree in nutritional sciences from Cornell University.

Dr. Michael F. Jacobson is co-founder and executive director of CSPI, and a former "Nader's Raider." Dr. Jacobson received his Ph.D. in Microbiology from the Massachusetts Institute of Technology.

Dr. Jacobson initiated actions that led to improved sodium labeling of packaged foods and to restrictions in the use of sulfites, nitrite, and other food additives. He also worked to halt deceptive food and beverage advertising, improve the nutritional quality of fast foods, and to achieve other consumer protections. The author of several books and numerous magazine articles and scientific papers, Jacobson is a frequent guest on national television programs including *Phil Donahue, The MacNeil/Lehrer News Hour* and *Sixty Minutes*.

Greg Moyer, a graduate of Bucknell University, was Director of Publications and Marketing at CSPI for five years, where he edited *Nutrition Action Health Letter,* the highly acclaimed monthly publication of CSPI.

PART ONE

THE
SALT STORY

CHAPTER ONE

The Loss of Innocence

Salt. The beautiful white crystals that glisten in shakers on dining tables across the land look as innocent as newly fallen snow. But for millions of us, appearances can be deadly. Medical researchers have convincingly linked this humble seasoning, so abundant in the American diet, to hypertension, or high blood pressure, a condition that raises the risk of a debilitating disease or premature death for one out of three Americans.

There are sobering statistics. Thirty-five million Americans with "definite" high blood pressure are:

- at eight times the average risk of suffering a stroke;
- at three times the average risk of suffering a heart attack; and
- at five times the average risk of suffering congestive heart failure.

Another 25 million people with "borderline" hypertension run about twice the average risk of suffering from these life-threatening diseases. And researchers are poised to classify still another 15 million as being not quite hypertensive, but still at slightly higher than normal, "intermediate," risk of getting cardiovascular disease.

The vast majority of these 60 million people have "essential hypertension," which means that their high blood pressure is not related to some other disease. Those with hypertension caused by kidney failure or some other medical problem number well under 10 percent of all hypertensives.

In nearly all cases, hypertension is painless. Most people with high blood pressure lead normal lives, oblivious to the increasing force with which their blood is pressing against blood vessel walls. It is only when an artery gets blocked or ruptures that victims of hypertension discover the gravity of their situation. Tragically, for some the knowledge comes too late.

3

PROFILE 1:

Steven Broyle*

Steven Broyle is very lucky: he is still alive, and he can read, think, and talk clearly. But Broyle, once a sales representative for Hitachi Corporation, is physically just a shadow of his former self. He can "walk," but only if someone assists. Basically he is confined to the wheelchairs he keeps on both the first and second floors of his suburban Washington home. He is paralyzed on the right side of his body, and must take care not to exert himself too strenuously.

At age 55, Broyle suffered a stroke. It was so massive that the doctors at the hospital told Broyle's wife that she would never be able to take care of her husband herself. They suggested she find a nursing home. By this standard, Broyle's partial recovery during the past eighteen months has made a tragic situation slightly less so.

But the terror of the day Steven was stricken remains with his wife, Beth. She remembers waking up that morning to see Steven's arm hanging limply at his side. He first dismissed the trouble he had getting his arm into a shirt, but Beth insisted that he see a doctor and she got a neighbor to come help. Little did Steven know as he walked out the front door toward the family car that this simple daily ritual would soon be beyond his grasp. But by the time he and the neighbor reached the curb in front of the house, Steven's right side had frozen. Beth helped drag him back to the front porch before she called an ambulance.

Meanwhile, Steven's mouth began sagging, and he started vomiting. He drooled, and his right eye rolled. The doctors said a blood clot somewhere in the left side of his brain led to the stroke that morning. Steven spent the next seven months in a rehabilitative hospital in New York. Now Beth is grateful for simple things: Steven lives at home, and she can take care of him without outside help.

---

*Pseudonyms have been given to those individuals whose case histories are described.

When a blocked or ruptured artery cuts off blood to the brain, the person suffers a *stroke*. Hypertension is the most important factor contributing to over 1,300 cases of stroke each day in the U.S. Of the 500,000 people who have strokes each year, about 154,000 die soon after.

When a clogged artery blocks off the blood to the heart, the victim has a *heart attack*. High blood pressure is a major factor in triggering a large proportion of the 1,250,000 heart attacks suffered annually—more than two every minute. Almost half of these attacks lead to immediate death.

Hypertension is a killer. More than a third of all deaths in the United States each year are due to either stroke or heart attack; by comparison, only one death in five is caused by cancer.

But the medical complications springing from hypertension do not stop at stroke and heart attack. Blood pumping through the circulatory system at dangerously high pressures strains the heart muscles. The result is often an "enlarged heart" leading to *congestive heart failure*—a condition that afflicts roughly 1,500,000 Americans today.

The direct and indirect costs of high blood pressure—encompassing everything from medical care to the loss of the victims' productivity—amount to a staggering $8 billion a year. Add to that the suffering of the patient and his or her family, and it becomes clear why society must do all in its power to curb high blood pressure. Fortunately we may have the means to curb this hemorrhage of misery and death.

The Culprit Revealed

As early as 1904, two scientists suspected that salt, composed of sodium and chloride, possessed the power to raise blood pressure when eaten in excessive amounts. Today medical authorities strongly suspect that America's hypertension epidemic is intimately tied to our high-salt diet. To be more precise, they have discovered the real culprit to be sodium, one part of the salt molecule. The other part, chloride, is harmless.

The Surgeon General, the U.S. Food and Drug Administration, the U.S. Department of Agriculture, the U.S. Department of Health and Human Services, the National Academy of Sciences, the American Heart Association, and other authorities have studied the problem and reached the same conclusion:

Americans should cut back their intake of sodium. Over 90 percent of the sodium we ingest comes from salt.

This advice applies foremost to the 60 million who already have high blood pressure. A diet moderately low in sodium may be all that it takes to restore the blood pressure of millions to normal levels. For millions more whose blood pressure requires control by drugs, a low-sodium diet can actually decrease the necessary dose of the medication and thereby reduce its side effects.

But low-salt diets are not only for those already suffering from high blood pressure. Public health officials are urging *all* Americans to eat less salt. They are convinced that a low-sodium diet may prevent the onset of hypertension. Studies show that one of every two people aged 65 or older has hypertension; this means that half the population of *any* age group is susceptible to high blood pressure. We know that people who are overweight and people whose parents are hypertensive have a greater-than-average chance of developing the condition. But medical science has no way of predicting precisely which individuals may be sensitive to salt and will eventually suffer from high blood pressure. The only prudent course for all healthy people is to cut back.

How Is Hypertension Measured?

Doctors and nurses measure blood pressure with a sphygmomanometer, a device that looks like a thermometer connected by tubes to an inflatable cuff. When your blood pressure is taken, the doctor or nurse records two numbers. The first is the pressure read from the gauge when the heartbeat is first heard through the stethoscope. It represents the pressure exerted on the blood vessel walls when the heart is beating, and is called the systolic pressure. The second number, the lower of the two, is the pressure read from the gauge when the sound of heartbeat disappears. It represents the pressure on the arteries between heartbeats, and is called the diastolic pressure.

Use this reference to evaluate your blood pressure:

Normal: 120/80 or less
Borderline: 140/90 to 160/95
Definite: 160/95 or more

A National Institutes of Health task force is currently considering a new category called "intermediate." Roughly 15 million people have blood pressures between 120/80 and 140/90 millimeters of mercury. Doctors would not treat these people with drugs or even label them hypertensive, but they would advise them to eat less salt and, if necessary, to lose weight. Statistics show that even this slight elevation in blood pressure doubles an individual's chances of dying within eight years, from 1 in 20 to 1 in 10.

These numbers apply to adults only. Children and adolescents rarely have blood pressures that fall within these hypertensive ranges. However, a child can have an elevated pressure for his or her height and weight. Because such children have an increased risk of developing hypertension in adulthood, most doctors will advise them to avoid salt and, in most cases, to lose weight. However, the American Academy of Pediatrics encourages physicians to use the term "high normal blood pressure" rather than "high blood pressure" to avoid labeling the child as sick.

Blacks have particular reason to heed the warning. For some undetermined reason, hypertension affects blacks twice as often as whites. The death rates due to high blood pressure are more than three times higher in black women under the age of 54 than among white women of the same age. Death rates are more than twice as high in black men under the age of 55 than among white men of the same age.

Despite these grim statistics, the government and health authorities have only recently told people how much salt, or sodium, is too much. Studies of people in other cultures and of people on restricted diets have shown that the normally active body needs only 200 milligrams of sodium a day to function properly. That amounts to 1/10 teaspoon of salt. Doctors can confidently tell people to drastically reduce their salt intake because nature makes the chances of eating too little sodium almost nil. A diet with as few as 200 milligrams of sodium is hard to imagine, because sodium occurs naturally in many foods. As Chart 1A (page 8) shows, a typical diet comprised mostly of wholesome

foods with little added salt yields 1,461 milligrams of sodium—
over seven times the daily need.

Overdoing Our Intake

But the average American eats 4,000 to 6,000 milligrams of sodium
per day—twenty to thirty times the daily need. Why?

Look for a moment at the whole picture. Chart 1B (page 10)
depicts the diet of someone who relies heavily on processed
foods. Of the 6,678 total milligrams of sodium, 6,419 milligrams—
over 95 percent—come from processed foods.

Most people fall somewhere between the processed foods
eaters and the natural foods folks. We estimate that the average
American gets about 15 percent of his or her sodium from the salt
shaker. Another 10 percent occurs in foods naturally. But 75
percent comes from processed foods. Some of these foods, such
as pickles and cured meat, have been popular for centuries. But
in the past few decades, Americans have been "treated" to
thousands of new high-sodium processed foods.

One and a half ounces of Kraft pasteurized processed cheese
supply nearly four times the day's necessary sodium, one
McDonald's Quarter Pounder with Cheese weighs in with six
times the minimum requirement, and one Swanson's Hungry
Man Turkey Pie is loaded with nearly nine times the body's daily
need. In fact, a diet with as few as 200 milligrams of sodium is
hardly possible, given the amount of sodium in processed foods.

CHART 1A

Sodium Content of an
Unprocessed Foods Diet

*(Approximately 2,000 calories; if no brand name is
mentioned the figures were supplied by the USDA.)*

	PORTION	SODIUM (MG.)
Breakfast		
Shredded Wheat	1 biscuit	10
w/strawberries	1 cup	2

	PORTION	SODIUM (MG.)
w/sugar	1 tsp	0
w/low-fat milk	½ cup	60
Blueberry Muffin		
(Betty Crocker)	1	150
w/margarine, unsalted	1 pat	2
Orange juice	6 oz.	2

Lunch

Sliced chicken sandwich	3 oz. chicken	54
on Whole Wheat Bread		
(Arnold/Orowheat/		
Brownberry)	2 slices	220
w/mayonnaise	1 tbl.	80
Spinach salad	1 cup	47
w/oil & vinegar	2 tbl.	0
Grapes	1 cup	6
Apple juice	6 oz.	2

Dinner

Flounder w/lemon		
and butter	4 oz.	268
Green beans w/dill	1 cup	5
Baked potato	1 large	14
w/salt	1/10 tsp	200
w/margarine, unsalted	1 pat	2
Carrot Cake (Sara Lee)	1.5 oz.	125
Tea	1 cup	2

Snack

Low-fat milk	8 oz.	120
Oatmeal Cookies	2½	90
(Nabisco)		

                                                        ———————
                                                        1,473 mg.

Sources of Sodium:

Naturally occurring	596 mg.
Added at table	200 mg.
From processed foods	665 mg.

                        Total 1,461 mg.

CHART 1B

Sodium Content of a Processed Foods Diet

(Approximately 2,000 calories; if no brand name is mentioned the figures were supplied by the USDA.)

	PORTION	SODIUM (MG.)
Breakfast		
McDonald's Egg McMuffin	4.9 oz.	885
McDonald's Hash Browns	1.9 oz.	325
w/salt	⅒ tsp.	200
Coffee w/cream & sugar	1 cup	7
Lunch		
Campbell's Tomato Soup	10 oz.	938
Celeste Frozen Cheese Pizza	4.8 oz.	803
Canned Green Beans	½ cup	153
Tab Diet Soda	12 oz.	30
Dinner		
Hamburger Helper Cheeseburger Macaroni, w/hamburger	⅕ package	1025
Birds Eye Italian Vegetables	3.3 oz.	575
Jell-O Instant Chocolate Pudding	½ cup	515
Tea	1 cup	2
Snack		
Kraft Pasteurized Processed American Cheese	1.5 oz.	698
Nabisco Premium Saltines	1 oz. (10)	460
Diet Pepsi	12 oz.	62
		6,678 mg.

Sources of Sodium:

Naturally occurring	9 mg.
Added at table	200 mg.
From processed foods	6,469 mg.
Total	6,678 mg.

The National Academy of Sciences recommends that adults without hypertension consume between 1,100 and 3,300 milligrams of sodium per day. This translates into between ½ and 1½ teaspoons of salt. (The Academy set its recommended level higher than the 200 milligram minimum to be realistic. It knew that few people could ever reduce their sodium intake to a mere 200 milligrams.)

Yet manufacturers of processed foods dump so much sodium into their products that most people would be hard-pressed to stay within the Academy's realistic limits. What's more, few manufacturers tell us how much sodium they add to their processed foods. Less than half of all foods carry sodium labeling, rendering helpless those conscientious shoppers who are willing to take the time to read the fine print.

That's why the brand name guide to the sodium content of over 5,000 foods, which begins on page 143, becomes so important to concerned consumers like yourself. Unless the government orders companies to lower and label the sodium content of processed foods—a move that is unlikely, as you shall read—this guide is your main defense against foods that expose your family to an increased risk of hypertension and its potentially lethal consequences.

With this guide, you can note down the low-sodium brands of the foods on your shopping list before heading for the store, or you can carry this book with you while scanning the supermarket shelves.

Searching for low-sodium foods may seem to be just another nuisance in our already too complex world. But high blood pressure cannot be ignored. It renders the body vulnerable to heart attack, stroke, and congestive heart failure, as the next chapter explains in greater detail. As difficult as it may be, we must learn to eat less salt. It is a matter of life and death for millions.

CHAPTER TWO

The Silent Killer

The human body is made up of billions and billions of cells, each of which needs constant nourishment. Blood, carrying nutrients and oxygen to these cells, travels from the heart through a network of arteries, veins, and capillaries. Like water squirting through a garden hose or air inflating a tire, blood exerts pressure on the vessel walls as it flows through the body.

But over the years, the arteries of some people slowly constrict. No one knows the precise cause of the constriction, but the effect is clear: their blood pressure rises higher and higher. With each heartbeat, the blood pushes harder against their artery walls, weakening the thin muscular linings.

In some people, a weak spot develops in the wall of an artery feeding the brain. The artery balloons like a bubble in a worn-out tire. For a short while, the artery manages to keep delivering its vital supply of blood to the cells. But the ballooned wall is perilously thin. One day, without warning, it bursts. The person suffers a *cerebral hemorrhage*—a stroke.

Within five to ten minutes, brain cells starved for oxygen and blood lose their ability to function and die. If too many cells die, the brain and the person also die.

Atherosclerosis

In the United States, strokes are usually caused by a *blood clot* that lodges in a brain artery "clogged" by atherosclerosis. Atherosclerosis is the most common form of arteriosclerosis, also known as hardening (sclerosis) of the arteries (arterio).

Although most people who have the condition don't realize it until later in life, atherosclerosis actually begins quite early. Fat deposits (atheromas) accumulate in the artery walls during child-

---

PROFILE 2:
Edith Handy

Edith Handy might well have died. One day seven years ago, she was home, sitting at her dinner table, when all of a sudden she couldn't move her arm or use her hand. The last thing she remembers is being carried by her husband and son to another room. Edith suffered a massive hemorrhage that sent her to the hospital for months.

Edith is now 61. She has adapted to her limitations, but she isn't the person she used to be. Her speech is slurred, and everyday household chores pose problems. Her husband explains that the family paid for a nurse to stay with Edith during her first eight months out of the hospital, and that she learned to dress herself. But she "didn't learn as much as the doctors had hoped," he said.

---

hood and adolescence. By middle age, the deposits get covered with mounds of muscle cells and scar tissue known as plaque. The plaque protrudes inside the artery, partially blocking the flow of blood. At an advanced stage of the disease, even a small clot can cut off the flow of blood entirely.

Excess saturated fat and cholesterol in the typical American diet are primarily responsible for atherosclerosis. But high blood pressure also plays a major role. For some unknown reason, *hypertension speeds up atherosclerosis*.

The consequences of a blood clot are almost the same as a rupture. Blood is blocked from reaching a part of the brain, and cells die. When a blood clot, or thrombus, occurs, doctors attribute the death or paralysis to a thrombotic stroke.

Hypertension is as quiet and inconspicuous as a tiger stealthily eyeing its prey. The victim of a cerebral hemorrhage or a thrombotic stroke is no wiser to the dangers about to befall him or her than is the antelope about to be surprised by an agile cat.

In a few cases, high blood pressure causes headaches or dizziness. But most often it does not cause any symptoms. Millions of Americans don't know they have high blood pressure, because the only sure way of knowing is to have it checked at a clinic or doctor's office. In a 1976 to 1980 survey, doctors at the National Institutes of Health estimated that roughly one in four people with hypertension is totally unaware that his or her blood pressure is dangerously high.

About 500,000 Americans suffer strokes each year. Of the 346,000 who survive, only about 10 percent remain unimpaired. Roughly 40 percent have mild residual disabilities and 40 percent require special nursing care. The remaining 10 percent must be institutionalized for the rest of their lives.

PROFILE 3:
Jerry Rausch

Jerry Rausch, age 52, had suffered from angina attacks for twenty years. Angina pectoris, as doctors call it, is a severe pain in the chest, shoulder, or arm. The pain occurs when the heart muscle receives too little oxygen. Athero-sclerosis is the culprit. When coronary arteries are clogged, the heart muscle can't get enough oxygen-rich blood. As soon as exertion of any kind raises the demand for oxygen, the meager blood flow to the heart muscle is no longer adequate. The sharp pain ensues.

Whenever Jerry ate big meals or exercised, he felt the pain in his arms. Then years ago, doctors told Jerry to follow a low-salt, low-cholesterol, low-calorie diet for the rest of his life. Initially he kept to the diet, but over time he abandoned it.

Then, during the 1982 Christmas holidays, Jerry began to feel sluggish and lethargic. He vowed to himself to "slow down" after the excitement of the season passed. One morning he woke early with pains in his arm and shoulder. He took some nitroglycerin pills, and, "like a dummy," went to his job at a supermarket. By now a burning pain gripped his chest and his breath was "shorter than it had ever been."

He drove himself to the hospital, where blood tests confirmed that he had suffered a heart attack. The doctors then found that five of Jerry's coronary arteries were partially or completely blocked. Today, after quintuple bypass surgery, Jerry is back on his diet. There is little doubt that his high blood pressure, first noted over a decade ago, hastened his rendezvous with America's most common killer disease.

Heart Problems

The twin evils of high blood pressure and atherosclerosis connive to block arteries at other sites in the body, too. Blood vessels feeding the heart muscle—the coronary arteries—get clogged even more frequently than those leading to the brain. If a fresh supply of blood fails to reach the heart muscle, a part of the muscle dies and the victim suffers a *heart attack*. If too much of the muscle dies, the heart stops beating and the victim becomes one of the 536,000 who die each year from heart attacks. Of the 714,000 people who do survive their attack, 38 percent can resume their normal lives. But 42 percent are limited in activity and 20 percent remain incapacitated.

High blood pressure is also the primary cause of *congestive heart failure*, an affliction that plagues 1.5 million Americans. The hypertensive's "tightened" arteries put a strain on the heart, which has to work harder to pump the same amount of blood through a constricted circulatory system. At first, the heart compensates for the increased resistance by growing larger. But once the muscles stretch beyond a certain point, they become less rather than more effective at circulating blood throughout the body. Soon the heart begins to fail.

The person with a failing heart feels short of breath just from walking up a flight of stairs. Even such mild exertion, which increases the muscles' need for oxygen only slightly, overwhelms the heart. It can no longer pump enough oxygen-rich blood to the muscles in time.

People with congestive heart failure often develop *edema*, or fluid retention. Without a vigorous heart to keep it moving, blood accumulates in the tiny arteries and veins of the legs. Pressure in these overloaded vessels rises, allowing the clear, watery part of the blood to seep out into the body's tissues. As a result, edema victims have swollen legs and ankles.

In the most serious cases, the weakened heart has so much trouble pumping fresh blood into the circulatory system that some backs up into blood vessels in the lungs. High pressure again causes the watery part of the blood to seep out of the vessels and this time collect in the lungs. This person has developed *acute pulmonary edema* and is in danger of drowning in his or her own blood. Acute pulmonary edema is so dangerous that even if one survives the immediate crisis, the chances of dying within five years are 50 percent.

How a high-sodium diet causes blood vessels to constrict in the first place is still a mystery. Scientists are pursuing several theories, but so far they have no clear-cut explanations.

Researchers do understand how eating less sodium can reduce blood pressure in people who already have hypertension. The body retains a given amount of blood and other fluids for every milligram of sodium it holds. Rid the body of excess sodium, and you rid the body of excess fluid. Less fluid means lower pressure in the blood vessels, just as letting some air out of a tire lowers the pressure in it.

Tackling High Blood Pressure

Strokes, heart attacks, congestive heart failure . . . the toll of death and disability caused by hypertension reaches staggering heights. Yet the battle against this silent killer and crippler is still in its infancy. As recently as 1960, physicians thought all but extreme high blood pressure was a basically harmless condition. By the late 1950s, a massive study in Framingham, Massachusetts, had begun to show that hypertension was associated with

---

PROFILE 4:

Arlene Bennett

A strange set of circumstances led Arlene Bennett, a 28-year-old clothes buyer for a chic department store, to discover that she had dangerously high blood pressure.

While standing on the curb hailing a cab, she and a friend were grazed by a car driven by a drunk driver. They were both taken to an emergency room, where a doctor treated their immediate needs. In checking for other injuries, he took Arlene's blood pressure. That stopped him cold. "You're not coming back here until you see a hypertension specialist," he said.

Arlene has never had a weight problem, averaging 130 pounds for her 5' 10" frame. But her job demanded that she travel, when for days on end her diet became a succession of fast-food lunches and dinners. Holding to low-salt foods would have been difficult, even if Arlene had known enough to try.

The doctor found no obvious explanation for Arlene's high blood pressure. Her kidneys were normal. She did complain of frequent headaches, and when he asked about her family history, she told him her father had suffered a stroke in his 50s. The specialist surmised that Arlene's family may be genetically predisposed to developing hypertension. He put her on medication that she subsequently took for several years.

During this time, the doctor never said a word about diet. But a friend did speak out. She told Arlene of press accounts linking salt and hypertension. This spurred Arlene to learn more. Over the course of four years, she changed her eating habits markedly.

A career change allowed Arlene to eat more meals at home. There she avoided high-sodium convenience foods and cooked with a minimum of added salt. After marrying a fitness enthusiast, she also started exercising regularly. Gradually her hypertension diminished. In Arlene's case, it was probably the combination of factors—a low-sodium diet, exercise, and the medication—that brought her blood pressure to normal.

Arlene is like millions of other Americans who have a special sensitivity to high blood pressure. And just like her, most of these millions do not know they are susceptible to hypertension. It is for their sake, in particular, that food manufacturers should reduce the amount of sodium they add to foods. Arlene's hypertension is not "cured." But the lifestyle changes are now enough to keep her blood pressure down, and she no longer needs antihypertension drugs.

Arlene had good reason to want to be free of hypertension: she knew that blood pressure has a tendency to rise during pregnancy. Just recently, she gave birth to a healthy baby boy, and her blood pressure remained normal throughout the nine months of pregnancy and through delivery.

heart attacks, strokes, and congestive heart failure. But doctors were still not convinced that lowering blood pressure would necessarily reduce the incidence of these deadly "complications."

Then, in 1972, Dr. Edward Freis, a renowned hypertension authority at the Veterans Administration Hospital in Washington, D.C., reported the results of a five-year study involving over 500 veterans with hypertension. Lowering blood pressure with drugs, Freis found, markedly reduced the risk of most "complications."

Freis's findings stirred public health officials. In 1972, the then Secretary of the U.S. Department of Health, Education, and Welfare, Elliot Richardson, established the National High Blood Pressure Education program (NHBPEP). Its purpose, he said, was to reach the millions of Americans "walking around with a time bomb ticking inside them."

NHBPEP set up conferences for doctors, nurses, and other health practitioners; produced public service announcements, pamphlets, and posters for the public; and provided technical assistance to local organizations involved in blood pressure control.

Treat It for Life

In approach, the education program amounted to a search-and-destroy mission. Its objectives were threefold: first, to identify everyone who had hypertension; second, to send them to doctors for high blood pressure pills; and third, to coax them to keep taking the pills for the rest of their lives. NHBPEP's two slogans were "See Your Doctor" and "Treat It for Life."

The program succeeded in many ways. In the early 1960s, one out of two hypertensives was unaware of his or her condition. By the late 1970s, the rate of undiagnosed hypertension had dropped to about one in four. By 1982, the number of visits patients paid doctors regarding hypertension had jumped 54 percent. And, due in part to NHBPEP, the number of Americans dying from heart attacks dropped 26 percent between 1972 and 1981. The number dying from stroke fell 41 percent—an amazing drop—over the same period.

But the "Treat It for Life" approach has serious drawbacks. It means that millions of people with high blood pressure must endure a lifetime of drugs, often with dangerous and disturbing side effects. Causing symptoms such as lethargy, frequent urination, nausea, and sexual dysfunction, the antihypertensive medicines have proved a bitter pill to swallow. Chart 2A (below), adapted from a National Institutes of Health publication, outlines some of the side effects. But even this distressing listing does not tell the whole story:

CHART 2A

Side Effects of Commonly Used Antihypertensive Drugs*

DRUGS	SIDE EFFECTS
Diuretics	
Thiazide, related sulfonamides, and loop diuretics	Decreased blood potassium levels, increased blood uric acid, cholesterol, and triglyceride levels, glucose intolerance, sexual dysfunction
Potassium-sparing agents	
Spironolactone	Increased blood potassium levels, enlarged breasts (in males), breast pain, sexual dysfunction
Triamterene	Increased blood potassium levels
Amiloride	Increased blood potassium levels, sexual dysfunction
Adrenergic Antagonists	
Clonidine, Guanabenz, Methyldopa	Drowsiness, fatigue, dry mouth, sexual dysfunction
Guanethidine, Guanadrel	Sexual dysfunction, abnormally low blood pressure, diarrhea, stuffed nose
Propranolol, Metropolol, Nadolol, Timolol, Atenolol, Pindolol, Oxyprenolol	Slowed heartbeat, bizarre dreams, insomnia, fatigue, sexual dysfunction, increased blood triglyceride levels,

*Source: The 1984 Report of the Joint National Committee on Detection, Evaluation, and Treatment of High Blood Pressure. *Archives of Internal Medicine 144*: 1045, 1984.

DRUGS	SIDE EFFECTS
	decreased blood high-density lipoprotein cholesterol levels
Prazosin	Palpitations, weakness, abnormally low blood pressure
Rauwolfia alkaloids, Reserpine	Stuffed nose, lethargy, mental depression, sexual dysfunction
Vasodilators	
Hydralazine	Headache, rapid heartbeat, fluid retention
Minoxidil	Rapid heartbeat, headache, fluid retention, excess hair growth
Angiotensin-Converting Enzyme Inhibitors	
Captopril	Rash, impaired taste sense (rare)

- In January 1980, Selacryn, an anti-hypertensive drug, was banned after it caused liver damage in 363 patients and killed 24.
- Again in 1980, reserpine, another drug that lowers blood pressure, was found to cause cancer in animals. In April 1983, the U.S. Food and Drug Administration announced that reserpine labels must inform physicians about the study.
- In 1982, the results of the Multiple Risk Factor Intervention Trial (MRFIT) raised new fears. This $115-million/ten-year study funded by the government suggested that diuretics, the most commonly prescribed anti-hypertensive medication, significantly increased the risk of death in about a third of the patients studied. For reasons scientists cannot explain, those with irregular heartbeats seem to be in danger.

Even if the drugs had no serious side effects, the "Treat It for Life" philosophy had other flaws. When NHBPEP was created in the early 1970s, doctors considered only a fraction of the 35 million Americans with "definite" hypertension as likely candidates for drug therapy. Then, in 1979, a major study called the Hypertension Detection and Follow-up Program concluded that all people with either "definite" or "borderline" hypertension would benefit from taking the pills. The implication of this study staggered the medical community. If all the borderline and definite hypertensives were to receive antihypertension medication, 60 million people—one out of every three adult Americans—would be taking drugs for the rest of their lives!

Dr. Freis estimates the sheer cost of medicating one out of every three adults to be $30 billion a year. But more importantly, researchers such as Dr. Ronald Prineas of the University of Minnesota consider it "obscene" to put a third of the adult population on drugs. The absurdity of medicating one fourth of the nation became clear to NHBPEP, too. When the results of the

---

IS LOW BLOOD PRESSURE A PROBLEM?

If your blood pressure is considerably lower than 120/80, you may wonder whether you can ignore all the fuss about salt. Unfortunately, there are no golden assurances. For most people in the U.S., blood pressure rises with age. So even if your pressure is far below 120/80 at age 45, it may be 140/95 or more in another twenty-five years.

According to Dr. William Castelli, director of the Framingham (Massachusetts) Heart Study, it is possible that the 10 percent of the population with the lowest blood pressures might be resistant to hypertension as they grow older. Theoretically, these people could eat a salty diet with no second thoughts. But Castelli says that as yet we don't have the scientific evidence to assert that these people will remain hypertension-free.

Some people with low blood pressure may wonder whether they should eat extra salt to raise their blood pressure to more "normal" levels. "Absolutely not," says Castelli. "Years ago, if a patient came into a hospital with very low blood pressure, a doctor might have given the patient drugs to raise it. The doctor should have left the poor guy alone. Such people are better off than the rest of us because their risk of cardiovascular disease is lower."

As for the idea that people with low blood pressure have cold hands and feet, Castelli says "that's an old wives' tale."

follow-up study were released, Dr. Robert Levy, then director of the National Heart, Lung, and Blood Institute (the parent of NHBPEP), recommended that patients with borderline hypertension try reducing sodium consumption or losing weight before resorting to drugs.

A Public Health Solution

When a health problem afflicts a sizable percentage of the population, as hypertension does, it demands a prevention-oriented "public health" solution rather than individual diagnoses and treatment. For example, in 1913, the U.S. Public Health Service discovered that each year 200,000 Southerners were developing pellagra, a vitamin deficiency disease characterized by diarrhea, skin lesions, and dementia (mental disorders). Health authorities realized that the problem was too large to solve with a traditional medical solution. It was foolish to wait for thousands of people to contract pellagra, and then send each to a doctor for pills containing the missing vitamin, niacin. Instead officials opted for a "public health" approach. They took steps to prevent pellagra by requiring bread and cereal manufacturers to add niacin to their products.

The hypertension epidemic cries out for a public health approach aimed at *preventing* the disease. The patient-by-patient medical approach is simply not appropriate when a public health solution exists. Fortunately, in the case of hypertension we have both: we can treat people who already have hypertension with diet and/or drugs, and we can lessen the occurrence of high blood pressure by reducing the amount of sodium added to foods. The seemingly innocent white crystal that seasons so much of our food is one of our most formidable public health enemies. The next chapter describes the medical evidence exposing the link between high-salt diets and high blood pressure.

Can You Get Too Little Sodium?

Should you ever be afraid of eliminating too much sodium from your diet?

Sodium is a necessary nutrient. It controls the amount of fluid the body retains by regulating how much fluid stays within individual cells. Without any sodium, the body would become dehydrated. Sodium keeps some fluid inside cells because of its ability to attract water. To guard against too much sodium causing the cells to swell and burst, each cell has a "sodium pump" that transports excess sodium out, carrying water with it.

Sodium also establishes an electrical balance across the cell wall. Most of the positively charged sodium ions are kept outside nerve cells, creating an electrical charge difference between the inside and outside of the cell. Nerve cells "fire" and muscle cells contract by allowing the sodium to temporarily enter the cells and eliminate the electrical difference. Without sodium, our muscle and nervous systems would not function.

Clearly, life would not exist without sodium. Yet the amount the body needs—about 200 milligrams (equivalent to $\frac{1}{10}$ teaspoon of salt) a day—is minuscule. That amount replenishes the sodium we excrete in urine and feces (23 milligrams) and the amount the average person loses in sweat (46 to 92 milligrams).

Keep in mind that no one has recommended that we cut down to 200 milligrams. The National Academy of Sciences has advised us to consume between 1,100 and 3,300 milligrams of sodium (about ½ to 1½ teaspoons of salt) each day.

But we currently consume twenty to thirty times the 200-milligram amount and two to five times the Academy's recommended levels—between 4,000 and 6,000 milligrams of sodium (two to three teaspoons of salt) a day.

Still, some defenders of salt continue to charge that people who exercise regularly, such as joggers or bicyclists, might deplete their store of sodium and risk dangerous consequences. Few studies have carefully examined the effect of sodium on athletic performance. However, we do know that exercisers eating the current American diet derive no benefit from extra sodium.

Dr. David Costill, a leading authority on exercise physiology at Ball State University, tested the value of Brake Time, one of the commercial soft drinks designed to replace the sodium and other substances lost in sweat. Costill subjected twelve people to 10 days of "dehydration sessions," causing them to lose more than two quarts of water weight through strenuous exercise. During five of the days they drank Brake Time; the rest of the time they drank water. His subjects retained enough sodium by drinking either water or the special beverage. Costill concluded that the soft drinks were unnecessary—Americans get enough

SALT AND PREGNANCY

One note of caution: Pregnant women should not drastically restrict their sodium intake. For many years, some doctors prescribed strict low-sodium diets and/or diuretics for pregnant women. Their aim was to prevent or control pre-eclampsia, a dangerous disorder of late pregnancy in which blood pressure rises steeply, the body retains excess sodium and water, and protein is lost in the urine.

Since the early 1970s, most experts have changed their minds, proscribing diuretics and allowing pregnant women a liberal salt intake. Researchers still don't know what causes pre-eclampsia, but some recent studies suggest that severe sodium restriction does not help and may actually worsen the situation.

Pregnancy does increase the body's need for sodium. A total of 12,000 to 21,000 milligrams are accumulated over the nine-month period, and most experts now agree that women should be allowed to salt food to taste during pregnancy.

Diuretics can deplete pregnant women of potassium, sodium, and uric acid, and can cause other physiological disturbances. Some are suspected of causing birth defects, either directly or as a result of other complications. Needless to say, diuretics should be avoided, except in women suffering from certain heart conditions.

sodium in their regular diets to offset even large losses of sweat.

But Costill's dozen were eating roughly 6,000 milligrams of sodium—quite a high-sodium diet. How might they have responded had Costill cut their intake to between 1,000 and 3,000 milligrams a day? No recent studies have tested this possibility. But back in 1944 researchers found that army recruits consuming a diet with 3,400 milligrams of sodium and sweating an astonishing 6 to 7 quarts a day were more likely to show symptoms of salt depletion than were other recruits eating 6,000 milligrams of sodium and enduring the same workout.

Still, this result does not mean that active people on a low-salt diet will show signs of salt depletion. The 1944 army experiment was certainly performed under severe and extreme conditions.

By way of comparison, Costill estimates that Alberto Salazar (winner of the 1980, 1981, and 1982 New York City marathons) loses 3 to 4 quarts of sweat during a major race. How many Americans ever work up anything close to a 3-to-4-quart, 26-mile sweat?

A person who consumes between 1,100 and 3,300 milligrams of sodium a day can work out fairly rigorously on a hot day and not worry about sodium depletion. Dr. Norman Kaplan, a cardio-vascular expert at the University of Texas Health Science Center in Dallas, says: "As long as we are eating a fourth of our present level [of sodium], we're sure to get all the sodium that is needed, even during August, pre-season, two-a-day football practices in Texas."

Costill agrees. His studies show that people must secrete an entire quart of sweat to lose 1,000 milligrams of sodium. And it seems that the body has a built-in insurance policy that helps prevent severe sodium losses during strenuous exercise. Once you've worked out in hot weather for eight days or more, the body begins excreting only a third to a half the normal amount of sodium through sweat, Costill says.

And what about suspicions that a low-sodium diet reduces the body's intake of other necessary nutrients? The concern is unfounded.

Dr. David McCarron, director of the hypertension program at the University of Oregon Health Sciences Center, warns that people trying to restrict their sodium intake may cut back on dairy products and, in so doing, eat too little calcium. But people can easily cut their salt intake by avoiding high-sodium processed foods such as Hamburger Helper, Big Macs, and Campbell's Chicken Noodle soup, each of which contains about 1,000 milligrams of sodium.

No one has advised the average person with normal or high blood pressure to cut down on calcium-rich dairy products, such as milk. After all, a glass of milk only contains 120 milligrams of sodium, an ounce of Swiss cheese contains only 75 milligrams, and cheddar cheese contains 190 milligrams.

CHAPTER THREE

The Case Against Salt

Pause for a moment, and imagine yourself in the tropical South Pacific. You are part of a medical research team studying the eating habits of people who live far from the fast-paced industrialized world from which you come. You are there to test a hypothesis that has emerged after years of research by globe-trotting scientists like yourself. Is the amount of salt people consume related to the incidence of hypertension in their society? You are here, just south of the equator, to find evidence of the sodium-hypertension connection.

The Aita are a mountain-dwelling people who live on one of the Solomon Islands. Their diet consists of taro root, sweet potatoes, green leafy vegetables, and fruit. Occasionally they trade for small quantities of rice and canned fish. The Aita do not cook with salt, nor do they eat food preserved with salt. On the average, the Aita's diet provides a remarkably low 500 milligrams of sodium, or ¼ teaspoon of salt, a day.

When Dr. Lot Page, professor of medicine at Tufts University School of Medicine, studied the Aita in 1974, he found no cases of hypertension in the tribe. Young and old, all the individuals he examined had blood pressures within the normal range.

The Aita are not alone. The Yanomamo Indians of Brazil, the Australian aborigines, the Kalahari bushmen of Africa, the Polynesian residents of the Cook Islands, the Melanesian tribes in New Guinea, and the people of several other cultures eat little salt. They, too, are virtually hypertension-free.

Obviously, other aspects of these peoples' lifestyles may protect them from hypertension. On the average, their bodies are leaner and better exercised than ours. Still, scientists must question whether it is low-salt diets—and not other factors—that protect these societies from the ravages of high blood pressure.

The answer in part rests with another Solomon Island tribe,

the Lau. They are lowland, coastal dwellers similar to the Aita in nearly all ways except one: the Lau cook their vegetables in seawater. This practice makes the Lau's diet almost as salty as ours: they consume 3,500 to 5,000 milligrams of sodium, or 1¾ to 2½ teaspoons of salt, per day. Significantly, one of every ten Lau suffers from high blood pressure.

Still, their 10 percent rate of hypertension pales next to the incidence of high blood pressure in some modern countries. But the similarity of the Lau and the Aita in nearly every regard except for the ways the tribes use salt convinced Lot Page and many other scientists that salt, or sodium, raises blood pressure.

The sodium-hypertension connection gains further credence when tested in societies that consume very large amounts of salt. The Japanese are by far the world's top sodium consumers. The general population averages an intake of 6,000 to 10,000 milligrams of sodium per person per day. In parts of northern Japan, residents consume as much as 20,000 milligrams of sodium, equivalent to 10 teaspoons of salt, per day.

With so much sodium, Japanese high-blood-pressure rates soar into the stratosphere. Roughly 60 of every 100 Japanese have high blood pressure, as compared to 33 of every 100 adult Americans. Their stroke rate is an alarming 169 percent higher than that of the United States. The sodium-rich Japanese diet, complete with soy sauce, miso soup, salted fish, and salt-preserved foods, appears to be the major culprit.

Our Dietary Heritage

The experiences of the Aita, Lau, Japanese, and numerous other societies may be explained by our understanding of human evolution. Our bodies may simply not be designed to handle large amounts of salt. By 50,000 B.C., the human body had evolved to what it is today. Biologically, we are identical to the people of that time, even though those men and women had a lifestyle very different from ours. The invention of agriculture, which eventually brought to an end lives of nomadic existence, was still 40,000 years off.

Instead of growing food, early humans ate plants and animal milk, marrow, blood, and meat. Dr. Henry Blackburn, professor of medicine and director of the Laboratory of Physiological Hygiene of the University of Minnesota, calculates that even a diet generous in the remains of animals killed during hunts would provide only 1,200 to 1,600 milligrams of sodium a day. This is the diet on which our bodies evolved for hundreds of thousands of years. We developed physiological mechanisms to assure survival on very little sodium. Now that we are eating much more salt—Americans average 4,000 to 6,000 milligrams of sodium a day—our bodies cannot cope.

Cultures such as the Aita, the Yanomamo Indians, the Australian aborigines, and others are living links to the past. These people eat no more sodium than our hunter-gatherer ancestors, and they are totally free of hypertension. This evidence suggests that Americans and Japanese develop high blood pressure because high-salt diets overload their bodies with sodium.

When scientists match lifestyle or dietary patterns to the disease rates of Americans, Japanese, Aita, Lau, or other cultures, they are performing a type of medical detective work known as epidemiology. Even though these experiments cannot "prove" a link between salt and hypertension, they are very suggestive. On these grounds alone, public health authorities might be justified in calling for across-the-board reductions in the amount of sodium Americans consume. But epidemiological studies are not the only evidence indicting salt. Laboratory experiments with animals make the case against sodium even stronger.

Forced Hypertension

In nearly every species tested, salty foods cause hypertension. Researchers have fed large amounts of salt to rats, chickens, Yorkshire pigs, and especially monkeys—the animals that most resemble humans physiologically—and all have developed high blood pressure after months or years on a high-sodium diet. Researchers fed the animals more salt than most humans eat. As in many experiments with animals, high doses of the suspected culprit are necessary in order to see results on a small number of animals in a short period of time.

These simple but telling experiments intrigued pioneering hypertension researcher Lewis K. Dahl, of the Brookhaven National Laboratory in Upton, New York. He noticed that within a given group of animals, most developed high blood pressure—but to different degrees. Some developed very high blood pressure; others moderately high blood pressure; and still others retained normal blood pressure.

To carry the experiment further, Dahl bred hypertensive rats with other hypertensive rats, and hypertension-resistant rats with other hypertension-resistant rats. After several generations, he had created two distinct strains of animals: one group that was clearly salt-sensitive, and the other clearly salt-resistant. The blood pressure of the salt-sensitive rats skyrocketed whenever they received a high-salt diet. The salt-resistant rats maintained normal blood pressure no matter how much salt they were fed. Dahl's experiments opened a new chapter in our understanding of hypertension. Dahl showed not only that hypertension is related to salt intake, but also that sensitivity to salt is rooted in one's genetic heritage.

Human beings are not specially bred rats, of course. Like Lewis Dahl's original rat population, humans show a wide range of sensitivities to salt and the high blood pressure it can cause. The very salt-sensitive individuals will develop high blood pressure on fairly small amounts of salt; the less salt-sensitive will succumb only to higher salt levels; and the salt-resistant will retain normal blood pressure no matter how much salt they ingest.

In a culture such as the Yanomamo Indians of South America or the Kalahari bushmen of Africa, the amount of salt consumed is so low that it never triggers a reaction among the salt-sensitive. By contrast, in northern Japan the salt intake is so high that even some of the least salt-sensitive people develop high blood pressure.

Learning about these experiments might prompt you to wonder about your own salt sensitivity. Unfortunately, your doctor cannot administer a salt-sensitivity test. So far, scientists have not found a way to detect this predisposition to hypertension. Therefore, health experts have advocated a public health solution: reduce the amount of salt in the entire nation's diet with the knowledge that many genetically salt-sensitive people will be saved from hypertension and its often fatal consequences.

Testing Difficulties

Those with an interest in discrediting the sodium-hypertension connection—such as producers of salty processed foods—scoff at the mounds of epidemiological and animal studies strongly suggesting a link. They imply that the only way to settle the controversy once and for all would be to conduct studies on human subjects. However, certain types of human studies, such as feeding people a large amount of salt over a long period of time, would be unethical, given the large amount of evidence indicting sodium. Doctors might just succeed in giving these subjects high blood pressure.

Alternatively, researchers at the University of Indiana have tried feeding humans large amounts of sodium for a short time—one to three days. They wanted to test whether "salt loading" would cause even a temporary rise in blood pressure.

Dr. Friedrich Luft and his associates found that very high sodium intakes do hike blood pressure levels in normal people. But Luft noticed something even more fascinating. In one study, high-sodium diets (14,000 milligrams a day) raised blood pressure higher in black subjects than in white subjects. In another experiment, blacks, people over age 40, and close relatives of people with high blood pressure took longer to get rid of the excess sodium than others. Hypertension is unusually prevalent among each of these three groups.

Says Luft: "People in populations at risk [for hypertension] handled salt differently. They were more sluggish, not as facile at excreting salt as others." That salt loading raises blood pressure higher in blacks than whites suggests that "these people can only excrete excess sodium at a higher pressure," he explains. In other words, high blood pressure might be the body's way of building up enough force to rid itself of excess salt.

While scientists must take precautions when feeding people large amounts of salt, giving them a low-sodium diet poses no risk. After all, if too much sodium raises blood pressure, one might expect very little sodium to reduce it.

Indeed, a low-sodium diet does reduce blood pressure in people with hypertension. The phenomenon was first reported in 1904 by two French physicians who found that a low-salt diet reduced their patients' blood pressures. A conclusive 1982 study by Dr. Graham MacGregor and colleagues of Charing Cross Hospital Medical School in London demonstrated beyond any doubt that reducing sodium lowers blood pressure in hypertensives.

MacGregor put two groups of hypertensives on a low-sodium diet for several weeks. Then one group received a capsule consisting of nothing but sugar, a placebo. The other group took a capsule consisting of moderate amounts of sodium. Neither group knew anything about the chemical composition of the capsules they were taking. As expected, people taking the placebo had no change in blood pressure, while those taking the sodium-laced capsule watched their blood pressures jump. The experiment was repeated with each group getting the opposite "prescription." Again, the sodium capsule caused high blood pressure, while the placebo had no apparent effect.

Today scientists agree that low-sodium diets can reduce blood pressure for many hypertensives. But industry trade associations, such as the Salt Institute, contend that "it usually takes rigid sodium restriction, often less than 1,000 milligrams per day, to control hypertension effectively without medication." The salt industry makes this claim despite experiments over the past decade showing that sodium reductions needn't be that severe to lower blood pressure.

In MacGregor's study, the 19 patients only cut their sodium consumption by little over half—from 4,400 to 1,900 milligrams. Their blood pressures dropped significantly.

Dr. Jan Parjis of the University of Louvain in Belgium recorded similar results among the seventeen hypertensives that he studied. Like MacGregor, Parjis found that cutting sodium consumption in half was enough to lower the blood pressure. But neither MacGregor nor Parjis monitored their subjects for more than several weeks, leaving unanswered the question of long-term prospects for managing hypertension with low-salt diets.

Dr. Trevor Morgan and colleagues at the University of Melbourne in Australia have investigated just this concern. The research team put thirty hypertensive patients on a reduced-sodium diet for two years. On the average, subjects cut their sodium intake by a mere 18 percent, from 4,400 to 3,600 milligrams a day. Blood pressures not only dropped significantly, but in a third of their patients, blood pressures fell to normal levels.

The extremely important experiments of Morgan, Parjis, and MacGregor dealt with people who already had high blood pressure. At the University of Indiana, researchers have asked whether cuts in sodium consumption would affect the blood pressures of normal people. Dr. Janet Miller and her colleagues convinced sixteen couples to cut their sodium intake for three months from 3,500 to 1,600 milligrams a day. Even in these healthy, nonhypertensive adults blood pressures dropped significantly.

The Indiana study was small, but it does suggest that healthy people who eat low-sodium diets over a lifetime may avoid high blood pressure. In fact, all the evidence described so far—the comparisons among high- and low-salt cultures, the experiments with animals, and the human feeding studies—points to the same conclusion: a high-sodium diet can cause hypertension and a low-sodium diet can prevent it.

Still, segments of the food industry and a handful of scientists continue to carp: "There is no absolute proof that salt causes high blood pressure!"

A Long Range Study

The definitive study they demand may be in the making. In 1982, Dr. Ronald Prineas, of the University of Minnesota, and cooperating scientists at the universities of Alabama, Mississippi, and California embarked on a two-year pilot study of 800 people. These individuals are being carefully selected to meet two criteria: their blood pressures have to be normal, and they have to be genetically predisposed to developing hypertension (someone in their family has to have been hypertensive). One group of participants is eating a low-sodium diet (1,600 milligrams a day). Another group is following a low-sodium, high-potassium regimen. The remaining two groups are on weight-loss diets, one with sodium restrictions and one without.

At the end of the two years, researchers will know which diets are most effective at keeping blood pressures low. This pilot study will determine the shape of a much larger experiment in which Prineas will test the most successful diets on 7,000 people. The object is to see whether one or two of these eating plans—three out of four are low in sodium—can protect people from hypertension.

The Prineas experiment promises to quiet the naysayers once and for all. But results are not due for another ten years. Should we wait a decade before deciding to eat less salt? Yes, say segments of the food industry and their academic consultants.

But Prineas says no. In public health issues, he explains, "by not taking action, you're actually adopting a conscious policy...of letting people at risk go on being at risk. I am not willing, given the current evidence, to make that decision."

The only tenable public health position calls for cutting the salt. The Surgeon General, the U.S. Departments of Health and Human Services and of Agriculture, the American Heart Association, and the National Academy of Sciences have said as much.

But admonitions, however appropriate, are not enough. The public needs to learn where salt is hidden in the American diet.

Forgoing the salt shaker is a start. But as the next chapter explains, that's not the real problem. Thanks to food manufacturers, few of us leave the supermarket or restaurant without having purchased a megadose of salt in some of the most unlikely foods.

The Role of Potassium

"A low-potassium intake can be considered as an unindicted co-conspirator in the hypertension connection," according to Dr. Herbert Langford, chief of the endocrinology and hypertension division at the University of Mississippi Medical Center. In other words, too little potassium may work in combination with too much sodium to set the stage for high blood pressure.

Like sodium, the potassium story goes back hundreds of thousands of years. Human beings evolved on a diet that was both low in sodium and high in potassium, says Dr. Henry Blackburn, of the University of Minnesota. Fifty thousand years ago, a meal might have consisted of potassium-rich, sodium-poor plant foods, such as roots or leafy green vegetables, beans, nuts, seeds, and berries. On a good hunting day, fresh meat, milk, and/or eggs might have supplemented the diet of early humans. A day's worth of such food would provide no more than 2,000 milligrams of sodium (equivalent to 1 teaspoon of salt) and 7,000 milligrams of potassium.

Today the tables are turned. The typical American diet is not only spiked with sodium, it is also comparatively low in potassium. Processing is partly responsible. Canned vegetables, for example, have only about two-thirds the potassium of fresh vegetables. We also eat less potassium than our prehistoric ancestors because breads, grains, fats, and sugar—all low in potassium—comprise a substantial part of our diet. Our ancestors of 50,000 years ago ate virtually none of these foods. The result: the average American male now consumes about 2,800 milligrams of potassium per day, according to Langford. The average female eats only 2,300 milligrams.

Knowing that our forebears ate far more potassium than we do is fascinating, but it alone does not prove that too little potassium contributes to hypertension. Instead, several clues from different kinds of experiments have convinced scientists such as Langford that potassium affects blood pressure.

In 1957, Dr. George Meneely, of the Vanderbilt School of

Medicine, found that extra potassium protected animals against some of the damage caused by a high-sodium diet. He fed rats a diet high in both potassium and sodium. The excess sodium still caused high blood pressure, but those rats lived longer than did rats fed a high-sodium diet without potassium.

Meanwhile Dr. Naosuke Sasaki, of the Hirosaki University in Japan, noticed that the people in two northern Japanese villages ate huge amounts of salt. Mysteriously, one village had a much higher rate of hypertension than the other. Sasaki looked closely at the diets of the villagers. He found that the people in the village with the lower blood pressures ate plenty of potassium-rich apples, while those in the other village didn't. Sasaki drew a conclusion similar to Meneely's: that potassium protects against high blood pressure.

At the University of Mississippi, Langford measured the sodium and potassium in the urine of 662 high-school girls to get a fix on how much sodium and potassium the girls consumed. He found that the girls with higher blood pressures had relatively more sodium and less potassium in their urine. Langford also found that blacks generally consume less potassium than whites— only about 1,600 to 2,000 milligrams a day. He speculates that this dietary habit may contribute to the higher rates of hypertension among blacks.

In 1982, Dr. Graham MacGregor performed a critical experiment on potassium similar to his experiment on sodium. He secretly added 2,300 milligrams of potassium to the diets of twenty-three people with hypertension. Their blood pressures dropped an average of 7 to 8 millimeters of mercury.

The evidence indicting sodium as a cause of hypertension is far more extensive than the evidence suggesting that potassium may prevent it. Still, the few experiments on potassium are compelling. Eating a high-potassium diet full of fresh fruits and vegetables, fish, beans, and poultry, along with low-fat milk and lean meat, cannot hurt and may prove to help control high blood pressure. (See Chart 3A, page 36.) According to Langford, 4,000 milligrams of potassium a day is a desirable goal.

The Role of Chloride

For over 40 years, researchers have stated with confidence that it's the sodium part of sodium chloride that raises blood pressure. Then in late 1983, two groups of scientists suggested that chloride was the real culprit.

In their studies, blood pressure rose in rats fed sodium chloride, but not in those fed mixtures made primarily of sodium bicarbonate. Only when sodium is combined with chloride, they concluded, does blood pressure rise.

But other evidence makes it highly unlikely that chloride is the hypertension culprit. For example, researchers can lower blood pressure by feeding potassium chloride. If chloride were a blood pressure-raiser, potassium would have to be a very potent blood pressure-reducer to counter chloride's effect. A variety of studies, however, indicate that potassium does not have a powerful effect.

What, then, might explain the new results? Possibly, sodium might raise blood pressure while bicarbonate lowers it. That's what the Church & Dwight Corporation hopes. The makers of Arm & Hammer baking soda (which is sodium bicarbonate) helped fund one of the studies, and immediately seized on the results. In May, 1984, the company petitioned FDA to reconsider its sodium labeling regulation, which treats sodium linked to bicarbonate and sodium hitched to chloride equally.

FDA refused, saying it had too little evidence to exonerate sodium bicarbonate. Indeed, it is too early to tell. Other researchers have had trouble reproducing the first two studies' results. Until otherwise notified, assume sodium raises blood pressure, regardless of its partner.

Other Risk Factors

Certain factors that raise the risks of developing hypertension are within our power to control, while others are not. People with a family history of high blood pressure have double the risk of experiencing this often fatal condition compared to those without such a history. And race makes a difference. As explained, blacks in the U.S. are twice as likely to develop hypertension as whites. Scientists aren't sure why.

While you can't change your race or your family tree, you do have the power to control other suspected factors. Listed below are some lifestyle and dietary choices, other than sodium, that researchers have suspected of causing or preventing hypertension. The evidence so far indicates that some clearly do, and others probably do not, increase your chances of suffering from high blood pressure.

Obesity. No doubt about it, losing weight can lower blood pressure. Physicians routinely advise hypertensive patients to achieve and maintain their ideal weight. Even a moderate, 10 to 30 percent, weight loss can make a difference.

Studies show that keeping thin can also prevent the onset of hypertension. In a Georgia community, overweight people were four times as likely to develop high blood pressure as lean people.

Lack of Exercise. Surprisingly, few studies show that exercise— as opposed to the weight loss it induces—lowers blood pressure. Compare a group of exercisers with sedentary people and you'll find that the exercisers have lower blood pressures. But then again, they also have lower weights. Studies that have tried to control for this weight-loss variable offer no clear indication of whether exercise itself will lower blood pressure.

CHART 3A:

Potassium Values of Sample Foods

DAIRY PRODUCTS	PORTION	MG.
Cheddar cheese	1.5 oz.	42
Cottage cheese	½ cup	95
Ice cream, hard	½ cup	123
Milk, low-fat 2%	1 cup	412
Milk, skim	1 cup	426
Milk, whole	1 cup	370
Yogurt, plain, low-fat	1 cup	531

FRUITS AND FRUIT JUICES

Apple	1 medium	159
Apple juice	6 oz.	222
Avocado	½	602
Banana	1 medium	451

	PORTION	**MG.**
Blueberries, raw	½ cup	65
Cantaloupe	½	825
Cherries, sweet, raw	10	152
Grapefruit	½	167
Grapefruit juice	6 oz.	252
Grapes	10	93
Grape juice, frozen	6 oz.	42
Orange	1	237
Orange juice, frozen	6 oz.	354
Peach	1	171
Peaches, canned in juice	½ cup	158
Pear	1	208
Pineapple, raw	½ cup	113
Plum	1	112
Raisins	½ cup	553
Strawberries	½ cup	123
Watermelon	⅟₁₆	560

GRAINS AND GRAIN PRODUCTS

Bread, white	2 slices	52
Bread, wheat	2 slices	144
Oatmeal, cooked	1 cup	146
Rice, brown	1 cup	41
Shredded wheat	1 cup	87

LEGUMES AND NUTS

Almonds, roasted	1 oz.	219
Beans, white navy, cooked	1 cup	790
Beans, red kidney, cooked	1 cup	629
Black-eyed peas, cooked	1 cup	573
Lentils, cooked	1 cup	498
Peanuts, roasted	1 oz.	191
Soybeans, cooked	1 cup	972
Split peas, cooked	1 cup	592
Walnuts, English, shelled	1 oz.	128

MEAT, FISH, POULTRY, AND EGGS

Chicken, dark meat, roasted	4 oz.	284
Chicken, light meat, roasted	4 oz.	458

MEAT, FISH, POULTRY, AND EGGS

	PORTION	MG.
Chuck, choice, lean, cooked	4 oz.	276
Cod, broiled w/butter or margarine	4 oz.	460
Egg	1 large	65
Flounder, broiled w/butter or margarine	4 oz.	460
Ground beef, lean, cooked	4 oz.	305
Halibut, broiled	4 oz.	596
Pork, loin chops, lean roasted	4 oz.	373
Shad, baked	4 oz.	428
Sirloin steak, lean, cooked	4 oz.	409
Tuna, in water, chunk (tuna with liquid)	3 oz.	257
Turkey, dark meat, roasted	4 oz.	450
Turkey, light meat, roasted	4 oz.	465

VEGETABLES AND JUICES

		MG.
Beet greens, cooked	1 cup	480
Beets, cooked, diced, drained	1 cup	354
Broccoli, cooked	1 cup	414
Brussels sprouts, cooked	1 cup	846
Cabbage, red, raw, shredded fine	1 cup	240
Carrot, raw	1	246
Cauliflower, cooked	1 cup	240
Celery, raw	1 stalk	136
Cole slaw (w/mayonnaise)	1 cup	239
Collards, cooked	1 cup	498
Corn, cooked	1 ear	151
Eggplant, cooked, diced	1 cup	300
Green beans, cooked	1 cup	190
Green pepper	1	349
Lettuce, iceberg	1 cup	96
Lettuce, romaine	1 cup	143
Lima beans, immature, cooked	1 cup	376
Mushrooms, sliced raw	1 cup	390
Okra, cooked	1 cup	268

	PORTION	MG.
Peas, cooked	1 cup	314
Potato, baked	1 long	782
Radishes	½ cup	185
Rutabaga, cooked, mashed	1 cup	1320
Spinach, cooked	1 cup	1166
Spinach, raw	1 cup	259
Squash, acorn, baked	1 cup	984
Squash, butternut, baked	1 cup	1248
Sweet potatoes, baked	1	342
Tomato	1 medium	300
Tomato juice, canned	6 oz.	413
Turnips, boiled, mashed	1 cup	432
Zucchini, cooked, sliced	1 cup	508

Sources: USDA Handbooks #456, 8-1, and 8-9

Stress. Many people think that if you have hypertension, you must be tense. They are wrong. A person may feel relaxed and serene but have dangerously high blood pressure.

Although hypertensives don't necessarily feel nervous, some researchers suspect that years of psychological or social stress might cause hypertension. After all, they've reasoned, a stressful situation does raise the blood pressure temporarily. But to date, scientists have found very little substantial evidence that stress causes high blood pressure.

On the other hand, preliminary studies do suggest that certain relaxation techniques may lower blood pressure in hypertensives. Relaxation methods are currently experimental, but a combination of diet and relaxation therapy may someday replace drugs as the primary treatment for mild hypertension.

Alcohol. A handful of respected studies all agree: the heaviest drinkers have the highest blood pressures. "Heavy" drinking meant two drinks or more per day in some studies, and five drinks or more in others. Whether it is alcohol or some other factor in the lives of heavy drinkers that raises their blood pressure is not clear.

Lack of Polyunsaturated Fats. A few preliminary studies suggest that a diet with a smaller amount of total fat or one with a higher ratio of polyunsaturated fat to saturated fat seems to lower blood pressure. When both these changes are made—less total fat, and a higher percentage of polyunsaturated fat—pressures have been reduced 8 to 10 percent.

Lack of Calcium. Several small studies have led a few researchers to conclude that low calcium diets contribute to high blood pressure. Most medical experts say it is far too early to tell whether a deficiency of this mineral actually induces hypertension.

As mentioned, scientists still have no way of knowing which factors have the greatest impact on which people.

The only solution, then, is to pay some attention to each, but most attention to those backed with the greatest amount of evidence. In other words, cut back on sodium and maintain normal weight, but also think about exercising, getting enough calcium, replacing some of the saturated fats in your diet with polyunsaturated fats, and avoiding stressful situations and excessive alcohol consumption. Luckily, you can't go wrong making these changes—each is healthful in its own right.

CHAPTER FOUR

Salt, Salt Everywhere

Stop and think about the last time you grabbed lunch at McDonald's. Your choices might have included a cheeseburger, a vanilla shake, french fries, and apple pie. Now which of these foods had the most sodium?

If you said "french fries," you couldn't be more wrong. According to McDonald's, a serving of the salt-coated fried potatoes contains about 109 milligrams of sodium, not that much really. But the cheeseburger packs in 767 milligrams of sodium—seven times the amount in the "salty" fries! Even the vanilla shake and apple pie are surprises; they contain two and four times as much sodium as the french fries, respectively.

In fact, your entire meal at McDonald's provided 1,475 milligrams of sodium, more than seven times the body's physiological need and within the range you should be consuming for an entire day!

McDonald's excesses are the rule, not the exception, in the restaurant world. Two slices of Pizza Hut's Thin 'n Crispy Supreme Pizza contain 1,200 milligrams of sodium, an Arby's Roast Beef sandwich supplies a hefty 880 milligrams, and Burger King's Whopper has a whopping 975 milligrams. That's before you add a couple of sprinkles of salt at the table.

Concern about sodium apparently falls on deaf ears at the fast food franchisers. In 1982 the centerfold of *Upfront,* a magazine circulated to young kitchen and counter workers at Burger King restaurants, proclaimed, "To make the best french fries better... add more salt."

Kitchen Cupboard Criminals

Clearly the food companies have been slow to accept the fact that salt can be a substance of abuse. But fast food retailers are no more to blame than their food-manufacturing brethren. We estimate that roughly 75 percent of America's sodium intake comes from processed foods. From breakfast cereals to soups to frozen dinners, there's no escaping salt.

For example, had you skipped McDonald's for a lunch of Campbell's Chicken Noodle Soup and a serving of Kraft Macaroni and Cheese, a Coke, and some Jell-O Instant Chocolate Pudding for dessert, you would actually have fared worse. The bowl of soup contains 1,169 milligrams of sodium, the macaroni and cheese 665 milligrams, the soda 15, and the pudding 515, for a total of 2,354 milligrams.

But maybe you shy away from processed concoctions and eat at McDonald's only occasionally. In fact, you may consider yourself a careful eater, someone who makes an effort to avoid the excess additives, calories, sugar, and fat so prevalent in processed foods. You may start your day with a bowl of General Mills' Total cereal, then brown-bag a lunch consisting of a tuna salad sandwich on whole wheat bread, a strawberry yogurt, a handful of unsalted peanuts, and a small can of V-8 juice.

These are wise food choices if you want to get enough vitamins and fiber while avoiding excess fat and sugar. But with regard to sodium, you might as well have driven through the "Golden Arches." Did you know that Total cereal contains 375 milligrams of sodium per serving? That's more salt than you'd get by crushing a biscuit of shredded wheat (10 milligrams) into your cereal bowl and dousing it with a teaspoon of La Choy soy sauce (325 milligrams).

As for lunch, two slices of Arnold/Orowheat Bran'nola bread contain 355 milligrams of sodium. Three ounces of Chicken of the Sea tuna packed in water add another 400 milligrams (tuna packed in oil supplies even more). Add a tablespoon of mayonnaise to hold your sandwich together, and chalk up another 80 milligrams. The strawberry yogurt contains about 125 milligrams of sodium. And the unsalted peanuts a mere 2. Then, wash down the sandwich with the 6 ounces of Campbell's V-8 and savor your final 625 milligrams. This "nutritious" alternative meal to the McDonald's fare packs even more of a hypertension-threatening wallop: 1,587 milligrams.

The point is: large amounts of sodium are hidden in most processed foods, even in foods that don't taste salty. Potato chips taste salty, but fifteen of them contain only 140 milligrams of sodium. Three medium-size Aunt Jemima pancakes do not taste very salty, but together they contain over four times as much sodium (643 milligrams) as the chips.

But It Doesn't Taste Salty

That people often can't taste the excess sodium makes the problem all the more insidious. In some foods the salt is masked by other flavors. For example, salt in spaghetti sauce has to compete with tomatoes, garlic, onions, olive oil, oregano, and often meat to impress your tastebuds. And you swallow much of the salt before you've had a chance to taste it. On the other hand, all the salt in a potato chip or french fry lies on the surface, dissolving into a solution that coats your tongue with salty fluid. The chips and fries thus taste saltier than many other foods that actually contain more salt.

Processed foods are the number one sodium villain in our diet. Just look at Chart 4A (page 45), which compares the amount of sodium in relatively unprocessed foods with the amount used in making the processed equivalent. The differences are striking: Chef Boy-ar-dee Beef Stew contains nearly four times more sodium than the homemade variety prepared and tested by the U.S. Department of Agriculture. A serving of regular oatmeal contains 1 milligram of sodium, but Quaker Instant Oatmeal packs 252 milligrams. Cooked rice contains a mere 1 milligram of sodium, but the maker of Minute Rice manages too sneak about 570 milligrams into one serving of its Long Grain & Wild Rice. A serving of this Minute Rice product alone accounts for about a third the sodium the average American should eat in a day.

Why Salt?

Motives for food companies' love affair with salt are hard to discern. Few explain their actions. In some cases, salt is an inexpensive flavor substitute for real food. It is cheaper for Campbell to serve 1,169 milligrams of sodium in 10 ounces of chicken noodle soup than to create an equally pleasing taste with herbs, spices, and more chicken.

Until recently, few companies tried to add as little salt as necessary to produce an acceptable taste. It almost seemed as if companies salted foods for the least sensitive tongue. For example, a spokesperson for Campbell Soup Company told us that his company recently discovered that it could add less salt and nobody would know the difference. Campbell food chemists are reportedly marching through the company's line of soup products, testing the palatability of lower sodium levels. The Campbell spokesperson said, "We can reduce sodium 20 to 25 percent in many flavors and consumers don't even know it."

Of course, some companies produce high sodium foods with recipes that call not only for salt, but also for other sodium-containing additives. Sodium phosphate, an emulsifier, contributes most of the sodium found in pasteurized processed American cheese. Sodium nitrite, a preservative, adds a sizable helping of sodium to processed meats. And sodium bicarbonate or baking soda contribute a substantial portion of the sodium in pancake mixes and baked goods, such as Pillsbury's Pipin' Hot cinnamon rolls. Overall, food additives contribute less than 10 percent of the sodium in processed foods. But, as in these examples, additives can sometimes account for a much higher percentage. (See Chart 4B, page 47, for a list of sodium-containing additives in food.)

Still other companies are driven by the force of habit. When many of the foods on our grocery shelves were first introduced, salt was considered a benign, inexpensive, and popular flavor enhancer. The makers of those products that have become American classics are understandably loath to tamper with success.

Fortunately, more and more companies are realizing that the success of their product has little to do with current levels of sodium used in their recipes. Gwaltney Meats, a division of Smithfield Foods, recently ran a public taste test to introduce its fifteen new reduced-salt, reduced-sugar ham, bacon, sausage, lunch meat, and hot dog products. Gwaltney officials placed three unmarked servings of meat—its own and two competing brands—before a pack of hungry journalists. The company asked members of the media to identify which of the three products contained between 30 and 47 percent less sodium. According to an article in the *Newport News Times Herald*, "When the results were in, the ratings were very close and most could not tell which was the low-salt brand."

CHART 4A:

Sodium in Fresh vs. Processed Foods

(in milligrams)

FRESH FOODS*	PROCESSED FOOD EQUIVALENT
Beef stew (1 cup): 290 mg.	Chef Boy-ar-dee Beef Stew (1 cup): 1120 mg.
Cheddar cheese, natural (1.5 oz.) 300 mg.	Kraft Pasteurized Processed Cheese (1.5 oz.): 698 mg.
Chicken breast, roasted (4 oz.) 75 mg.	Swanson Fried Chicken Breast (4 oz.): 769 mg.
Corn, cooked (½ cup): 1 mg.	Green Giant Canned corn (½ cup): 180 mg.
Green beans, cooked (½ cup) 5 mg.	Green Giant Canned green beans (½ cup): 190 mg.
Flounder, broiled w/butter (6 oz.): 285 mg.	Mrs. Paul's Light and Natural Flounder Fillet (6 oz.) 975 mg.
Hamburger, lean and cooked (5 oz.): 100 mg.	Banquet Cookin' Bag Meat Loaf (5 oz.): 894 mg.
Oatmeal, cooked (¾ cup): 1 mg.	Quaker Instant Oatmeal (¾ cup): 252 mg.
Peas, cooked (½ cup): 1 mg.	Green Giant Canned peas (½ cup): 255 mg.
Pork chops, broiled (3.2 oz.) 64 mg.	Oscar Mayer Wieners (3.2 oz.) 1028 mg.
Rice, cooked (½ cup): 1 mg.	Minute Rice Long Grain and Wild Rice (½ cup): 570 mg.

*Figures obtained from USDA Handbook #456

Ironically, taking out most of the sugar—another nutritional plus for Gwaltney—allowed the salt content to be reduced. It seems that sugar and salt counter one another. When the sugar was eliminated, the amount of salt needed to assure a pleasing taste fell markedly.

The guide to the sodium content of processed foods starting on page 143 confirms the Campbell and Gwaltney experiences. Some companies must be adding more salt than is necessary to preserve and flavor food, since other manufacturers of competing products are able to use much less. For example, B&M's Red Kidney Baked Beans offers 873 milligrams per cup, while Friends' brand contains 1,320 for an equivalent portion. And Gulden's Dijon Mustard has 120 milligrams of sodium per tablespoon, but Grey Poupon Dijon Mustard contains nearly four times that amount: 445 milligrams per tablespoon.

These are not comparisons between the low-sodium dietetic formulations and the "regular" brands, but differences between products that sit next to each other on the supermarket shelves. If one "regular" brand has less salt, why can't the other? Inertia and tradition must be part of the reason companies produce highly salted foods.

Manufacturers' Responsibility

There can be no better proof of the indifference to sodium than to find companies that do not even know how much sodium their products contain. When we wrote to companies for sodium information, Chun King (of Chinese food fame), Golden Grain (makers of Rice-A-Roni), Shakey's Pizza Parlors, and a number of other manufacturers sent us information on calories, protein, carbohydrates, fat, vitamins, and mineral composition—everything but sodium.

Irresponsible as it may be that companies do not know the sodium contents of their products, the real shame rests with the federal government for not demanding quantitative sodium labeling on foods in the first place. Salt and sodium-containing additives must be noted in the list of ingredients that appears on almost all food packages. But the government doesn't require companies to tell consumers how much sodium the product contains!

Merely listing salt in the ingredient statement can mislead consumers who know that manufacturers are required to list food contents in order of descending weight. These people may

CHART 4B:

Sodium-Containing Additives

ADDITIVE   **FUNCTION**

dioctyl sodium sulfosuccinate ....dissolving agent
disodium EDTA *(ethylene*
   diamine tetraacetic acid) ....sequestrant (traps metal
                                       impurities and retards
                                       discoloration or rancidity)
disodium guanylate ....flavor enhancer
disodium inosinate ....flavor enhancer
monosodium glutamate (MSG)....flavor enhancer
sodium acid pyrophosphate ......sequestrant
sodium alginate ....stabilizer, thickener
sodium aluminosilicate ....anti-caking agent
sodium aluminum phosphate ....acid in baking powder
sodium ascorbate (vitamin C) ....nutrient, antioxidant
sodium benzoate ....preservative
sodium bicarbonate ....baking soda
sodium bisulfite ....preservative, antioxidant
sodium calcium aluminosilicate ..anti-caking agent
sodium carbonate ....alkali (adjusts acidity)
sodium carboxymethylcellulose ..stabilizer, thickener
sodium caseinate ....whitener, protein
sodium citrate ....antioxidant
sodium erythrobate ....color preservative
sodium gluconate ....sequestrant
sodium hexametaphosphate ......sequestrant
sodium nitrite ....preservative
sodium pectinate ....thickener
sodium phosphate (mono-, di-,
  and tribasic) ....sequestrant
sodium propionate ....preservative
sodium saccharin ....artificial sweetener
sodium sorbate ....preservative
sodium stearoyl fumarate
  (or lactylate) ....dough conditioner

assume that a food is low in sodium when salt appears near the end of the list. Not true. With 1,236 milligrams of sodium, a McDonald's Quarter Pounder Sandwich with Cheese is a high-sodium food. As the name implies, it contains a quarter pound of meat, 4 ounces. With the bun and other ingredients, the sandwich weighs a total of 6.8 ounces, or 190,400 milligrams. The whopping 1,236 milligrams of sodium is only 0.65 percent of the total weight, so of course salt would fall low on the ingredient list if fast foods had labels!

Labels are useful only when they tell consumers how much sodium, in milligrams, the food contains. Currently the government requires a sodium declaration only if the product advertises itself as low in salt. For instance, Tillie Lewis No Salt Added Fancy Sweet Peas must list the number of milligrams of sodium in a serving. Apparently the government assumes that people who buy low-salt foods do so upon the advice of a physician and only they need to know precisely how much salt they are consuming. Now that government officials have recommended that everyone eat less salt, all foods should declare sodium content.

A Top Priority

Reducing sodium in processed foods must be the number one priority. Processed foods account for three-quarters of the nation's sodium intake, according to our estimates.* Another 10 percent occurs naturally in foods, and about 15 percent comes from salt added at the dinner table. But these figures are just averages. People who salt their food before tasting it get a higher percentage of their sodium from added salt. Those who frequent-

---

*The government has estimated that 25 to 50 percent of our sodium is added in cooking or at the table; about 25 percent occurs naturally in foods; and the remaining 25 to 50 percent comes from processed foods. We found that these estimates were unsubstantiated. Moreover, we found it nearly impossible to construct realistic sample diets based on the government's figures. Instead, our sample diets suggested that 75 percent of sodium comes from processing, 15 percent comes from the salt shaker, and 10 percent is naturally occurring. A small study recently conducted at Georgetown University in Washington, D.C., found that less than 10 percent of sodium is added at the table, suggesting that our estimates are reasonably accurate.

ly dine in restaurants get a larger fraction from processed foods. Some people derive significant amounts of sodium from two other surprising sources: drinking water and over-the-counter medicines.

Sodium from the Tap

The sodium content of tap water varies tremendously from one part of the country to another. In most cities, a glass of water provides far less than 10 milligrams of sodium. But in San Diego, for example, that same 8 ounces contains 30 milligrams.

People in coastal communities such as San Diego drink saltier water because the sea infiltrates the city water supply. In addition, Dr. Edward Calabrese, an environmental expert at the University of Massachusetts, says that if your water contains more than 2 to 3 milligrams of sodium per glass, a likely reason is that your city uses salt to de-ice roadways.

Road salting worries Calabrese because of a study he completed in 1977. It found that high-school students from a nearby community with a high-sodium water supply had higher blood pressure levels than students in a town where the water was virtually sodium-free. Calabrese doesn't know why an extra 25 milligrams of sodium per glass should make such a big difference, but his study has convinced him to recommend that municipalities use calcium chloride instead of salt to de-ice roads where run-off could contaminate reservoirs, wells, and aquifers. (Chart 4C, page 50, lists the sodium content of drinking water in major cities.)

Water softeners in homes further exacerbate the drinking water problem. The softener is designed to replace "hard" minerals, such as magnesium, cadmium, and lead, with "softer" minerals, such as sodium. "Hard" minerals form insoluble compounds when they combine with soap and detergents, making soap less sudsy.

The amount of sodium derived from the water softener depends on the "hardness" of the natural water. In Indianapolis the water is naturally hard. A spokesperson for the Water Quality Association says water softeners add about 29 milligrams of sodium to each 8-ounce glass. By contrast, Chicago has fairly soft water. The typical softener there adds only 14 milligrams of sodium to the same amount of water.

BOX 4C:

Sodium Content of Major City Water Supplies

(Unless otherwise noted, 100% of each city's water supply was tested)

STATE AND CITY	MG. PER 8 OZ.
Alabama	
Birmingham	2.1
Mobile	0.6
Montgomery	12.7
Arizona	
Phoenix (74%)	25.0
Tucson	10.4
California	
Fresno	4.0
Long Beach	29.3
Los Angeles	16.0
Oakland	2.8
Sacramento (85%)	2.8
San Diego (57%)	23.6
San Francisco	1.0
San Jose (80%)	6.8
Colorado	
Denver	3.5
Connecticut	
Bridgeport	0.8
Hartford	0.6
New Haven	0.9
District of Columbia	
Washington, D.C.	2.1
Florida	
Jacksonville	3.3
Miami	4.0

STATE AND CITY	MG. PER 8 OZ.
St. Petersburg	1.4
Tampa	2.0
Georgia	
Atlanta	0.5
Savannah	0.9
Hawaii	
Honolulu (57%)	9.6
Illinois	
Chicago	0.9
Rockford	0.8
Indiana	
Evansville	3.0
Fort Wayne	3.5
Gary	1.0
Indianapolis	2.6
South Bend (74%)	1.9
Iowa	
Des Moines (75%)	7.8
Kansas	
Kansas City	5.9
Topeka	26.2
Wichita	14.4
Kentucky	
Louisville	6.1
Louisiana	
Baton Rouge	17.7

STATE AND CITY	MG. PER 8 OZ.
New Orleans	4.2
Shreveport	5.6
Maryland	
Baltimore	0.8
Massachusetts	
Boston	0.5
Springfield	0.6
Worcester	0.7
Michigan	
Detroit	0.9
Flint	6.6
Grand Rapids	1.1
Minnesota	
Minneapolis	1.4
St. Paul (90%)	1.4
Mississippi	
Jackson	0.8
Missouri	
Kansas City	8.9
St. Louis	5.2
Nebraska	
Lincoln	5.9
Omaha	15.3
New Jersey	
Jersey City	1.0
Newark	0.8
Paterson	0.9
New Mexico	
Albuquerque	10.4
New York	
Albany	0.3
Buffalo	2.2
New York City	1.4
Rochester	1.4

STATE AND CITY	MG. PER 8 OZ.
Syracuse	4.0
Yonkers	1.8
North Carolina	
Charlotte	0.9
Greensboro	0.6
Ohio	
Akron	1.4
Cincinnati	4.2
Cleveland	2.6
Columbus	4.2
Dayton	4.0
Toledo	2.8
Youngstown	6.1
Oklahoma	
Oklahoma City	19.8
Tulsa	1.0
Oregon	
Portland	0.2
Pennsylvania	
Erie	2.8
Philadelphia	1.4
Pittsburgh	2.6
Rhode Island	
Providence (92%)	0.7
Tennessee	
Chattanooga	1.9
Memphis	2.6
Nashville	0.09
Texas	
Amarillo	5.4
Austin	7.8
Corpus Christi	14.6
Dallas	9.2
El Paso (97%)	28.3
Fort Worth	4.7

STATE AND CITY	MG. PER 8 OZ.	STATE AND CITY	MG. PER 8 OZ.
Houston (83%)	19.3	**Washington**	
Lubbock (82%)	10.4	Seattle	0.3
San Antonio (Unknown %)	1.8	Spokane (77%)	0.7
		Tacoma (94%)	0.6
Utah		**Wisconsin**	
Salt Lake City (92%)	8.5	Madison	0.7
		Milwaukee	0.9
Virginia			
Norfolk	2.1	Source: U.S. Geological Survey, "Public Water Supplies in the United States, 1962."	
Richmond	0.08		

Consumers who choose to eliminate this unnecessary source of sodium have two options. They can attach the water softener to the hot water line only, allowing water used for washing to be softened but water used for drinking to remain hard. Or they can install a water filter to the pipes under the sink to remove the sodium from the line.

Over-the-Counter Drugs

Antacids, laxatives, and over-the-counter medications are the last major source of sodium bedeviling consumers. Chart 4D (page 53) shows how much sodium many of these products contain. For most people, the total won't amount to much. But many elderly people take laxatives and antacids regularly. One popular medicine, Bromo-Seltzer, provides 717 milligrams per dose. We suggest looking for new low-sodium alternatives. For example, Riopan and Riopan Plus are antacids with considerably less sodium.

From processed foods to tap water to nonprescription drugs, high sodium intakes condemn Americans to an unnecessarily high risk of hypertension. The problem is real: millions of people are suffering heart attacks and strokes due to high blood pressure. The scientific evidence is compelling: epidemiological, animal, and human studies attest to a sodium-hypertension connection. The solution is apparent: reduce or eliminate sodium in foods and drugs whenever possible, and at the very least, label it. The institutions needed to protect the public are in place: the federal government has the authority to curb sodium. But what is being done? The next chapter shows that the government is coming late to the hypertension debate, but is making progress despite fervid industry opposition against raising the visibility of this pressing public health issue.

Please note that the data contained in Chart 4D, found on the following four pages, was obtained from the various manufacturers and *Handbook of Nonprescription Drugs, Ed. 7*, American Pharmaceutical Association, 1982.

Note: Some vitamin C tablets contain sodium ascorbate, which is sodium plus ascorbic acid (vitamin C). These pills contain 11 mg of sodium per 100 mg of vitamin C.

CHART 4D

Sodium Content of Nonprescription Drugs and Supplements

PRODUCT	DOSAGE	SODIUM (MG.)
Alka-Seltzer (gold box)	1 tablet	226
Alkets	1 tablet	1
Allbee with C	1 capsule	1
Allbee-C 800	1 tablet	1
Allbee-C 800 with Iron	1 tablet	1
Allbee-T	1 tablet	68
ALternaGEL (suspension)	1 teaspoon	2
Aludrox (suspension)	1 teaspoon	1
Aludrox (tablet)	1 tablet	2
Aluminum Hydroxide Gel USP	1 teaspoon	8
Amitone	1 tablet	2
Amphojel (suspension)	1 teaspoon	7
Amphojel (tablet)	1 tablet	1
A.M.T. (suspension)	1 teaspoon	5
A.M.T. (tablet)	1 tablet	4

PRODUCT	DOSAGE	SODIUM (MG.)
Arthralgen	1 tablet	1
Banacid	1 tablet	0
Basaljel (capsule)	1 capsule	3
Basaljel (suspension)	1 teaspoon	2
Basaljel (tablet)	1 tablet	2
Basaljel Extra Strength (suspension)	1 teaspoon	17
Bell-Ans.	1 tablet	142
BiSoDol (powder)	1 teaspoon	156
BiSoDol (tablet)	1 tablet	under 1
Brioschi (capsule)	1 capsule	770
Bromo-Seltzer (capsule)	1 capsule	761
Camalox (suspension)	1 teaspoon	3
Camalox (tablet)	1 tablet	3
Chooz (gum tablet)	1 tablet	3
Citrocarbonate (suspension)	1 teaspoon	700
Comfolax	1 capsule	10
Creamalin	1 tablet	41
Delcid (suspension)	1 teaspoon	15
Dialume	1 tablet	1
Dicarbosil	1 tablet	3
Di-gel (liquid)	1 teaspoon	9
Di-gel (tablet)	1 tablet	11
Dimacid	1 tablet	0
Dimacol (capsule)	1 capsule	1
Dimacol (liquid)	1 teaspoon	5
Dimetane Decongestant (elixir)	1 teaspoon	1
Dimetane Decongestant (tablet)	1 tablet	1
Dimetane Elixir	1 teaspoon	2
Dimetane Tablets, 4 mg.	1 tablet	1
Donnagel (suspension)	1 teaspoon	5
Doxidan	1 capsule	2
Doxinate (capsule)	1 capsule	12
Doxinate (solution)	1 teaspoon	14
Dulcolax	1 tablet	0
ENO Powder	1 teaspoon	780
Enzymet	1 tablet	0
Estomul-M (solution)	1 teaspoon	12
Estomul-M (tablet)	1 tablet	16
Ex-Lax (chocolate tablet)	1 tablet	1

PRODUCT	DOSAGE	SODIUM (MG.)
Ex-Lax (unflavored pill)	1 pill	1
Ferancee	1 tablet	25
Ferancee-HP	1 tablet	67
Fero-Grad-500	1 tablet	55
Ferro-Sequels	1 tablet	5
Festal	1 tablet	2
Fletcher's Castoria	1 teaspoon	3
Gaviscon	1 tablet	19
Gaviscon-2	1 tablet	39
Gelusil (suspension)	1 teaspoon	1
Gelusil (tablet)	1 tablet	2
Gelusil-II (suspension)	1 teaspoon	1
Gelusil-II (tablet)	1 tablet	3
Gelusil M (suspension)	1 teaspoon	1
Gelusil M (tablet)	1 tablet	3
Gustalac	1 tablet	1
Haley's M-O	1 teaspoon	0
Kessadrox	1 teaspoon	9
Kolantyl (gel)	1 teaspoon	under 5
Kolantyl (tablet)	1 tablet	15
Kudrox	1 teaspoon	15
Liquid Antacid (McKesson)	1 teaspoon	9
Maalox (suspension)	1 teaspoon	3
Maalox #1 (tablet)	1 tablet	1
Maalox #2 (tablet)	1 tablet	2
Maalox Plus (suspension)	1 teaspoon	3
Maalox Plus (tablet)	1 tablet	1
Maalox TC	1 teaspoon	1
Magnesia and Alumina Oral Suspension	1 teaspoon	8
Marblen (suspension)	1 teaspoon	3
Metamucil	1 teaspoon	10
Metamucil	1 packet	1
Milk of Magnesia (concentrated)	1 teaspoon	10
Milk of Magnesia USP	3 teaspoons	1
Mitrolan	1 tablet	1
Modane (liquid)	1 teaspoon	0
Modane (tablet)	1 tablet	0
Mylanta (suspension)	3 teaspoons	12

PRODUCT	DOSAGE	SODIUM (MG.)
Mylanta (tablet)	1 tablet	1
Mylanta-II (suspension)	1 teaspoon	4
Mylanta-II (tablet)	1 tablet	2
Nephrox	1 teaspoon	3
Niferex (capsule)	1 capsule	5
Niferex (elixir)	1 teaspoon	15
Niferex (tablet)	1 tablet	2
Niferex with Vitamin C	1 tablet	21
Novafed Liquid	1 teaspoon	14
Novafed A Liquid	1 teaspoon	13
Novahistine Cough Formula	1 teaspoon	14
Novahistine Cough/Cold	1 teaspoon	12
Novahistine Elixir	1 teaspoon	5
Peri-Colace (capsule)	1 capsule	5
Peri-Colace (syrup)	1 teaspoon	26
Pertussin Cough Syrup	1 teaspoon	10
Phenaphen	1 capsule	1
Philip's Milk of Magnesia (liquid)	1 teaspoon	1
Philip's Milk of Magnesia (tablet)	1 tablet	3
Phospho-Soda	4 teaspoons	2217
Ratio	1 tablet	1
Riopan (suspension)	1 teaspoon	1
Riopan (tablet)	1 tablet	1
Riopan Plus (suspension)	1 teaspoon	1
Riopan Plus (tablet)	1 tablet	1
Robalate	1 tablet	1
Robitussin	1 teaspoon	3
Robitussin-CF	1 teaspoon	1
Robitussin-DM (liquid)	1 teaspoon	6
Robitussin-DM Cough Calmers Lozenges	1 lozenge	1
Robitussin-PE	1 teaspoon	5
Rolaids (tablet)	1 tablet	53
Senokot-S	1 tablet	5
Siblin	1 teaspoon	130
Silain (tablet)	1 tablet	1
Silain-Gel	1 teaspoon	4
Simeco	1 teaspoon	10

PRODUCT	DOSAGE	SODIUM (MG.)
Soda Mint	1 tablet	90
Sodium Phosphate and Biphosphate Oral Solution USP	6 teaspoons	3340
Stimulax	1 capsule	12
Syntrogel	1 tablet	8
Tempo	1 drop	2
Titralac (suspension)	1 teaspoon	11
Titralac (tablet)	1 tablet	under 1
Tralmag	1 teaspoon	0
Trisogel (suspension)	1 teaspoon	1
Trisogel (tablet)	1 capsule	1
Tums (tablet)	1 tablet	3
2G	1 teaspoon	14
2G-DM	1 teaspoon	13
WinGel (suspension)	1 teaspoon	3
WinGel (tablet)	1 tablet	3
Z-Bec	1 tablet	1

CHAPTER FIVE

The Politics of Sodium

The television lighting accentuated the boyish good looks of Rep. Albert Gore, Jr., Democrat from Tennessee, as he leaned into the microphone before him. There, from the chair at the center of the raised mahogany dais of the Science and Technology Committee hearing room, he surveyed the invited representatives of medicine, government, industry, and the public who had come to examine government policies concerning sodium and hypertension. It was April 13, 1981.

They were there, Gore reminded his guests, "to probe the largest single contributor to the number one cause of death, heart disease, and...a significant cause of stroke."

Nearly all the witnesses agreed that the salt-rich American diet leads to hypertension. High blood pressure experts from the American Medical Association, the National Heart, Lung, and Blood Institute, and several major universities testified that labeling the sodium content of processed foods was an important first step in enabling consumers to avoid excess sodium.

Gore's hearing had a promising outcome: in June 1981, he introduced a bill calling on the U.S. Food and Drug Administration (FDA) to require companies to label the sodium content of processed foods. For the next six months, over 100 representatives signed on as co-sponsors of the proposed legislation.

In September, Rep. Henry Waxman, a Democrat from California and chairman of the Subcommittee on Health and the Environment, held hearings on the bill, a necessary hurdle before it could be considered by the full House of Representatives. To Rep. Gore's delight, the American College of Cardiology, the American Association of Retired Persons, the American College of Preventive Medicine, and the American Heart Association added their enthusiastic support to the medical and public interest groups already pushing the measure.

But there was one ominous sign. Food industry trade associations, such as the Grocery Manufacturers Association, the National Food Processors Association, and others who had refused to testify at Gore's hearings, were still not supporting the bill. This troubled the congressional staffers charged with shepherding the proposed legislation into law. "Industry opposition in this Congress [the 97th] is the kiss of death," lamented one. Still, with such unanimous support from the medical community, the bill's backers thought they had convinced a majority of the subcommittee members to endorse the sodium-labeling initiative.

The Food Industry Steps In

They were wrong. On February 23, 1982, the day scheduled for the subcommittee vote, Reps. Waxman and Gore struck the bill from the agenda. It was an oblique admission that the proposed legislation had lost support and would not be approved by the subcommittee.

What had happened? The two congressmen later learned that the food industry had persuaded at least one and possibly two colleagues to change their votes.

Rep. James Florio, a liberal Democrat from southern New Jersey, had signaled his support for sodium labeling as recently as three weeks before the subcommittee session. But, not twenty-four hours before the vote, this generally progressive politician attended a breakfast meeting with forty food producers from his district, organized by the Grocery Manufacturers of America. Chief among the breakfast guests was the Campbell Soup Company, one of the largest employers in Camden, South Jersey's largest city. In addition to preparing its well-known soups, Campbell owns Mrs. Paul's Kitchens, Swanson, Franco-American, and Pepperidge Farm, all makers of sodium-laden processed foods.

According to *The Washington Post*, the topic of the day was sodium, and by the time Florio returned to Washington later that morning, he had withdrawn his support of the labeling initiative.

Rep. Thomas Luken, Democrat from Ohio, also made an eleventh-hour about-face. We do not know whether the food industry influenced Luken, but suffice it to say that Luken's district includes the headquarters of Procter & Gamble, makers of Duncan Hines cake mixes, Pringles potato chips, Jif peanut butter, and other foods.

Florio and Luken are not completely to blame for killing the labeling measure. The American Medical Association, one of the most powerful lobbies in Washington, also changed horses in midstream. The AMA had testified at the Gore hearing in support of mandatory labeling and promised to lobby for the bill when it was introduced. But by November 1981, the AMA was nowhere to be seen. "The food industry reminded the doctors that the AMA didn't like being regulated any more than it did," commented one congressional aide.

Another aide added that the physicians' group thought forceful support for the labeling bill might harm its relationship with the administration. "The AMA simply wasn't willing to burn up any political capital on sodium labeling."

This scenario is not unusual. When public-spirited legislation steps on corporate toes, industry lobbyists move into top gear to block the bill.

CSPI States the Case

It was our public health advocacy group, the Center for Science in the Public Interest (CSPI), that first petitioned the FDA to require food companies to both label and limit the sodium content of processed foods. Back in 1978, we argued that millions of Americans with high blood pressure had to follow doctors' orders to eat less sodium. Millions more just wanted to avoid unnecessary salt, as health experts had advised. It should be obvious, we said, that there is no way of learning the sodium content of processed foods without labeling.

Our group pushed the case even further. We said labeling alone would not solve the problem. Millions of Americans do not read food labels—they're in a rush or "can't be bothered"—a teenager with above-normal blood pressure for his age is unlikely to scrutinize the tiny print on food packages. Others cannot read the labels—individuals with language problems or poor eyesight, for instance. And still others, forced to exist on convenience foods—the traveling salesman, the trucker on the road—have little choice in their diet. The only way to lower the risk for these people would be to limit the amount of sodium that manufacturers add to processed foods. The law already forbids food companies from using dangerously high levels of other harmful food additives, we said. Salt does not deserve special treatment. We were not talking about putting the entire nation on a bland, salt-free diet. The limits we proposed would not ban salt, but reduce the amount of salt that companies could use in certain foods.

At the 1981 congressional hearing, Rep. Robert Walker, Republican from Pennsylvania, voiced opposition to CSPI's proposal. "I'm a little concerned about a process that has a Big Brother approach," he said. "Why should we tell the companies precisely how much [sodium] to put in foods, then tell consumers to salt to taste?"

Dr. Michael Jacobson, the executive director of CSPI, replied, "I don't like Big Brother any more than you do. But I find when I want to eat processed foods that the salt content is dictated by another Big Brother—or Big Aunt—Betty Crocker."

In other words, the public needs more control over its food supply, and only the federal government can provide it. CSPI petitioned the FDA because that agency regulates 80 percent of all processed foods. We could have laid the same proposals before the U.S. Department of Agriculture (USDA) since it regulates meat, poultry, and egg products, the remaining 20 percent of processed foods.

A Thumbnail Account

The FDA did not act quickly on sodium. Here is a review of the government's response to the sodium-hypertension connection:

July 1978: CSPI petitions the FDA to label and limit sodium in processed foods. Those calling for sodium labeling on all processed foods include 24 senators, 50 representatives, 22 health organizations, and numerous hypertension experts.

January 1979: The FDA says any change in labeling must await action on the agency's overhaul of the entire food label. Any move to limit sodium must follow a report on salt by the FDA advisory panel, the Select Committee on GRAS (Generally Recognized as Safe) Substances (SCOGS).

July 1979: SCOGS concludes that salt is not "generally recognized as safe," and recommends that the FDA develop "guidelines for restricting the amount of salt in processed foods."

August 1979: The Surgeon General of the United States issues a report, *Healthy People*, that states, "Americans would be healthier, as a whole, if they consumed . . . less salt."

February 1980: The National Academy of Sciences publishes the 1980 edition of *Recommended Dietary Allowances*, advising the public to reduce sodium intake to between 1,100 milligrams and 3,300 milligrams per day.

February 1980: The U.S. Department of Health and Human Services and the U.S. Department of Agriculture jointly release *Dietary Guidelines for Healthy Americans*. It advises people to "avoid excess sodium."

February 1981: CSPI submits to the FDA petitions from 5,769 health professionals, including a number of noted hypertension experts, calling on the agency to limit and label the sodium content of processed foods.

April 1981: The House Subcommittee on Investigations and Oversight, chaired by Rep. Albert Gore, holds hearings to explore the problem of dietary sodium and high blood pressure.

September 1981: The House Subcommittee on Health and the Environment, chaired by Rep. Henry Waxman, holds hearings on Rep. Gore's bill requiring sodium labeling on all foods regulated by the FDA.

February 1982: Reps. Waxman and Gore withdraw the sodium-labeling bill from the agenda of the House Subcommittee on Health and the Environment, fearing it would not pass the subcommittee.

August 1982: The FDA finally responds to CSPI's original 1978 petitions. It formally denies them, rejecting our request for the mandatory labeling and limiting of the sodium content of processed foods. But all the effort was not in vain.

Some Progress

Dr. Arthur Hull Hayes, Jr., the Reagan appointee who headed the FDA between April 1981 and September 1983, happens to be a hypertension expert. For the nine years prior to his move to Washington, he was director of the hypertension clinic at the Hershey Medical Center in Hershey, Pennsylvania. This experience sensitized Hayes to the undeniable link between sodium and hypertension. "Those of us who have treated hypertension," explains the FDA head, "understand how effective low-sodium diets can be."

Early in his tenure, Hayes made sodium his major public health challenge and, considering the anti-regulatory constraints of the Reagan Administration, was reasonably effective. At a March 1982 conference sponsored by the American Medical Association, Hayes said, "It is clear . . . that we as a society and we as health professionals must address the sodium issue, and we must do it now. . . . Sodium reduction must remain a general health goal for our nation and indeed for all developed countries."

The FDA's objectives, said Hayes, are twofold: educate the public about the sodium-hypertension connection, and lower sodium consumption in the U.S.

To implement these goals, he has met with food companies and asked them to label and lower the sodium content of their products. He has appeared in a television public service announcement encouraging consumers to read sodium labeling. And he has established a task force on sodium education that coordinates the publication of pamphlets and articles by the FDA, USDA, National Institutes of Health, industry, and other groups.

But the core of the FDA's anti-sodium campaign is its labeling plan, officially proposed in June 1982. According to the plan, the FDA would require sodium labeling on foods that already have nutrition labeling, which include roughly half of processed foods. Those products already listing calories, protein, carbohydrate, fat, and at least seven vitamins and minerals designated by the FDA would also have to include sodium content. (Companies that fortify or make a nutrition claim for their product, such as "High in Vitamin C," must use a nutrition label. But in most cases, the company chooses to list nutrients on its labels voluntarily.)

Other foods, those without nutrition labeling, would carry sodium labeling only if manufacturers chose to do so. The FDA plan is headed in the right direction, but it limits the agency's role to pleading with these companies to label sodium. The final decision on whether to label sodium for at least half the food supply would remain with the manufacturers. In all likelihood, these foods will remain unlabeled. The FDA admits as much. "The agency believes that [these] sodium labeling proposals...will provide the consumer with information on the sodium content of about half of all processed food," declares the FDA's June 1982 document.

The Allies

CSPI and a host of other organizations say half is not good enough. We have been joined by over 100 members of Congress and the following groups in calling for the sodium labeling of all processed foods:

American Heart Association
American Public Health Association
American College of Preventive Medicine
American Association of Retired Persons
Citizens for the Treatment of High Blood Pressure
National Center on Black Aging
National Consumers League
National Council of Senior Citizens
National Kidney Foundation
National Urban Coalition
Consumer Federation of America
American College of Cardiology
National Institutes of Health, Hypertension Task Force

What You Can Do

You can help decrease the hypertension rate in this country by taking direct action.

● *Ask your congressional representatives and senators to support mandatory sodium labeling. Write: (Name of sena-*

tor), Washington, D.C. 20510, or (Name of representative), Washington, D.C. 20515.

● *Urge the FDA to require sodium labeling on all foods it regulates. Write: Commissioner, Food and Drug Administration, 5600 Fishers Lane, Rockville, Maryland 20857.*

● *Urge the USDA to require sodium labeling on meat, egg, and poultry products. Write: Secretary of Agriculture, U.S. Department of Agriculture, 14th and Independence Avenues S.W., Washington, D.C. 20250.*

● *Buy no-salt-added or reduced-sodium products. If sales soar, more companies will jump on the low-sodium bandwagon.*

● *Ask your supermarket to order low-sodium items that you've seen advertised or sold elsewhere.*

● *Ask your supermarket manager to use colored posters or shelf tags to direct shoppers to low-sodium foods. Such signs will make low-sodium shopping easier for you and your neighbors.*

● *Ask restaurant owners or waiters for low-sodium foods. If enough people ask, the restaurant should get the hint.*

● *When you see grocery products without sodium labeling, write to the manufacturer for sodium information. (The address is on the label.) You might also urge the company to list the sodium content of the food on the label.*

● *When a favorite food is high in sodium, write and ask the manufacturer to reduce the sodium level.*

To push the FDA beyond its half-hearted response, CSPI sued the agency in March 1983 to require sodium labeling on all processed foods. The lawsuit also charged that the FDA had acted illegally when it decided in June 1982 to postpone indefinitely a response to the 1979 report of the FDA's own advisory panel,

SCOGS. That report said salt was not "generally recognized as safe" and should be limited in processed foods.

In June 1984, Judge June Greene handed down her decision. FDA, she ruled, has the discretion to embark on a voluntary labeling scheme before considering mandatory measures. However, Judge Greene did conclude that FDA could not defer *indefinitely* its decision on the safety of salt, leaving CSPI the opportunity to file a future lawsuit should FDA continue to procrastinate.

Meanwhile, in April 1984, FDA announced that its voluntary proposal would finally become a regulation on July 1, 1985. By that date, FDA projected, roughly half of all foods would carry sodium labeling.

However, in July 1985, the agency waffled, deciding to postpone the deadline one full year. Delaying the deadline, said FDA, was easier than responding to the numerous requests it had received from companies wanting extensions.

Although the FDA has been willing to go halfway, the USDA, in control of meat, poultry, and egg products, is willing to go nowhere. In May 1982, the Agriculture Department announced that it would not require sodium labeling even on meat and poultry products like hot dogs and frozen dinners that already carry nutrition labels.

By late 1984, the results of USDA's toothless non-policy had become evident. While 40 percent of foods regulated by FDA had sodium labeling, the percentage was only 18 for USDA's foods. Still, the department refused to change its rules. The public's health, one can only conclude, is less important to USDA than the food industry's desire to withhold information.

Labeling aside, neither USDA nor FDA has accomplished much in a second, crucial area—lowering the sodium content of foods. True, some food processors such as Libby, Hunt-Wesson, and Del Monte have created new, no-salt-added versions of their products. But few have reduced the sodium in their mainstream foods eaten by the majority of Americans.

In early 1985, CSPI released its 1984 Sodium Index, a system set up to track changes in the salt content of over 1,700 commonly eaten foods. Comparing 1984 with earlier years, the findings were grim.

Roughly two-thirds of the products hadn't changed at all. Of the remainder, almost as many sodium levels went up as down. While Pepperidge Farm cut the sodium in their English muffins,

Pillsbury hiked the sodium in its Figurines. While Kraft lowered the sodium in its processed cheeses, Green Giant started adding more to its pouch frozen entrees.

Food manufacturers may argue against sodium labeling and reductions on ideological or cost grounds, but the bottom line is this: millions of people are dying prematurely due to hypertension-related diseases.

Nearly seven years after CSPI first petitioned FDA to label and limit sodium in processed foods, this is how the situation stands:

● Health experts have expressed nearly unanimous support for sodium labeling on all processed foods.

● The FDA responded with a modest educational effort and a labeling plan that should ensure sodium labeling on roughly half of all processed foods.

● The USDA has published a brief pamphlet on sodium, but decided not to require sodium declarations on foods containing meat or poultry.

● While a few companies are adding less or no salt to their foods, an almost equal number raised the sodium content, and the majority of corporate chefs have not changed their cooking habits.

● Sodium labeling has gained fresh support in Congress. In early 1985, Sen. Howard Metzenbaum and Rep. Neal Smith each introduced legislation that would require sodium labeling on all foods.

The nation's hypertension epidemic should provoke stronger governmental action than has yet been proposed. Many foods will remain unlabeled, and even if they were labeled, too many people will not take the time to learn the sodium content of the foods they eat. But you, as a concerned consumer, do not have to wait for the federal turtle to get tough with food manufacturers. The brand name listing of over 5,000 foods begins on page 143. Use it in conjunction with existing labeling and with the tips for cooking with less salt outlined in the next chapter to help protect your family.

CHAPTER SIX

Tips on Cutting Back Salt

We've told you about the brain damage and premature deaths linked to hypertension, and we've said that salt aggravates high blood pressure. We've exposed hidden sodium in the food supply, and we've warned that the government will not protect you by setting limits on the amount of sodium in processed foods. By now you've guessed that your only defense is to take matters into your own hands. Still, you resist taking the initiative to change your diet. You're just like everyone else who has ever wondered, "Will I miss my salt?"

The answer is "probably not." Well-known food writers such as Craig Claiborne of *The New York Times* confirm the experiences of others who have taken the low-sodium road. Using less salt allows subtle new flavors to emerge—flavors once lost by dousing food with the white stuff. Cutting the salt can lead to new culinary treats. In fact, it may be one of the easiest of your eating habits to change.

While most assertions about taste fall outside the realm of science, some preliminary studies conducted by the Monell Chemical Senses Center, a leading research laboratory in this field, tell us that people's desire for salt falls off as they use less of it. Dr. Gary Beauchamp recruited nine college students for one of his taste tests at Monell. He ordered the students to avoid high-sodium foods for five months. On the average, the students cut their sodium intake by 40 percent. All the while, Beauchamp tested his student volunteers. For example, he often prepared nine bowls of soup salted to different degrees. He would ask the students to rate the soups in order of their saltiness and to tell him which of the nine they found most pleasing.

Beauchamp's results were dramatic: after three months, the

students said soups that once tasted normal now seemed salty. Moreover, they preferred less and less salty versions of the soup as time went on.

Beauchamp's study and the experiences of Claiborne and thousands of others suggest that you can overcome a craving for salt over the long haul. This chapter should help start you thinking strategically about cutting back on salt, recognizing that well-entrenched habits cannot be changed overnight. Take it gradually and victory will be yours.

For total success, you'll have to control sodium from these three sources: processed foods, the salt shaker, and restaurant foods. Processed foods eaten at home or in restaurants probably account for 75 percent of your sodium intake. The salt shaker may contribute another 15 percent. (The rest of the sodium occurs naturally in foods.) Let's map out the difficulties and discuss how to overcome the challenges posed by each.

Processed Foods

To get a handle on salt in processed foods, you have to know how much sodium each food contains. The brand name sodium listing on page 143 is an indispensable tool. You will also find a sodium declaration on many food labels. Once you've zeroed in on the sodium content of the food, determine into which of these categories the food falls.

1. *If the product contains less than 135 milligrams of sodium per serving, consider it "safe."* Most fruit, for example, contains less than 10 milligrams per serving. Feel free to consume any canned, dried, or frozen fruit without fear of overloading on sodium. Other processed foods in the very low sodium category include pasta (without the sauce), ice cream, soft drinks, alcoholic beverages, and frozen vegetables (sauceless).*

2. *If the product contains more than 135 milligrams per serving, check the grocery shelves for a "no-salt-added" version.* Many Libby, Del Monte, Hunt, and store-brand canned vegeta-

---

*You should, of course, consider the nutritional aspects of these foods as well. A low-salt food may be loaded with refined sugar, fat, or alcohol.

bles now come without added salt. As Chart 6A (page 71) shows, the list of other no-salt-added foods is growing.

Curiously, companies sometimes charge more for a product that actually contains less of an ingredient. That is the case with many no-salt-added foods. In a few cases, manufacturers may have substituted more expensive items for the missing salt. More likely, it costs more per individual unit to produce smaller quantities of a less popular food. You might write a letter to the company complaining about the price but we still suggest buying the more expensive product. It may seem costly, but it's far less expensive than suffering a heart attack or stroke.

Aside from the price, you may have trouble adjusting to the taste of some no-salt-added products. Salt-free club soda, canned vegetables, tomato sauce, peanut butter, and potato chips taste fine. But other foods may taste flat without salt. Salt-free cottage cheese struck us as bland. Try adding spices to give the food better flavor. This remedy works best for foods such as canned beef stew, canned soups, hot cereal, canned vegetables, and spaghetti sauce. (See Chart 6B, page 79.)

3. *If a no-salt-added version is not available, check for competing brands with lower sodium values.* Use the brand name guide to look for the compatible product with the least sodium. For example, Star-Kist makes a tuna packed in water with about half the sodium of most other brands. Weight Watchers low-fat cottage cheese has roughly half the sodium content of most other brands.

4. *If you can't switch to a lower-sodium competitor, consider substituting another food.* Often, high-sodium foods can be replaced with similar foods containing far less.

5. *If none of these options work, consider making your own dish.* Spaghetti sauce, mashed potatoes, vegetables in sauce, pancakes, macaroni and cheese, and soups take much less time to prepare than most people think. Make your own, and salt according to your taste. Your final product will almost certainly be cheaper, fresher, and tastier.

6. *If you don't want to make the food yourself, and dread the thought of giving up a favorite food, make allowances for this extra sodium by cutting back elsewhere in your diet.* Unless you

CHART 6A

Low- or Reduced-Sodium Foods

BAKING POWDER
Featherweight Low Sodium
  Baking Powder

BEVERAGES
Campbell's
Low Sodium V-8 Vegetable
  Cocktail Juice

Canada Dry
Unsalted Club Soda

Featherweight
Tomato Juice

Hunt's
No Salt Added Tomato Juice

Schweppes
Unsalted Club Soda

BREAD
Mrs. Wright's (Safeway)
Low Sodium Raisin Bread
Low Sodium Wheat Bread
Low Sodium White Bread

Stop & Shop
100% Whole Wheat, No Salt
  Added Bread
Cracked Wheat, No Salt
  Added Bread
Daisy Loaf White, No Salt
  Added Bread
Oatmeal, No Salt Added
  Bread
Premium White, No Salt
  Added Bread

Wonder
Low Sodium White Bread

**BUTTER AND
MARGARINE**
Fleischmann's
Unsalted Parve Margarine
Unsalted Stick Margarine

Land O Lakes
Lite Salt Sweet Cream
  Whipped Butter
Unsalted Sweet Cream Butter
Unsalted Sweet Cream
  Whipped Butter

Mazola
Unsalted Margarine

**CAKES, COOKIES, AND
PASTRY**
Featherweight*
Cake & Cookie Mix
Cookies (8 varieties)
Low Sodium Cake

Health Valley*
Cookies (12 varieties)

Stella D'Oro
Cookies (5 varieties)
Pastry (4 varieties)

---

*Usually only available in health
food stores

CANNED ENTREES
Featherweight*
Beef Ravioli
Beef Stew
Chicken Stew
Chili with Beans
Spaghetti with Meat Balls
Spanish Rice
Stuffed Dumplings with
    Chicken

Health Valley*
Chili

CEREAL
Featherweight*
Corn Flakes
Crisp Rice

Health Valley*
Granolas (6 varieties)

Kellogg
No Salt Added Corn Flakes
No Salt Added Rice Krispies

CHEESE
Breakstone
Dry Curd Cottage Cheese with
    Added Skim Milk

Dorman's
No Salt Added Muenster
    Cheese
No Salt Added Swiss Type
    Cheese

Featherweight*
Cheddar Cheese
Colby Cheese

Health Valley*
Cheddar Cheese

Jack Cheese
Muenster Cheese

Heluva Good Cheese
Unsalted Cheddar Cheese

Lite-line
Sodium Lite Pasteurized
    Process Cheese Product**
Reduced Sodium Pasteurized
    Process Cheese Product**

Pauly
Low Sodium Colby Cheese

Sargento
50% Reduced Sodium
    Cheddar Cheese
75% Reduced Sodium Swiss
Pot Cheese

Tillamook
Low Sodium Cheddar Cheese

Tuttle
Dry Curd Cottage Cheese

CONDIMENTS
Del Monte
No Salt Added Catsup

Featherweight*
Catsup
Chili Sauce
Mayonnaise
Mustard
Pickles

Hunt's
Tomato Ketchup

Kikkoman
Lite Soy Sauce**

**Reduced- but not low-sodium foods

CRACKERS AND BREAD PRODUCTS

Devonsheer
Unsalted Melba Rounds
Unsalted Seasoned Croutons

Featherweight
Melba Toast
Unsalted Crackers
Whole Rice Crackers

Health Valley*
Herb Crackers, Unsalted
Sesame Crackers, Unsalted
Stoned Wheat Crackers,
    Unsalted

Ideal
No Salt Whole Grain
    Flatbread

Manischewitz
(5 matzo varieties)
Wheat Crackers

Old London
Unsalted White Melba Toast

Stella D'Oro
Diet Breadsticks

Venus
Bite-Size Wheat Wafers, No
    Salt Added
Bran Wafers, No Salt Added
Whole Wheat Wafers, No Salt
    Added

FISH

Chicken of the Sea
Dietetic, Albacore Chunk
    White Tuna in Water

Featherweight*
Salmon
Sardines (3 varieties)
Tuna

Health Valley*
Dietetic Tuna
Pink Salmon

Starkist
60% Less Salt Chunk White
    Tuna in Spring Water
Diet Pack Chunk Light Tuna
    in Water

NUTS AND PEANUT BUTTER

Fisher
Raw Almonds
Raw Black Walnuts
Raw English Walnuts
Raw Pecans
Unsalted Peanuts

Health Valley*
Chunky Peanut Butter,
    Unsalted
Creamy Peanut Butter,
    Unsalted

Koeze
Old Fashioned, No Salt Added
    Peanut Butter

Laura Scudder
Goobers, Roasted in Shell
Pistachios, Roasted in Shell
Sunflower Nuts, Roasted in
    Shell

Peter Pan
Low Sodium Peanut Butter

Planters
Unsalted Cashews
Unsalted Mixed Nuts
Unsalted Peanuts
Unsalted Sunflower Seeds

PANCAKE MIX
Featherweight*
Pancake Mix

Soy-o*
Low Sodium Pancake Mix

PROCESSED MEATS
Featherweight*
Cooked Ham
Corned Beef Loaf
Pork Franks

Olde Smithfield**
Bacon
Bologna (2 varieties)
Franks (2 varieties)
Ham (4 varieties)
Luncheon Meat
Salami
Sausage (2 varieties)

SALAD DRESSINGS
Aristocrat
French
Onion and Garlic

Featherweight*
2 Calorie Low Sodium
  Dressing
Imitation French Dressing

Health Valley*
Unsalted Avocado Dressing
Unsalted Herb Dressing
Unsalted Thousand Island
  Dressing

SNACKS
Charles Chips
Potato Chips

Featherweight*
Unsalted Popcorn
Unsalted Pretzels

Granny Goose*
100% Natural Unsalted Potato
  Chips

Health Valley*
Unsalted Corn Chips
Unsalted Potato Chips
Unsalted Sesame Whole
  Wheat Bavarian Pretzel

Wise
No Salt Added Potato Chips

SOUPS AND CHILI
Campbell's
Canned Low Sodium Soups
  (9 varieties)

Featherweight*
Canned Soups (4 varieties)
Dry Soups (4 varieties)

Health Valley*
Canned, Unsalted Soups (11
  varieties)
No Salt, No Meat, Mild Chili

Herb-Ox
Low Sodium Beef Broth &
  Seasoning
Low Sodium Chicken Broth &
  Seasoning

---

*Usually only available in health
food stores
**Reduced- but not low-sodium foods

Low Sodium Vegetable Broth
    & Seasoning

Lite-Line
Low Sodium Instant Bouillon
    Beef Flavor
Low Sodium Instant Bouillon
    Chicken Flavor

TOMATO PRODUCTS
Contadina
No Salt Added Tomato Paste

Del Monte
No Salt Added Stewed
    Tomatoes
No Salt Added Tomato Sauce

Featherweight*
Spaghetti Sauce
Tomato Juice
Tomato Sauce

Health Valley*
No Salt Added Bellissimo
    Sauce
No Salt Natural Tomato Sauce

Hunt's
No Salt Added Whole Peeled
    Tomatoes
No Salt Added Spaghetti
    Sauce
No Salt Added Stewed
    Tomatoes
No Salt Added Tomato Paste
No Salt Added Tomato Sauce

VEGETABLES, CANNED
Del Monte
(8 varieties)

Diet Delight*
(6 varieties)

Featherweight*
(7 varieties)

Libby's
(19 varieties)

Stop & Shop
No Salt Added Vegetables
    (6 varieties)

---

must follow a strict low-sodium regimen, you can probably
afford to eat a few high-sodium items on occasion. If the thought
of life without salted cottage cheese, Jewish rye bread, or frozen
pizza is too much to bear, avoid other high-sodium foods.

Salt from the Salt Shaker

The salt shaker is the one source of sodium over which you have
complete control. The only trick to cutting back is this: do it
gradually. Unless you are under strict doctor's orders, there is no
reason to go cold turkey. People adjust to the taste of low-sodium
foods over a period of time. Give yourself from two to three
months to retune your tastebuds for low-sodium dining.

At the dinner table we suggest you follow these steps adapted from the Stanford Heart Disease Prevention Program to wean your family off the salt shaker.

Stage One: Taste food before salting it. Some people are so accustomed to the shaker habit, they don't realize how many foods taste delicious without additional salt.

Stage Two: When you do reach for salt, use as little as possible. One shake is often enough.

Stage Three: Put an herb shaker on the table. Fill your salt shaker with a mixture of herbs from the recipe below, or choose a few herbs that mix well with the meal of the evening. Use Chart 6B (page 79) for suggestions for seasoning food without salt. We recommend this herbal substitute:

HERBAL SEASONING

1   teaspoon celery seed, ground
2½ teaspoons marjoram, crushed
2½ teaspoons summer savory, crushed
1½ teaspoons thyme, crushed
1½ teaspoons dried basil, crushed

Stage Four: Remove the salt shaker from the table. Most hungry eaters won't bother walking back into the kitchen to get it.

Stage Five: Don't make a fuss if someone in your family continues to add salt. Children, especially, react poorly to heavy pressure. Some will continue to add salt simply to be rebellious. Others may take longer to adjust to new flavors. Be patient. It won't help if kids start to consider salt a forbidden treat. The best way to convert recalcitrant children is to let them know you are enjoying the lower sodium foods.

Salt in Cooking

As a nation, Americans are not very adventurous when it comes to seasoning food. Ask a friend which spices he or she uses to season food and you are likely to hear "salt and pepper." Most traditional cookbooks give similar instructions. You would think salt and pepper were the only seasonings available.

On the other hand, most "average Americans" are delighted to discover that herbs and spices add dozens of new, distinctive flavors to our vocabulary of good tastes. Salted baked potatoes are the standard, but potatoes seasoned with curry, paprika, nutmeg, garlic and onions, dill, or parsley are terrific. Use Chart 6B, page 79, for mixing and matching foods with suitable spices and herbs.

Another pitfall comes in following too literally an old family recipe or a traditional cookbook. (We've compiled a tasty collection of low-sodium recipes; see page 90. A list of low-sodium cookbooks appears on page 429.) Here are general tips for modifying any old-fashioned, salt-laden recipe:

- When making casseroles, baked goods, or other dishes to which salt is ordinarily added before cooking, try halving the amount. If the dish tastes fine, halve the amount again the next time you make the recipe. Keep reducing the salt until you reach the point where you think it really needs the seasoning. Often recipes call for lots of salt when very little or none is really needed.
- When cooking soups, meats, stews, sauces, vegetables, or other foods where salt can be added at the last minute, hold the salt. Add other spices and taste the dish. Does it really need salt? If so, add a pinch. Keep adding pinches until you are satisfied with the taste.
- Add more of the other spices in a recipe to "cover" for the salt you omit. Often, extra onions, oregano, mushrooms, paprika, and many other spices add a pleasant zip to tired dishes. On the other hand, be careful with the amount of additional garlic, cayenne, vinegar, lemon juice, cloves, and chili powder you include. Too much of some of these seasonings can zap your food right into the garbage can. And a word of warning about curries: these recipes call for exact amounts of spices. If you increase one spice, increase them all proportionately.
- Use fresh ingredients to get flavors so robust that you won't notice the "missing" salt. Fresh garlic is better than garlic powder; fresh-squeezed lemons are better than bottled lemon juice; fresh tomatoes or mushrooms win hands down over canned varieties. Consider growing your own herbs and spices in a window pot or garden. Once you've tasted fresh-dried

Shaker Hints

The easiest way to cut down on the salt your family adds at the table may be to buy a salt shaker with smaller holes—so says a recent study at the University of New South Wales in Australia. Three researchers measured the amount of salt used by over 1,600 people dining in a department store cafeteria, an airline complex, and in four restaurants. Sure enough, the researchers found that the larger the holes in single-holed salt shakers, the more salt people consumed.

The researchers also wondered whether proximity to the salt shaker might affect people's salt intake. They compared the amount of salt used by 377 employees at five different canteens at a large public utility. The first canteen placed salt shakers on the tables; three others kept them at the end of the food line; and a fifth placed not salt shakers but salt packets at the end of the line.

The results: people forced to quickly use the shaker while standing at the end of the line averaged only 530 milligrams per person. Those who could add salt at their leisure while sitting at their tables used 730 milligrams each. And people given the opportunity to grab a packet and bring it back to their tables ended up using the entire contents, raising their total to between 1,000 and 1,100 milligrams.

The two studies offer further proof that people use salt shakers as much out of habit as for taste. Indeed, the researchers found that less than 25 percent of salt users tasted food before salting it.

basil, mint, thyme, oregano, or bay leaves, you will never be satisfied with the store-bought variety.

● Besides salt, reduce the amount of other sodium-containing ingredients. Use low-sodium baking powder (less than 2 milligrams per teaspoon) rather than regular baking powder (339 milligrams per teaspoon) or baking soda (821 milligrams per teaspoon). Use 1½ teaspoons of Featherweight low-sodium powder for every 1 teaspoon of baking powder or soda the

CHART 6B

How to Add Spice to Your Life

FOOD	RECOMMENDED SPICES AND OTHER NATURAL FLAVORINGS
Beans	Cayenne, curry, cumin, chili powder, garlic, onions
Beverages	Cinnamon, nutmeg, ginger, mace, mint
Breads	Anise, caraway seeds, poppy seeds, sesame seeds, fennel seeds, raisins
Cakes, cookies, desserts, pastries	Anise, cardamom, cinnamon, cloves, nutmeg, ginger, allspice
Cottage cheese and cheese dips	Caraway seeds, chives, dill, cayenne, marjoram, coriander, thyme
Eggs	Basil, oregano, black pepper, chives, mustard, parsley, tarragon
Fruits, fruit pies	Allspice, basil, anise, cardamom, cinnamon, cloves, cumin, mint, ginger
Indian dishes	Cardamom, cayenne, coriander, cumin, curry, ginger, garlic, turmeric, mustard seeds
Italian dishes	Basil, oregano, garlic, thyme, bay leaf, onions, cayenne, wine
Mexican dishes	Cayenne, chili powder, oregano, cumin, garlic, coriander
Oriental dishes	Cumin, garlic, ginger, sesame seeds, sherry, cayenne, green pepper
Popcorn	Garlic powder, curry, onion powder
Rice	Saffron, cumin, curry, turmeric, green pepper, basil, oregano
Salads	Basil, chervil, dill, parsley, tarragon, vinegar, lemon juice

FOOD	RECOMMENDED SPICES AND OTHER NATURAL FLAVORINGS
Salad dressing	Caraway, chervil, chives, ginger, mustard, turmeric, pepper, celery seeds
Soups (creamed)	Peppercorns, bay leaf, dill, paprika, marjoram, tarragon
Soups (vegetable)	Thyme, dill, curry, garlic, onion, oregano, wine, bay leaf, basil

Meats, Chicken, and Fish

Beef	Cumin, ginger, thyme, sage, nutmeg, curry, coriander, mustard
Beef, ground	Onions, pepper, mustard, green pepper, oregano, nutmeg
Chicken	Paprika, oregano, sesame, curry, ginger, marjoram, sage, parsley
Fish	Lemon juice, wine, dill, basil, paprika, parsley, curry
Lamb	Curry, rosemary, mint, tarragon, pepper, savory, sage, cumin
Pork	Apple, cumin, rosemary, savory, fennel, onion, garlic
Tuna	Mustard powder, celery, green pepper, scallions
Veal	Ginger, sage, curry, mustard, mint, oregano, tarragon, dill

Vegetables

Artichokes	Savory, thyme, lemon juice, rosemary, vinegar, garlic
Asparagus	Garlic, lemon juice, vinegar, marjoram, savory, dill, pepper
Beets	Anise, caraway, fennel, ginger, savory, thyme
Broccoli	Tarragon, lemon juice, vinegar, black pepper, oregano, chervil
Brussels sprouts	Sage, lemon juice, caraway seeds, garlic, pepper

FOOD	RECOMMENDED SPICES AND OTHER NATURAL FLAVORINGS
Cabbage	Caraway seeds, curry, cumin, apple, anise, savory, fennel seeds
Carrots	Ginger, sage, thyme, mint, dill, parsley
Cauliflower	Nutmeg, curry, black pepper, tarragon, onion
Corn	Green pepper, fresh tomato, scallions, mushrooms, basil
Cucumbers	Dill, vinegar, garlic, chives, mint
Eggplant	Sage, garlic, oregano, marjoram, pepper, cumin
Green beans	Dill, lemon juice, marjoram, mushrooms, curry
Kale	Lemon juice, onion, pepper, vinegar, nutmeg
Parsnips	Mace, ginger, chervil, dill
Peas	Savory, sage, onion, mushrooms, pepper
Potatoes	Dill, parsley, paprika, black pepper, caraway, saffron, curry
Rutabaga	Mace, nutmeg, ginger, apples, cardamom
Summer squash, zucchini	Savory, paprika, dill, basil, oregano, pepper, thyme
Sweet potatoes, pumpkin, winter squash	Allspice, cardamom, cinnamon, cloves, fennel, ginger, apples
Tomatoes	Basil, marjoram, oregano, thyme, dill, savory
Turnips	Allspice, caraway, garlic, pepper, scallions

---

recipe calls for. Use soy sauce sparingly; it is far from salt-free.

● If you are using canned foods, you can actually rinse away the salt in some. Dieticians at Duke University Medical Center found that just one minute under the tap washed away 75 to 80 percent of the sodium in canned tuna fish. Forty percent of the sodium in canned green beans went down the drain after a one-minute rinse followed by heating in tap water rather than the can's liquid. Hold small amounts of food under the tap for different lengths of time, until you learn how much rinsing yields the right flavor.

● If you "Kosher" meat and poultry by soaking it in salt water before cooking, you're roughly doubling the sodium in beef and veal, and tripling the sodium in chicken. The solution: after soaking the meat in saltwater, immerse for one hour longer in tap water. That will remove the excess sodium from beef and veal. The chicken, unfortunately, will hold on to most of its newly acquired salt. That means a half chicken breast (3.5 ounces) will contain about 210 mg. not the usual 70 mg. of sodium.

Are Salt Substitutes Safe?

There are two kinds of salt substitutes: those consisting entirely of potassium chloride (NoSalt, CoSalt, and Morton's Salt Substitute), and those consisting of equal parts potassium and sodium chloride (Morton's Lite Salt).

Potassium chloride looks like regular salt, and to some people it tastes very much like salt. Others find that it has a bitter taste, though the bitterness is much less noticeable when the potassium chloride is mixed with sodium chloride.

Potassium chloride is safe for healthy people. The kidney excretes any the body can't use. According to Dr. Herschel Jick, of the Boston Collaborative Drug Surveillance Program, which monitors adverse effects of drugs throughout the Boston area, "We've kept track of hundreds of thousands of patients, and we have never seen anyone get into trouble because of salt substitutes."

Potassium can cause problems only if it builds up in the blood, says Jick, and that occurs under two conditions. The buildup can occur if the kidneys are impaired, or it can occur when the individual is taking the kind of diuretic that causes the kidneys to retain potassium.

Unfortunately, if potassium does build up in your blood, you can be in extreme danger. Too much (or too little) potassium in the blood can disrupt the heart's regular beat, leading to cardiac arrest. For this reason, the labels on some salt substitutes warn certain people to check with a doctor before switching to the product. The label of the highly advertised NoSalt reads:

For normal, healthy people. Persons having diabetes, heart or kidney disease, or persons receiving medical treatment should consult a physician before using a salt alternative or substitute.

Missing from this notice is any warning that people taking certain diuretics should avoid NoSalt. To make sure that labels on all salt substitutes warn against all possible dangers, the FDA should specify a precise notice that would be required on all labels.

Some doctors worry that people whose kidneys are functioning improperly might not know it. Dr. Allen Forbes, associate director for nutrition and food science at the FDA, disagrees. "Potassium intoxication can occur in people with extremely far advanced kidney or liver disease," Forbes says. "But by the time one reaches that stage, it is virtually certain that they are under medical care."

Dr. Herbert Langford, of the University of Mississippi, suggests that healthy people use salt substitutes to boost their potassium intake as long as they don't go overboard. "Up to one teaspoon a day may offer a modest benefit to millions," he says.

One teaspoon of NoSalt provides 2,730 milligrams of potassium. You can use the chart on page 36 to estimate the amount you would receive from other foods. Langford recommends a total of 4,000 milligrams of potassium a day.

Restaurant Food

Dining out on a low-sodium diet is not easy. If you go to restaurants only occasionally and are trying to prevent—rather than control—high blood pressure, you may want to throw caution to the wind. Occasional indulgences will not cancel the benefit of an otherwise sensible dietary routine. Nevertheless, you can take steps to minimize the salt you eat in restaurants. Some of the following tips are simple alternatives; others may require more sacrifice. Choose the solutions that best meet your needs.

- Stay with fruit or fruit juice for an appetizer. Or ask the waiter to serve you a salad while the others are slurping their salty soup.
- Choose oil and vinegar over other salad dressings. If you're itching for French, Russian, or Italian, have it served "on the side." Then use only a small amount. (You'll also avoid unnecessary calories and fat that way.) Consider stowing your own low-sodium dressing in a coat pocket or handbag.
- Order fish, poultry, meat, or vegetables to be cooked without salt or sauce. Ask for a lemon wedge as substitute seasoning.
- Rely on baked potatoes for an excellent low-sodium side dish. Have the butter or sour cream served on the side, and use them sparingly. Better yet, moisten the potato with yogurt or milk and add a grind of pepper.
- For lunch, consider the fresh fruit platter. If your diet is too strict to handle cottage cheese at 250 to 450 milligrams of sodium per ½ cup, order plain yogurt at 58 milligrams sodium per ½ cup instead.
- If you are ordering pasta and sauce, have them served separately. Mix only a small amount of the high-sodium sauce with the low-sodium noodles.
- When it comes to dessert, pies are usually worse than cakes or cookies. Other than fruit, ice cream and sherbet are probably lowest in sodium of most common desserts. Still, these foods are often high in sugar and/or fat.
- Take your herb shaker to the restaurant. Many people carry artificial sweeteners to drop in their coffee, so don't feel embarrassed about bringing your own healthful additive. Suggest to the waiter that the restaurant provide alternatives to table salt.

Remember to be bold in asking the restaurant for adjustments or special favors. The waiters and chefs are there to serve you. If enough people ask for alterations, the restaurant should get the message.

Natural Foods

We offer no suggestions for avoiding sodium that occurs naturally in foods. This source doesn't provide nearly enough sodium to worry about. For example, some people refer to celery (50 milligrams per stalk) and milk (120 milligrams per serving) as high-sodium natural foods. Yet these foods hardly qualify as high in sodium compared to processed foods such as frozen pizza, canned soup, or frozen TV dinners. Such delights from the corporate kitchen usually contain more than 1,000 milligrams per serving.

There is one exception: if your doctor has ordered an extremely low sodium diet of less than 1,000 milligrams a day, you may need to limit a few natural foods. Check our brand name guide to sodium carefully—it includes the sodium contents of natural foods.

There you have it: an entire strategy for lowering the sodium content of your diet by one-half to two-thirds with a minimum of sacrifice. Some scientific studies linking sodium and hypertension suggest that such modest reductions in dietary salt yield significant reductions in blood pressure. Keep these findings in mind as you sift through the thousands of foods in the brand name listing at the back of the book. Surely the effort it takes to make small dietary changes is more than offset by the sense of security that good health provides.

Other Dietary Concerns

In the past decade a revolution has taken place in the once quiet backwater of nutrition science. It began when researchers discovered that some of the nation's most frequently fatal diseases are linked to the American diet. Nutritionists with a public health perspective started to redirect their worries away from making sure that people got enough protein, vitamins, and minerals and toward the overabundance of fat, salt, sugar, and the lack of dietary fiber in most people's diets. The nutritionists noted that Americans were no longer dying of scurvy or beriberi, diseases caused by vitamin deficiencies. Rather, they were dying mainly from heart attacks, strokes, and cancer, diseases related to the overconsumption of fat and salt in foods.

In this book, we've focused on one of these major concerns: the sodium-hypertension connection. But the others are just as serious. Here are several dietary rules of thumb that you can follow to protect your health.

Sea Salt

In many people's minds sea salt is thought of as healthier than regular land salt. It isn't.

Sea salt is just as likely to raise blood pressures as is regular salt—both are loaded with sodium. Moreover, sea salt's reputation as a rich source of minerals is unfounded. While sea water is well endowed with minerals such as magnesium, calcium, and iodine, and trace elements such as fluorine, selenium, and molybdenum, sea salt has insignificant amounts of these nutrients.

Why the difference? Before the sea water is evaporated, it must be treated to remove impurities. As the Hain Pure Food Company explains:

> *Our sea salt is scraped from the bottom of elevated ponds and piled for drying. Prior to filling into containers, the sea salt is redissolved in pure drinking water and treated with lime and soda ash to remove foreign matter, dirt, fish, and other matter that is picked up when the salt is scraped from the bottom of ponds. This must be done to make it fit for human consumption. After that, the salt is again elevated under vacuum, ground for granulation and packed.*

Apparently, most of the minerals are lost in this purification process.

Some people say they prefer the taste of sea salt. They insist that the improved flavor means they can use less in cooking. You may choose to pay the higher price for sea salt's taste, but unless you do use less, don't expect your choice to benefit your health.

Eat less fat. Why? To reduce the risk of heart attack, stroke, cancers of the breast and colon, and to help prevent obesity. Avoiding excess fat is especially important for hypertensives, because high blood pressure already increases their risk of heart attack and stroke.

How? Eat less fatty meat, butter, margarine, oils, fried foods, whole milk products, hard cheeses, creamed foods, chips, pies, and pastries. Instead, eat more breads, potatoes, pasta, cereals, fruit, vegetables, broiled or baked fish and poultry, low-fat milk, yogurt, low-fat cottage cheese, and legumes such as pinto, kidney, and garbanzo beans.

Eat less sugar. Why? To reduce the risk of tooth decay, obesity, and nutrient deficiencies.

How? Drink fewer soft drinks, and eat less candy, ice cream, pastries, puddings, cakes, cookies, and pies.

Eat more fiber. Why? To reduce the risk or severity of constipation, diverticulosis, diabetes, obesity, and cancer of the colon.

How? Eat more legumes, whole grain breads and cereals, vegetables, and fresh fruits.

Eat fruits and vegetables rich in vitamins A and C. Why? To reduce the risk of developing certain cancers.

How? Eat more broccoli, Brussels sprouts, bananas, carrots, sweet potatoes, cantaloupe, grapefruit, kale, collards, spinach, peaches, oranges, cabbage, watermelon, cauliflower, green pepper, winter squash, and other fruits and vegetables.

Eat foods rich in calcium. Why? To reduce the risk of osteoporosis (brittle bones).

How? Eat low-fat milk products, kale, collards, broccoli, rutabaga, okra, tofu, sardines, canned salmon, and soybeans.

Avoid too much alcohol. Why? To prevent cirrhosis of the liver, birth defects, obesity, brain damage, and cancer of the mouth, throat, and esophagus.

How? Drink more water, seltzer, fruit juices, apple cider, and herb teas.

PART TWO

LOW-SODIUM RECIPES

A Low-Sodium Diet

When it comes to food, talk is cheap, We could rave on for pages about the health benefits of low-sodium foods, but without specific suggestions about cooking with less salt, we would convince nary a soul to change a single meal. So in this section, we offer recipes to prove that eating less salt can be not only healthful, but positively delightful.

The appetizers, soups, entrees, side dishes, and desserts offered here are designed for people—whether healthy or hypertensive—who want to eat about 2,000 milligrams of sodium a day. That number is roughly one-half to one-third of the average American's current intake and falls within the 1,100 to 3,300 milligram range recommended by the National Academy of Sciences' Committee on Dietary Allowances. These recipes are not intended for someone on a *strict* low-sodium diet. Such people may want to eliminate some of the higher-sodium ingredients to cut back even further.

You may be surprised to see that salt is not completely missing from these recipes. Sodium-rich tamari soy sauce is also used. We aren't ignoring our own advice by recommending a quarter teaspoon of salt or a tablespoon of tamari here and there. In fact, when you eat a diet consisting largely of unprocessed foods, such as those used in these recipes, there is room for a bit of salty seasoning. Even with the added salt or tamari, few of these dishes contain more than 350 milligrams of sodium per serving, and most contain only about 200 milligrams.

Indeed, avoiding salt-laden processed foods was the key to cutting the sodium in these recipes. We didn't have to eliminate any natural foods. Processed foods, such as cottage cheese or canned tomato sauce, were used sparingly and only if their sodium content was not excessive.

Glancing through the recipes, you may notice that meals

comprised of these dishes won't constitute your standard meat-and-potatoes dinner. Cauliflower Leek Soup, Chick-Pea and Vegetable Curry, Bulgur-Wheat Pilaf, and Poached Bananas are not exactly regulars on most American dinner tables, but they should be. They are designed to protect more than your blood pressure. The recipes are low in fat, saturated fat and cholesterol, to reduce the risk of atherosclerosis—which also hikes your chances of suffering a heart attack or stroke. In keeping with the recommendations of the National Academy of Sciences' Diet, Nutrition, and Cancer Committee, the recipes are also low in total fat; and they include both whole grains and vegetables rich in vitamins A and C, to lower the risk of cancer. That's why you'll see so little butter, margarine, oil, meat, cream, cheese, whole milk, and eggs. Beans, fresh fruits and vegetables, pasta, bulgur, rice, yogurt, low-fat milk, cottage cheese, chicken, and seafood have taken their place.

Anyone can simply "leave the salt out," but the result is often a tasteless meal. As a cooking teacher, Robin Rifkin is concerned about flavor, texture, color, and aroma as well as nutrition. Her recipes are delectable.

Please note the following regarding milligrams of sodium given for the recipes:

● The sodium levels were based on the sodium present in each ingredient, not by analysis of the finished product.

● The sodium content of processed ingredients will vary somewhat depending on which brand is used. We used USDA figures, which are averages of most brands on the market.

● Some recipes allow for alternative ingredients. In those cases, we used the ingredient with the highest sodium content, and therefore some levels we give may be overestimates.

● Some recipes called for rinsing an ingredient. Because the sodium content of rinsed foods is difficult to estimate, we used the sodium content for the unrinsed product. The sodium content of some dishes may therefore be lower than the level indicated.

A closed-minded individual once commented, "No fat, no salt, no sugar, no taste." That may be true for those with no imagination, no curiosity. For those with a little more gumption, low-salt cooking can be an adventure. These recipes are for people who would rather gamble on a new culinary experience than take a chance on damaging their health.

Appetizers, Dips, and Dressings

Antipasto Salad

1 small head romaine or red leaf or green leaf lettuce
Marinated Mushrooms
1 cup cooked, drained garbanzo beans (chick-peas)
2 roasted green or red pepper, sliced (see Note) or 2 whole
*   pimientos, rinsed well and thinly sliced*
2 ounces provolone cheese, very thinly sliced
Finely chopped parsley for garnish

1. Wash the lettuce and arrange 2 leaves on each plate to hold the salad.
2. Spoon some marinated mushrooms into the center of each plate. Spread the garbanzo beans around the mushrooms, and then place the strips of pepper or pimiento across the mushrooms and beans.
3. Roll up the slices of cheese and add them to the salads. Garnish with the parsley.

Makes 4 servings; 212 mg. sodium per serving.

Note: To roast peppers, place the oven rack in its highest position and set the temperature to broil. Broil the peppers, turning them as they blacken. When they are charred all over, remove them from the oven and run them under cold water. Drain, then peel off the charred skin, then seed and slice them into long, thin strips.

Marinated Mushrooms

1 onion, white or yellow, chopped fine
1½ cups vegetable stock (see page 108)
1 sprig fresh tarragon, chopped, or ½ tablespoon dried
1 tablespoon dry sherry
1 pound mushrooms with stems, rinsed, dried, and cut in half
4 teaspoons white miso (optional; see note)
¼ teaspoon freshly ground black pepper

1. In a large skillet, cook the onion in ½ cup of the stock until it is translucent.

2. Add the tarragon, sherry, and mushrooms, and a little more of the stock. Simmer for 2 to 3 minutes.

3. Add the miso, and stir well. Add the remaining stock and the pepper, and cook for 3 minutes more. Set aside to cool.

4. Refrigerate for 1 hour or more. Serve cold, or reheat to serve warm.

Makes 6 servings; 128 mg. sodium per serving with miso.

Note: Miso (fermented soy bean paste) is found in oriental markets or health-food stores. Like tamari and soy sauce, it is high in sodium, and should be used sparingly.

Sweet and Sour Dressing

¼ cup rice vinegar* or ⅛ cup cider vinegar
¼ cup frozen apple juice concentrate mixed with 1 cup water
⅓ onion, chopped
2 tablespoons chopped celery
1 clove garlic, peeled and minced
¼ teaspoon dry mustard or 1 teaspoon Dijon or stone-ground prepared mustard
2 tablespoons minced parsley
Pinch of cayenne pepper (optional)

Combine all the ingredients in a blender or food processor and blend well. Serve chilled (shake before using).

Makes 1¾ cups; 28 mg. sodium per ¼ cup.

*Available in oriental food stores.

Tofu Dressing and Dip

½ pound tofu (bean curd), cut into 1-inch cubes
1 small clove garlic, peeled and minced
2 tablespoons finely chopped onion
2 tablespoons chopped fresh parsley
1 cup water
1 teaspoon chopped fresh dill or ½ teaspoon dried
1 tablespoon stone-ground or Dijon mustard
2 tablespoons lemon juice
1 teaspoon tamari soy sauce
¼ teaspoon crushed dried basil
¼ teaspoon freshly ground black pepper (optional)

Place all the ingredients in a blender or food processor and purée until smooth. Chill and serve over salad.

To use as a dip, omit the water entirely if processing; add up to ¼ cup if puréeing in a blender.

Makes 2 cups; 184 mg. sodium per ¼ cup.

Curried Yogurt Dip

This is good with crisp fresh vegetables.

1 cup nonfat plain yogurt
¾ tablespoon curry powder

Blend the curry powder and yogurt thoroughly and chill. Serve as a dip.

Makes 1 cup; 41 mg. sodium per ¼ cup.

Yogurt Cheese Spread

2 cups plain low-fat yogurt
1 tablespoon chopped fresh dill
1 tablespoon finely grated carrot
1 tablespoon finely chopped chives or scallions
¼ teaspoon curry powder
Freshly ground black pepper to taste
1 tablespoon chopped fresh basil (optional)

1. Line a colander with cheesecloth and place it over a bowl. Add the yogurt and strain for 4 hours or overnight.
2. Combine the remaining ingredients with the strained yogurt and serve as a dip or spread.

Makes about ¾ cup; 108 mg. sodium per ¼ cup.

Hummus

Hummus is delicious as a dip with pita bread.

¾ cup dried chick-peas (garbanzo beans) cooked until soft
*   enough to mash between your fingers (2 cups cooked) or*
*   one 20-ounce can chick-peas, rinsed*
Juice of 1 lemon
1 to 2 cloves garlic, peeled
¼ small onion
1 tablespoon tahini (sesame seed paste) (optional)
2 teaspoons tamari, if dried chick-peas are used

2 tablespoons chopped fresh parsley
Vegetable stock (see page 108)

1. Place the chick-peas, lemon juice, and garlic in a food processor or blender and purée until smooth.
2. Add the onion, tahini, tamari (if you are using it), and parsley, and continue to purée. If the mixture is too dry, add a few tablespoons of stock.

Makes 4 servings; 151 mg. sodium per serving.

Tortilla Chips

Here's a way of baking corn tortillas, instead of frying them, to make crispy chips. They'll keep well if stored in a plastic bag.

12 corn tortillas
¼ cup tamari soy sauce
¼ cup water
Chili powder to taste

1. Cut each tortilla into 8 wedges.
2. Mix the tamari and water in a medium-size bowl. Marinate the tortilla wedges in the tamari mixture for 15 minutes, and then lay them out on a cookie sheet. Sprinkle them with chili powder.
3. Preheat the oven to 325° F.
4. Bake wedges for 10 to 15 minutes, until crisp. Remove from the oven and set aside for 2 minutes.

Makes about 100 chips; sodium level not available.

Nachos

3 fresh corn tortillas, each cut into 8 wedges
¼ cup shredded Monterey jack cheese
2 tablespoons chopped fresh green chilies, stems and seeds removed, or 1 canned whole green chili, rinsed and chopped

1. Preheat the oven to 400° F. Lay the cut tortillas on a cookie sheet. Bake for 15 minutes or until the tortillas are crisp and slightly browned.
2. Remove the cookie sheet from the oven. Spread the cheese evenly over the tortillas, then spread the chilies over the cheese. Bake 3 to 4 minutes or until the cheese melts.

Serve with Guacamole, Mock Sour Cream, or Salsa (all recipes on pages 97-98).

Makes appetizers for 2 or serves 1 as dinner; 144 mg. sodium per half recipe.

Guacamole

This can be made in a food processor. It's great on Tortilla Chips.

1 large ripe avocado, peeled and pitted
1 clove garlic, minced
¼ small onion, chopped fine
⅛ teaspoon chili powder, or more to taste
½ cup yogurt or Mock Sour Cream
½ medium tomato, peeled, seeded, and chopped fine
Juice of ½ lemon

In a bowl, mash the avocado until smooth. Add the garlic, onion, chili powder, yogurt or mock sour cream, tomato, and lemon juice and mix well. Chill briefly before serving.

Makes 1¾ cups; 11 mg. sodium per ¼ cup.

Mock Sour Cream

This is an excellent substitute for sour cream in all dishes. It has only 180 calories per cup; regular sour cream has 485.

1 cup no-salt cottage cheese
Nonfat milk or buttermilk (optional)
1 to 2 teaspoons lemon juice (optional)

Put the cottage cheese in a food processor or blender and purée, adding milk or buttermilk if necessary, to make a smooth sour-cream-like consistency. Taste for sourness, adding lemon juice, a teaspoon at a time, if you wish a more sour taste.

This freezes well. Stir well after thawing.

Note: If you use a dry-curd cottage cheese, you will need some liquid in order to purée; if your cheese is creamy, you won't. Also note that no-salt-added cottage cheese has a tart flavor and usually does not need lemon juice.

Makes 1 cup; 21 mg. sodium per ¼ cup.

Salsa
(Mexican Hot Sauce)

*1 28-ounce can unsalted Italian plum tomatoes, chopped, with
    liquid, or one 28-ounce can unsalted diced tomatoes, with
    liquid, or 5 ripe medium tomatoes, chopped, with their
    juice*
½ medium onion, finely chopped
*1 tablespoon diced canned green chilies, rinsed to remove salt
    (optional)*
2 teaspoons lemon juice
Pinch of crushed dried oregano
Pinch of ground cumin (optional)
2 tablespoons chopped scallions
2 cloves garlic, peeled and minced

Combine all the ingredients and mix well. Chill. Serve as a
tortilla chip dip or on salads, on baked potatoes, on rice, or on
cooked vegetables.

Makes 3½ cups; 4 mg. sodium per ¼ cup.

Bean Dip

Good with Tortilla Chips and as a filling in bean burritos.

1½ cups cooked pinto beans
1 clove garlic, peeled and minced
1 tablespoon chopped canned green chili
¼ onion, chopped
½ cup chopped fresh tomato (peeled and seeded first)
1 teaspoon ground cumin
½ teaspoon ground coriander
⅛ teaspoon salt
Freshly ground black pepper to taste

1. Place the beans in a blender or food processor and purée, or
mash them well with a fork.

2. Add the remaining ingredients and blend well.

Makes 2 cups; 31 mg. sodium per ¼ cup.

Sauces

Cranberry Sauce

¾ cup water
1 cup frozen apple juice concentrate
3 tablespoons frozen orange juice concentrate
3 cups fresh cranberries

1. In a large pot combine the water with the juices and bring to a boil over medium heat, uncovered.
2. Add the cranberries and cook approximately 5 minutes, or until the skins are popped. Serve hot or cold.

Makes 6 servings; 12 mg. sodium per serving.

Italian-Style Tomato Sauce

1 cup finely chopped onion
2 cloves garlic, peeled and minced
½ cup finely chopped celery
1 medium carrot, chopped or grated
¼ cup dry sherry or red wine
1 bay leaf
3 large ripe tomatoes, peeled and chopped (see Note)
1 30-ounce can plum tomatoes, drained
1 6-ounce can tomato paste
¼ pound mushrooms, rinsed, dried, and sliced
1 teaspoon dried oregano or 2 teaspoons fresh
1 teaspoon dried marjoram
1 teaspoon dried thyme or 2 teaspoons fresh
¼ teaspoon freshly ground pepper
⅓ teaspoon salt (optional)

1. In a 2-quart saucepan, cook the onion, garlic, celery, and carrot in sherry or wine until onion is translucent.
2. Add the rest of the ingredients and simmer, uncovered, 30 minutes to 1 hour, until the sauce is as thick as you want it. Stir occasionally.

Makes 1½ quarts; 101 mg. sodium per ½ cup.

Note: To peel tomatoes, first cut an X in the top and bottom of each tomato. Then drop them in a pot of boiling water to cover, and cook for 1 minute. Rinse under cold water and peel.

Tomato Basil Sauce

1 large onion, chopped
¼ cup dry white wine
7 large ripe tomatoes, peeled, cored, and chopped fine
15 fresh basil leaves
⅓ teaspoon salt
Freshly ground black pepper to taste

1. In a large skillet, cook the onion in the wine over medium heat until the onion is translucent. Add the tomatoes and cook another 5 to 10 minutes.

2. Pour the onion-tomato mixture into a blender, add the basil leaves, and purée until smooth.

3. Return the purée to the skillet and simmer for 10 to 15 minutes. Add the tamari and pepper to taste, and serve over hot pasta.

Makes 3 cups; 152 mg. sodium per ½ cup.

Oil-Free Pesto

1 cup chopped fresh basil leaves (rinsed and patted dry first)
1 cup chopped fresh spinach leaves (thoroughly rinsed and patted dry first)
1 tablespoon tamari soy sauce
3 cloves garlic, peeled
Freshly ground black pepper to taste
1 tablespoon grated Parmesan cheese
2 tablespoons plain yogurt (optional)
3 tablespoons pine nuts

1. Place all the ingredients except the pine nuts in a blender or (preferably) a food processor and process until the pesto is well blended and smooth. Stir in the pine nuts.

2. Serve as a sauce for spaghetti.

Makes ¾ cup; 678 mg. sodium per ½ cup.

Soups

Minestrone

1 tablespoon oil
1 medium onion, chopped

3 cloves garlic, peeled and minced
2 stalks celery with leaves, chopped
2 medium carrots, peeled and sliced in ½-inch rounds
1 medium zucchini, diced in ½-inch pieces
4 new potatoes, cut in quarters.
1½ cups bean stock (liquid from cooking beans)
1½ cups juice from unsalted canned plum tomatoes, or one
   6-ounce can unsalted tomato paste mixed with 1 cup water
2 ripe tomatoes, peeled, seeded, and cut in thin wedges
¼ pound fresh green beans, trimmed and cut in thirds
1 cup cooked kidney beans (see Note)
1 cup water
1½ cups elbow macaroni, cooked
2 tablespoons chopped fresh basil or 1 tablespoon dried
¾ teaspoon crushed dried marjoram
½ teaspoon crushed dried thyme
½ teaspoon crushed dried oregano
Freshly ground black pepper to taste
1 tablespoon tamari soy sauce (optional)

1. Heat the oil in a large soup pot and add the onion and garlic. Sauté on medium-high heat for 2 to 3 minutes, until soft.

2. Add the celery, carrots, zucchini, potatoes, and bean stock. Stir, cover, and cook for 15 minutes over medium heat.

3. Add the tomato juice or tomato paste mixture, the tomato wedges, green beans, kidney beans, and water. Cook, uncovered, for 20 minutes.

4. Add the remaining ingredients and cook another 15 minutes. Taste, and correct seasoning if you wish.

Makes six 8-ounce servings; 192 mg. sodium per serving.

Note: If you want to use canned beans, rinse them first; and substitute vegetable stock for the bean stock.

Black Bean Soup

2 cups dried black beans
1 bay leaf
3 cloves garlic, peeled and minced
2 cups chopped onion
1 cup chopped celery
1 tablespoon Sovex yeast (optional; see Note)
½ cup dry sherry

¾ teaspoon ground cumin
1 tablespoon tamari soy sauce
3 tablespoons chopped canned or fresh green chilies, rinsed
  well and seeded
2 tablespoons chopped fresh parsley
¼ cup fresh-squeezed lemon juice
¼ cup chopped scallions (bulb and 1 inch of green stem)
  and 8 lemon slices, for garnish

1. Pick over the beans to remove any pebbles. In a 4-quart pot bring the beans to a boil in 5 cups water. Boil for 3 minutes, then turn off the heat and allow the beans to soak for 1 hour.

2. Drain the beans and discard the soaking water. Return the beans to the pot, add 2 quarts fresh water, and bring to a boil. Lower the heat and simmer the beans, covered, for at least 2 hours, until tender.

3. Add the bay leaf, garlic, onion, celery, optional Sovex yeast, and sherry, and cook for 15 minutes. Add the cumin, tamari, green chilies, parsley and lemon juice, and continue cooking for a few minutes.

4. Remove the soup from the heat and purée it in batches in a food processor or blender. If you wish, purée only half and use it to thicken the remaining soup.

5. Serve hot, with a garnish of chopped scallions and a small slice of lemon on each serving.

Makes 8 servings; 145 mg. sodium per serving (excluding the yeast).

Note: Sovex yeast is a brand name for a smoke-flavored yeast found in health food stores. It is used to give foods a smoky bacon-like flavor.

Curried Lentil Soup

2 cups dried lentils
1 tablespoon oil
1 medium onion, chopped
2 to 3 cloves garlic, peeled and minced
1 dried red Szechuan chili pepper
1½ teaspoons curry powder
1 tablespoon dry sherry
1 bay leaf
1 stick cinnamon

5 cups vegetable stock (see page 108) or water
½ teaspoon crushed dried marjoram
¼ teaspoon crushed dried basil
1½ tablespoons tamari soy sauce or ¾ teaspoon salt

1. In a large pot, bring the lentils to a boil in 4 cups water. Boil for 3 minutes, then turn off the heat and allow the lentils to soak for 1 hour.

2. Drain the lentils, discard the soaking water, and rinse the lentils. Set aside.

3. Heat the oil in the large pot. Add the onion, garlic, and Szechuan pepper, and sauté over medium heat for 2 minutes. Add the curry powder, sherry, bay leaf, and cinnamon, and cook for another 2 minutes. Add the lentils and stock or water and bring to a boil. Reduce the heat and simmer, covered, for 30 minutes.

4. Add the marjoram, basil, and tamari or salt, and cook, covered, for 10 to 15 minutes more.

5. Serve the soup as is, or purée it in batches in a food processor or blender. If you wish, purée only half, using it to thicken the remaining soup.

Makes eight 1-cup or ten 6-ounce servings; 172 mg. sodium per 6-ounce serving, 229 mg. per 1-cup serving.

Lima Bean Soup

This chunky soup is almost a stew.

1 tablespoon oil
1 onion, chopped
2 cloves garlic, minced
2 carrots, peeled and sliced
2 stalks celery, sliced
4 cups vegetable stock (see page 108)
4 new red potatoes, quartered
1 10-ounce package frozen baby lima beans
5 ounces fresh spinach, destemmed and well rinsed
1 tablespoon tamari soy sauce
Freshly ground black pepper to taste

1. Heat the oil in a soup pot. Add the onion, garlic, carrots, and celery, and sauté for 3 to 5 minutes, until the onion is translucent.

2. Add the vegetable stock, potatoes, and lima beans. Cover and cook over medium heat for 20 minutes. Add the spinach, cover, and cook for another 25 minutes.

3. Ladle approximately 2 cups of the soup into a blender, and process until well puréed. Return the purée to the soup and stir thoroughly. Season with tamari and pepper to taste.

Makes six 8-ounce servings; 225 mg. sodium per serving.

Split Pea Soup

1 cup dried split peas
1 tablespoon oil
1 onion, chopped
3 cloves garlic, peeled
4 medium carrots, cut in ½-inch slices
7 cups chicken stock or vegetable stock (see page 108)
1 teaspoon Sovex yeast (optional; see page 102)
½ teaspoon crushed dried basil
1 tablespoon tamari soy sauce (optional)

1. In a large pot bring the peas to a boil in 3 cups water. Boil for 3 minutes, then turn off the heat and allow the peas to soak for 1 hour.

2. Drain the peas, discard the soaking water, and rinse the peas. Set aside.

3. Heat the oil in the large pot. Add the onion and garlic and sauté over medium heat for 2 to 3 minutes. Add the peas, carrots, stock, and Sovex yeast. Bring to a boil, reduce the heat, and simmer, covered, at least for 30 minutes, until peas are tender.

4. Add the basil and tamari (if you are using it) and cook 5 minutes more.

5. Remove the soup from the heat and purée it in batches in a food processor or blender. If you wish, purée only half, using it to thicken the remaining soup.

Makes six 6-ounce servings; 167 mg. sodium per serving (excluding the yeast; including the tamari).

Zucchini Barley Soup

1 tablespoon oil
1 onion, chopped
2 cloves garlic, peeled and minced

5 cups grated zucchini
1 cup mushrooms, rinsed, dried, and sliced
¼ cup dry sherry
½ cup barley, rinsed well and drained
4 cups chicken stock or vegetable stock (see page 108), or
    water
1½ teaspoons fresh chopped dill
1½ tablespoons tamari soy sauce
1 teaspoon dried crushed basil
Freshly ground black pepper to taste

1. Place the oil in a soup pot and sauté the onions and garlic in it over medium heat for 5 minutes, until the onion is translucent. Add 4 cups of the zucchini, all the mushrooms, sherry, barley, and stock or water. Bring to a boil, then reduce the heat and simmer, covered, for 50 minutes.

2. Add the dill, tamari, basil, and the rest of the zucchini, and cook for 5 minutes. Taste and add pepper.

Makes eight 6-ounce servings; 174 mg. sodium per serving.

Navy Bean Soup

1½ cups dried navy beans, rinsed and picked over
1 bay leaf
6 cups (1½ quarts) vegetable stock (see page 108) or water
2 garlic cloves, peeled and minced
1 onion, chopped
2 stalks celery, chopped
1 medium potato, diced
4 carrots, peeled and diced
1 teaspoon curry powder
1 teaspoon dry mustard
1½ teaspoons crushed dried basil
¼ teaspoon freshly ground black pepper
1 tablespoon lemon juice
1 teaspoon crushed dried dill
1 tablespoon tamari soy sauce

1. Rinse the beans under cold water, then place them in a medium-size saucepan. Cover with 8 cups of water—*do not* use the vegetable stock at this point—and bring to a boil. Let the beans boil for 3 minutes. Turn off the heat, and let the beans soak for 1 hour.

2. Drain the beans and discard the soaking water. Refill the pot with the bay leaf, the vegetable stock or fresh water, and the soaked beans. Bring the stock to a boil, lower the heat, and cook the beans for 30 minutes, covered. Check every so often, as beans will foam and may boil over.

3. Add the garlic, onion, celery, potato, and carrots. Cook, uncovered, for an additional 30 minutes or until the beans and vegetables are tender.

4. Add the seasonings and simmer for 10 to 15 minutes. Add more stock or water if the soup is too thick. Serve hot.

Makes about 10 servings; 123 mg. sodium per serving.

Italian Vegetable-Pasta Soup

1 cup Italian-Style Tomato Sauce, preferably homemade (see page 99)
¾ cup juice from canned unsalted plum tomatoes
1 small zucchini, cut in half lengthwise, then cut in ¼-inch slices
1 cup cooked kidney beans
1 cup cooked elbow macaroni or other small noodles
1½ tablespoons chopped fresh basil or ¾ tablespoon dried
2 tablespoons chopped fresh parsley
¼ teaspoon dried crushed marjoram
¼ teaspoon dried crushed thyme
1 teaspoon tamari soy sauce (optional)
3 tablespoons grated Parmesan cheese

1. Place the tomato sauce, tomato juice, and zucchini in a saucepan and cook, covered, over medium heat for 5 minutes.

2. Add the kidney beans and macaroni and cook for an additional 10 minutes. Add the seasonings and cook for 5 minutes more. Serve with Parmesan cheese.

Makes four ¾-cup servings; 198 mg. sodium per serving.

Cauliflower Leek Soup

1½ tablespoons oil
2 large leeks, green tops removed, white stems split, washed well, and patted dry
2 teaspoons peeled and minced or crushed garlic
¼ cup dry sherry
2 stalks celery, sliced thin

1 canned mild green chili pepper, rinsed and chopped
1 cauliflower, broken into small flowerets
1 bay leaf
3¼ cups vegetable stock or chicken stock (see page 108)
½ teaspoon Sovex yeast (optional; see page 102)
1 tablespoon prepared stone-ground mustard
½ teaspoon paprika
4 tablespoons chopped scallions
1 teaspoon crushed dried marjoram leaves
1 tablespoon tamari soy sauce
2 cups evaporated skim milk
Freshly ground black pepper to taste

1. Heat the oil in a large pot, then add the leeks and garlic, and sauté for a few minutes. Add the sherry, celery, chili pepper, cauliflower, bay leaf, and 1 cup of the stock. Cook for 10 minutes over medium heat, stirring occasionally.

2. Add the rest of the stock, the optional yeast, mustard, paprika, and half the scallions. Cook for 5 minutes.

3. Purée 4 cups of the soup in a processor or blender. Pour the purée back into the soup and stir. Add the rest of the ingredients and cook for 10 more minutes.

Makes seven 8-ounce servings; 320 mg. per serving (excluding the yeast).

Fish Stock

Use this for fish chowder, for poaching fish, or for making a sauce for a fish dish.

1 pound fish bones (heads, etc.)
Several large sprigs parsley
2 carrots, peeled and cut into thick slices
1 onion, cut in quarters or 1 leek, well washed and cut in pieces
1 clove garlic, peeled
2 tablespoons crushed dried dill or 1 sprig fresh
1 gallon water
2 peppercorns

1. Tie the fish bones up in a cheesecloth bag.

2. Combine all the ingredients in a large soup pot and bring to a boil. Reduce heat and simmer for at least 1 hour, uncovered, skimming off any foam until the stock looks clear.

3. Place a strainer over a large bowl and pour the stock through it.

4. Refrigerate the strained stock. Skim off any fat that rises to the top after cooling. Stock will stay fresh for a few days in the refrigerator, and it freezes well.

Makes about 2 quarts; sodium level not available.

Vegetable Stock

When you are cooking during the week, don't throw out the vegetable peels and odds and ends . . . (but avoid any that have been waxed).

> *Vegetable peels and trimmings, such as:*
>   *carrot peels and ends*
>   *onion skins*
>   *celery leaves*
>   *lettuce leftovers*
>   *cabbage hearts and leaves*
>   *tomato skins*
>   *green bean trimmings*
>   *asparagus stalks*
>   *broccoli stalks*
>   *cauliflower leaves*

1. Rinse whatever vegetable peels and trimmings you have on hand and put them in a soup pot. Add at least double the amount of water, and bring to a boil. Reduce the heat, cover, and simmer for about 1 hour.

2. Place a strainer over a large bowl and pour the stock through it. Strained stock will keep for several days in the refrigerator, and it freezes well.

Yield and sodium level not available.

Chicken Stock

Stock adds wonderful flavor to soups and grains.

> *1 chicken, or the skin and bones from deboning 1 chicken*
>   *or from deboning 5 whole chicken breasts*
> *2 carrots, peeled and cut into thick slices*
> *1 medium onion, peeled, plus outside skins from several onions*
> *1 bay leaf*
> *2 stalks celery, cut into large pieces*

Several peppercorns
Any vegetable odds and ends on hand

1. Wash the chicken, remove and discard the skin, and place the chicken in a large soup pot. Add all the remaining ingredients, then add water to cover, and bring to a boil. Reduce the heat and simmer, uncovered, 1 to 1½ hours, until the chicken has fallen apart.

2. Place a colander over a large bowl and strain the soup through it. Return the stock to the soup pot and continue cooking until it is reduced by about 25 percent—this makes it more flavorful. Then set the stock aside to cool.

3. Skim any fat from chilled stock, and strain the stock into a bowl through a strainer lined with cheesecloth. Use immediately or freeze for future use.

Note: If you use canned broth instead of fresh stock, try to find a salt-free brand.

Makes about 2 quarts; sodium level not available.

Beef Stock

6 pounds beef marrow and shin bones
4 quarts water
5 peppercorns
2 stalks celery, cut into large pieces
2 medium carrots, peeled and cut into thick slices
Several sprigs of parsley
1 medium onion, quartered, plus outside skins from several
*   onions*
1 bay leaf
1 teaspoon crushed dried thyme
6 whole cloves
3 star anise (optional)
Any other vegetable according to taste (optional)

1. Place all the ingredients in a large pot and bring to a boil. Reduce the heat and simmer, uncovered, for 2 hours. As the soup is cooking, skim off any scum that rises to the surface or clings to the inside of the pot.

2. Place a colander over a large bowl and strain the soup through it. Set the strained stock aside to cool, then refrigerate.

3. Skim any fat from the chilled stock, and strain the stock

into a bowl through a strainer lined with cheesecloth. Use immediately or freeze for future use.

Note: If you use canned broth instead of fresh stock, try to find a salt-free brand.

Makes about 2 quarts; sodium level not available.

Main Dishes

Chick-Pea and Vegetable Curry

1 large onion, diced
1 to 2 cloves garlic, peeled and chopped or pressed
1 slice fresh ginger, about 1 by ½ inch
1 to 2 teaspoons curry powder
⅓ cup vegetable stock or chicken stock (see page 108)
½ large potato, cut in 1-inch dice
2 carrots, peeled and sliced
¼ pound green beans
1 cup cooked chick-peas (garbanzo beans)
Juice of ½ lemon
2 tablespoons frozen apple juice concentrate
1 teaspoon tamari soy sauce
Freshly ground black pepper to taste

1. In a large skillet or sauté pan, cook the onion, garlic, ginger, and curry powder in the stock over medium heat, until onion is translucent.

2. Add the potato and carrots, and cook, covered, until partially soft, about 5 minutes. Add the green beans and chick-peas, and continue cooking until all the vegetables are tender but not mushy.

3. Mix in the lemon and apple juices, tamari, and pepper. Cook another few minutes and serve.

Makes 4 servings; 103 mg. sodium per serving.

Pasta Primavera

⅓ medium-size sweet potato, sliced in ¼-inch diagonal slices
½ cup fresh peas or frozen peas or lima beans
1 red bell pepper, cored, seeded, and sliced in thin strips
½ leek, cut in half and rinsed well, dried, and sliced
* diagonally*

*1 cup fresh spinach leaves torn in ½-inch pieces (rinsed well
    and dried first)*
¼ pound snow peas, stems removed
1 small zucchini, cut in ½-inch slices
½ pound whole-wheat fettuccine
1½ cups nonfat yogurt
1½ tablespoons cornstarch or arrowroot
1½ teaspoons unsalted butter
3 tablespoons grated Parmesan cheese

1. Place the sweet potato and peas or lima beans in a basket steamer in a 1-quart saucepan with 1 inch of water. Bring the water to a boil and steam the vegetables for 4 minutes, covered, over medium heat. Add the rest of the vegetables and steam for 4 more minutes. Rinse the vegetables under cold water to stop the cooking process. Drain, and set aside.

2. Bring a large pot of water to a boil, add the pasta, and cook (without salt) until *al dente* (soft but still firm, not mushy). Rinse under cold water, drain, and set aside.

3. While the pasta is cooking, combine the yogurt with the cornstarch or arrowroot and cook 5 minutes over medium heat, whisking to blend until smooth.

4. In a large skillet, melt the butter over low heat. Add the vegetables, pasta, and the yogurt mixture. Toss to combine and serve immediately, topped with the Parmesan cheese. If you wish, this dish is also delicious served cold.

Makes 4 servings; 167 mg. sodium per serving.

Zucchini Stuffed with Apples and Walnuts

4 small to medium zucchini or 2 large
½ cup minced onion
1 tablespoon oil
1 apple, cored, peeled, and chopped fine
½ cup chopped walnuts
½ teaspoon ground cinnamon
⅛ teaspoon ground cardamom (optional)
2 teaspoons lemon juice
½ cup cooked brown rice
Paprika for garnish

1. Rinse the zucchini, then trim off the ends. Cut the zucchini in half lengthwise, and steam over medium heat for 8 minutes.

Rinse under cold water to stop the cooking process and set aside to cool.

2. Preheat the oven to 350° F. Carefully scoop out and reserve the zucchini pulp, leaving a shell ¼-inch thick. Set aside.

3. In a skillet sauté the onion in the oil for 2 to 3 minutes over medium heat, until the onion is translucent. Add the apple, walnuts, cinnamon, cardamom (if you are using it), lemon juice, and zucchini pulp. Cook for 5 minutes, then add the rice. Continue cooking for an additional 3 minutes.

4. Lay the zucchini shells in a non-stick baking pan or one lined with cooking parchment. Fill each shell with the cooked mixture. Sprinkle with paprika and bake for 30 minutes.

Makes 4 servings; 6 mg. sodium per serving.

Shells Stuffed with Cheese and Vegetables

12 large white or whole-wheat pasta stuffing shells
¼ onion, chopped
½ teaspoon peeled chopped garlic
1 medium zucchini, shredded
8 large mushrooms, rinsed, dried, and finely chopped
Water or vegetable stock (see page 108)
¾ cup nonfat cottage cheese, puréed until almost smooth
  (see Note)
2 tablespoons grated Parmesan cheese (optional)
1 tablespoon chopped fresh parsley
Pinch each of crushed dried tarragon and basil
2½ cups Italian-Style Tomato Sauce (see page 99)

1. Place the shells in a large pot of boiling water to cover, and boil for 10 to 12 minutes or until the shells are tender but not mushy. Rinse the shells under cold water to stop the cooking process, then drain and set aside.

2. In a large skillet or sauté pan cook the onion, garlic, zucchini, and mushrooms in a few tablespoons of water or stock for a few minutes. Spoon off any excess liquid.

3. In a large bowl, mix the cottage cheese with the vegetables. Add the Parmesan cheese (if you are using it). Add the parsley, tarragon, and basil.

4. Preheat the oven to 350° F. Spoon some tomato sauce on the bottom of a baking dish large enough to hold the shells in one layer. Fill each shell with 2 tablespoons of the vegetable-cheese mixture. Place the shells in the baking dish as you fill them, and

cover with the remainder of the sauce. Cover with aluminum foil and bake for 30 minutes.

Makes 4 servings; 185 mg. sodium per serving.

Note: You may make this without the cottage cheese, but double the amount of vegetables and add 1 carrot, peeled and shredded. The amount of Parmesan and spices remains the same.

Eggplant Parmesan

4½ cups Italian-Style Tomato Sauce (see page 99)
1 medium to large eggplant, cut in ½-inch rounds
2 egg whites
1½ cups whole-wheat or unbleached white flour
2 teaspoons paprika
⅛ teaspoon freshly ground black pepper
6 ounces part-skim-milk mozzarella cheese, grated
4 tablespoons grated Parmesan cheese

1. Preheat the oven to 350° F. Place the egg whites in a bowl large enough to hold an eggplant slice. In a medium-size bowl, mix together the flour, paprika, and pepper.

2. Dip each eggplant slice first in the egg white, then in the flour. Shake off the excess and lay the slices on a non-stick baking pan. Bake for 30 minutes or until slices are lightly browned.

3. Ladle 1½ cups tomato sauce into a casserole or baking pan approximately 6 x 8 inches. Lay ⅓ of the eggplant slices on the sauce, then add 1 cup more sauce, ⅓ of the mozzarella cheese, and 1½ tablespoons Parmesan cheese. Repeat the layering, beginning with the eggplant, and continue until all the ingredients are used. End with Parmesan cheese.

4. Cover the casserole with aluminum foil and bake for 25 minutes. Remove the foil and bake 10 minutes, uncovered.

Makes 5 servings; 437 mg. sodium per serving.

Coquilles St. Jacques

A lovely rendition of an old favorite—scallops cooked with wine and herbs.

½ cup finely chopped yellow or white onion
2 tablespoons chopped shallots
2 cloves garlic, peeled and minced or pressed

¾ cup dry white wine
10 mushrooms, rinsed, dried, and sliced
1 bay leaf
⅛ teaspoon crushed dried thyme
⅛ teaspoon crushed dried rosemary or marjoram
Juice of ½ lemon or lime
1 pound bay or sea scallops
½ cup whole-wheat or unbleached white flour
½ teaspoon paprika
2 tablespoons unsalted butter or oil
¼ cup grated Swiss cheese

1. In a skillet cook the onion, shallots, and garlic in a few tablespoons of white wine for 3 to 4 minutes over medium heat, until the onion is translucent. Add the mushrooms, herbs, lemon or lime juice, and the remaining wine. Continue cooking for approximately 5 minutes more. Set the mixture aside in a small bowl.

2. Wash the scallops and pat dry. Combine the flour and paprika in a bowl and add the scallops to the flour mixture. Coat lightly, remove the scallops, and discard any extra flour.

3. Melt the butter or oil in a skillet over medium-low heat. Add the scallops and cook 2 to 3 minutes, stirring to keep them from sticking. Add the wine mixture to the scallops and continue cooking a few minutes more. Be careful not to overcook—just long enough for the flavors to blend.

4. Preheat the broiler. Divide the mixture into 4 scallop shells or au gratin dishes. Sprinkle 1 tablespoon grated Swiss cheese on each one. Place under the broiler and broil 1 minute or until the cheese melts.

Makes four 4-ounce servings; 219 mg. sodium per serving.

Salmon with Carrot Leek Sauce

1 carrot, peeled and cut into 1-inch chunks
1 leek, bulb only, rinsed well, dried, and chopped
2 tablespoons dry sherry
2 tablespoons water
1 to 1½ pounds salmon steaks, 1 to 1½ inches thick
¾ cup plain low-fat yogurt
½ teaspoon tamari soy sauce
2 teaspoons chopped fresh dill or 1 teaspoon dried

1. Preheat the broiler.

2. In a small saucepan, cook the carrot and leek in the sherry and water over medium heat, covered, for 15 minutes. Make sure the liquid does not evaporate. Set aside.

3. Lay the salmon on a broiler pan and broil for 15 minutes, turning once.

4. Meanwhile, in a food processor or blender, purée the carrot-leek mixture with its liquid, plus the yogurt, tamari, and dill.

5. Return the purée to the saucepan and heat, gently, for a few minutes.

6. Place the salmon on serving plates and spoon the sauce over.

Makes 4 servings; 240 mg. sodium per serving.

Poached Red Snapper

> 2 cups (approximately) fish stock or vegetable stock (see pages 107 and 108)
> 1/4 cup dry sherry
> 2 slices fresh ginger, cut in thin strips
> 1 pound red snapper fillets
> 1/4 pound snow peas, stems removed
> 2 scallions, cut in half
> 1/2 lemon, sliced

1. In a large skillet or sauté pan, combine 1½ to 2 cups fish or vegetable stock with the sherry and ginger and bring to a simmer. Add the fish, snow peas, and scallions and cook over medium heat for 8 minutes, turning the fish carefully after 4 minutes.

2. Carefully lift the fillets from the pan and serve accompanied by the vegetables and lemon slices.

Makes four 4-ounce servings; 113 mg. sodium per serving.

Shrimp Stir Fry with Chinese Mushrooms and Rice Noodles

> 5 dried shiitake mushrooms*
> 5 dried wood ear mushrooms*
> 5 dried lily flowers*
> 1 tablespoon peanut or safflower oil
> 2 cloves garlic, peeled

1 thin slice ginger
1 dried red Szechuan chili pepper*
1 pound shrimp, shelled and deveined
2 scallions, chopped (bulbs and 1 inch of green stem)
1 red bell pepper, cored, seeded, and sliced in julienne strips
4 ounces rice noodles,* soaked in hot water for 5 minutes
3 tablespoons dry sherry
1 tablespoon tamari soy sauce
1 to 2 tablespoons vegetable stock (see page 108) or water

1. Soak the mushrooms and lily flowers in 1 cup water for 30 minutes. Drain and pat dry. Shred each of the lily flowers into 4 to 5 pieces, and set all aside.

2. Heat the oil in a wok or skillet over medium heat. Add the garlic cloves, ginger, and Szechuan pepper, and cook for 2 minutes. Remove the spices from the oil and add the shrimp. Fry for 2 minutes, stirring constantly, then move the shrimp to the sides of the wok.

3. Add the mushrooms, lily flowers, and scallions, cook for 1 minute, stirring constantly, then move to the sides of the wok.

4. Add the bell pepper and rice noodles, stir in the sherry and tamari, and cook for 1 minute. Toss all the ingredients together and cook 1 minute more. Add 1 to 2 tablespoons of stock or water if the mixture seems too dry.

Makes 4 servings; sodium level not available.

*Available in oriental markets.

Hot Peppers

The oils in hot peppers can cause sensitive skin around the eyes, mouth, nose, and ears to burn. When working with them, never rub your eyes or any other sensitive area with your fingers. When you are finished cleaning and chopping the peppers, be sure to wash your hands thoroughly to remove all the oil.

Sole Baked in Paper

1 pound thin sole fillets
- 4 sheets cooking parchment, 8 x 8 inches each, or aluminum foil

1 teaspoon unsalted butter
Paprika to taste
Ground ginger to taste
Freshly ground black pepper to taste

1. Preheat the oven to 350° F. Divide the fish into 4 portions.
2. Fold each sheet of parchment in half and cut it into a heart shape. Open the paper, put a portion of fish on one side close to the center, and dot it with ¼ teaspoon of the butter. Sprinkle with paprika, ginger, and pepper. Fold the top over and crimp the edges together to form a tight seal.
3. Lay the packets on a cookie sheet and bake for 10 minutes. Makes 4 servings; 106 mg. sodium per serving.

Moussaka

1½ cups unbleached white or whole wheat flour
1 teaspoon paprika
½ teaspoon ground cinnamon
2 egg whites
1 large eggplant, unpeeled and sliced in ½-inch rounds
4 tomatoes, peeled, cored, seeded, and halved
Half of a 6-ounce can of unsalted tomato paste
1½ pounds lean ground beef
1 cup chopped onion
Freshly ground black pepper to taste
½ teaspoon freshly ground nutmeg
2 teaspoons chopped fresh thyme or 1½ teaspoons dried, crushed
8 ounces unsalted, low-fat cottage cheese
2 teaspoons cornstarch or arrowroot
6 tablespoons grated Parmesan cheese

1. Preheat the oven to 350° F.
2. Mix together the flour, paprika, and cinnamon in a flat dish. Put the egg whites in a separate bowl.
3. Dip the eggplant slices first in the egg whites and then in the flour mixture. Lay the coated slices on a non-stick baking sheet and bake for 30 minutes. (If you don't have a non-stick baking sheet, line a pan with parchment paper or oil the pan.)
4. Meanwhile, put the tomatoes in a processor or blender. Add the tomato paste and purée until smooth. Set aside.
5. Sauté the beef in a large skillet over medium heat until it is

browned. Add the onion, tomato sauce, pepper, nutmeg, and thyme. Cook, uncovered, over low heat for 15 minutes. Drain and set aside.

6. In a blender or food processor, purée the cottage cheese, cornstarch or arrowroot, and 2 tablespoons of Parmesan cheese until smooth. Set aside.

7. Spoon some of the tomato sauce over the bottom of an 8 x 8-inch casserole or baking dish. Lay half of the eggplant slices on top of the sauce. Then spread half of the meat mixture over the eggplant and sprinkle some of the remaining Parmesan over it. Repeat the eggplant and meat layers, and finish with the rest of the Parmesan cheese. Bake for 25 minutes, uncovered.

8. Remove the dish from the oven, spoon the puréed cottage cheese mixture over the top, and bake an additional 10 minutes.

Makes 4 to 6 servings; 256 mg. sodium per ⅕ recipe.

Mexican Chicken with Green Chilies

1½ tablespoons oil
1 large onion, halved, then cut into ¼-inch slices
2 cloves garlic, peeled and minced
1 pound chicken breasts, split, skinned, boned, and cut into
  4 pieces
1 4-ounce can whole green chilies, rinsed, seeded, and
  chopped
1 cup Mock Sour Cream (see page 97)
1 cup grated sharp cheddar or Monterey jack cheese
3 whole scallions, chopped

1. Preheat the oven to 350° F. Put the oil in a non-stick skillet or sauté pan and sauté the onion and garlic for 5 minutes, until the onion is translucent. Remove mixture from the pan and set aside.

2. Place the chicken breasts in the pan and cook for 3 minutes on each side.

3. Spread half the onion-garlic mixture in a 6- x 8-inch baking pan. Place the chicken on the mixture, and add half of the chilies and half of the mock sour cream. Beginning with half of the cheddar cheese, layer again, adding the rest of the onion-garlic mixture, sour cream, chilies, and cheese. Top with the scallions.

4. Cover the pan with aluminum foil and bake for 20 minutes. Uncover, then bake 5 more minutes.

Makes four 4-ounce servings; 261 mg. sodium per serving.

Chicken Enchiladas

For the Sauce

½ medium onion, chopped
2 cloves garlic, peeled and minced
5 or 6 medium tomatoes, puréed in a blender with ½ cup
   water
1 bay leaf
2 tablespoons chili powder
½ teaspoon ground cumin
1 teaspoon paprika
½ teaspoon crushed dried basil
½ teaspoon crushed dried oregano
1 green pepper, seeded and chopped
2 canned whole green chilies, rinsed and chopped or 1 fresh
   jalapeño chili, seeded and chopped

For the Enchiladas

1 tablespoon oil
1 pound boneless, skinless chicken breasts, cut in small pieces
   or shredded in a food processor
12 corn tortillas
1 cup shredded Monterey jack cheese
8 pitted jumbo black olives, rinsed well and sliced
Mock Sour Cream (see page 97)
Guacamole (see page 97)

1. Place a few tablespoons of water in a saucepan and add the onion and garlic. Cook 5 minutes or until tender.

2. Add the remaining ingredients and cook, covered, for 20 minutes over medium heat. Set aside.

3. Preheat the oven to 350° F.

4. Heat the oil in a medium-size skillet and add the chicken. Sauté over medium heat for 2 minutes. Add 1 cup of the sauce, cook for 1 more minute, and remove from the heat.

5. Dip the tortillas, one by one, in the remaining sauce to soften them. Spoon some of the sauce on the bottom of an 8- x 8-inch baking dish.

6. Spoon some chicken into the middle of each tortilla, and roll the tortilla into a cylinder shape. Place the rolled tortillas in the baking pan and spoon the remaining sauce over them.

7. Sprinkle the cheese over the tortillas and sauce, then the

olives. Cover the baking pan with foil and bake for 15 minutes, or until the ingredients are bubbling. (You can also broil the tortillas for 5 minutes, if you prefer.)

8. Serve with a dollop of Mock Sour Cream and Guacamole on each enchilada.

Makes 6 servings; 335 mg. sodium per serving.

Szechuan Chicken with Peanuts

2 whole chicken breasts, split, skinned and boned
1 tablespoon oil
1 thin slice fresh ginger
1 clove garlic, peeled
3 dried red Szechuan chili peppers, halved
⅓ cup unsalted peanuts
½ pound snow peas, stems removed
1 tablespoon tamari soy sauce
1 tablespoon dry sherry
1 teaspoon arrowroot or cornstarch
3 tablespoons cold water
2 scallions, cut in 1-inch pieces

1. Trim the fat from the chicken breasts, and cut each into long, ½-inch-wide strips.

2. Heat the oil in a wok, skillet, or sauté pan, add the ginger, garlic, and chili pepper, and cook for 2 minutes over medium heat.

3. Add the chicken and sauté for 2 minutes, stirring constantly.

4. Add the peanuts, snow peas, tamari, and sherry, stir, and continue cooking another minute or two.

5. Mix the arrowroot or cornstarch and water, until smooth. Add the mixture to the skillet and stir until the sauce thickens. Add the scallions, reduce the heat to low, and cook for 30 seconds. Remove the garlic, pepper, and ginger.

Makes 4 servings; 302 mg. sodium per serving.

Chicken with Prunes and Apricots

1 tablespoon safflower oil
1 cup sliced onion
1 pound chicken breasts, split, boned, and skinned
Paprika to taste
½ cup dried apricots

½ cup dried pitted prunes
1½ cups chicken stock (see page 108)
Juice of ½ lemon
1 tablespoon fresh chopped dill
1 teaspoon ground cinnamon
1½ teaspoons arrowroot
¼ cup cold water

1. Heat the oil in a large skillet or sauté pan. Add the onion and sauté over medium heat until translucent.

2. Add the chicken breasts to the skillet, sprinkle with paprika, and cook for 3 minutes on each side.

3. Add the fruit, stock, lemon juice, dill, and cinnamon, and simmer, uncovered, for 20 minutes.

4. Mix the arrowroot with cold water, add it to the skillet, and stir gently until the sauce thickens. Serve chicken with sauce spooned over.

Makes 4 servings; 24 mg. sodium per serving.

Triathlon Chicken (Coq au Vin)

¼ cup tamari soy sauce
1 cup water
Freshly ground black pepper to taste
1 pound chicken breasts, split, skinned, and boned
1½ small onions, chopped (about 1 cup)
1 clove garlic, peeled and minced
2 tablespoons dry sherry
½ cup red burgundy
1 tablespoon crushed dried marjoram
6 baby carrots, peeled and cut in half lengthwise
1 red potato, cut in ¼-inch slices
½ pound mushrooms, rinsed, dried, and cut in half lengthwise
Juice of 1 lemon
2 scallions, chopped (bulb plus 1 inch of green stem)
1 tablespoon arrowroot or cornstarch, mixed with ¼ cup cold water
¼ cup chopped fresh parsley

1. Place the tamari, water, and pepper in a large mixing bowl. Add the chicken, and coat it well with the liquid. Marinate for 1 hour, turning the pieces once. Pat the chicken dry, and discard the marinade.

2. In a skillet or sauté pan, cook the onions and garlic in the sherry over medium heat until the onion is translucent. Add the chicken and cook it for 3 minutes on one side, then turn and continue cooking for 3 minutes. If the pan becomes dry, add a little more sherry.

3. Add the wine, marjoram, carrots, and potato. Stir to make sure the chicken isn't sticking to the skillet. Cover, and cook for 10 minutes.

4. Add the mushrooms, lemon juice, and scallions, and continue cooking, uncovered, for another 10 minutes. Remove the skillet from the heat.

5. Spoon half the sauce from the skillet into a small saucepan. Place it over medium heat and add the arrowroot or cornstarch mixture. Stir until the sauce is thickened.

6. Return the skillet to the burner, and with the heat on low, pour the thickened sauce over the chicken. Stir to mix well. Serve garnished with fresh parsley.

Makes four 4-ounce servings; 231 mg. sodium per serving.

Veal Scallopini

½ cup unbleached flour
1½ teaspoons paprika
⅓ teaspoon (approximately) freshly ground black pepper
1 pound veal scallops, pounded thin
2 tablespoons unsalted butter
2 tablespoons chopped shallots
½ cup dry white wine
1 teaspoon lemon juice
½ pound mushrooms, rinsed, dried, and sliced
1½ teaspoons white miso (see Note) or 1½ teaspoons tamari soy sauce
3 tablespoons chopped watercress or fresh parsley, for garnish
Lemon slices, for garnish

1. In a bowl mix the flour with the paprika and pepper. Dredge the veal in the flour mixture, shake off the excess, and set aside.

2. Melt the butter in a large skillet or sauté pan over medium-low heat. Add the shallots and sauté for 1 minute. Raise the heat slightly and add the veal to the skillet. Cook for 2 minutes on each side, then add the wine, lemon juice, and mushrooms, and continue cooking for 2 minutes more. Turn off the heat.

3. Spoon 2 tablespoons of liquid from the skillet and mix with

the miso or tamari in a cup. Add this mixture to the pan and stir well.

4. Serve the veal on a platter sprinkled with watercress or parsley and lemon slices.

Makes 3 to 4 servings; 197 mg. sodium per ¼ recipe.

Note: Miso (fermented soy bean paste) is found in oriental markets or health-food stores. Like tamari and soy sauce, it is high in sodium, and should be used sparingly.

Tuna Fish Salad

1 6½-ounce can low-salt tuna packed in water, rinsed
¼ cup finely chopped celery
4 tablespoons unsweetened crushed pineapple, drained
1 teaspoon dried dill
4 tablespoons nonfat or low-fat plain yogurt
Freshly ground black pepper to taste

1. Drain the tuna well and place it in a medium-size bowl. Break it into small pieces.
2. Add the rest of the ingredients and mix well.

Makes 4 to 5 servings; 118 mg. sodium per ¼ serving.

Whole-Wheat Pancakes

Combine pancakes with fruit salad (recipe follows) or steamed vegetables for a change-of-pace main dish.

1½ cups whole-wheat flour
1½ teaspoons low sodium baking powder
2 teaspoons sugar or honey or frozen apple juice concentrate
1 egg, beaten
1½ cups skim milk
2 tablespoons unsalted butter, melted, or oil

1. In a bowl, mix together the flour and the baking powder.
2. In a separate bowl, mix together the sugar (or honey or apple juice), egg, and milk.
3. Add the wet ingredients to the flour mixture and stir until the batter is lump-free.
4. Place a non-stick griddle over medium-high heat and when it is hot, grease it with a small amount of the melted butter or oil. Ladle out a small amount of the batter and carefully pour it onto the griddle to form a pancake. If it is under 3 inches wide,

add more batter; if it is larger, ladle on less batter for the other pancakes. When the batter bubbles lightly and the edges seem firm, flip the pancakes over with a spatula. Cook for another minute or so.

5. Continue cooking the pancakes, greasing the griddle if it gets too dry, and adjusting the heat if the pancakes seem to be cooking too quickly.

Makes 16 to 18 pancakes, 3 inches wide; 14 mg. sodium per pancake.

Fruit Salad

½ cup seedless grapes
1 cup strawberries, hulled and cut in half
1 tangerine, peeled and divided into wedges
1 banana, peeled and sliced
1 apple, cored and cut into wedges
1 pear, cored and cubed
¼ cup raisins
1 cup plain lowfat yogurt
2 tablespoons chopped walnuts

Combine all the ingredients in a large bowl, toss thoroughly, and serve.

Makes four 1½-cup servings; 42 mg. sodium per serving.

Yummy Whole-Grain Pizza

For the Dough

⅔ cup warm water
1 package yeast
2⅓ cups whole-wheat flour
¼ cup frozen apple juice concentrate
2 tablespoons nonfat dry milk
1 teaspoon tamari soy sauce (optional)

For the Sauce

1 15-ounce can unsalted tomato sauce
1 6-ounce can unsalted tomato paste
¼ teaspoon freshly ground black pepper
¼ teaspoon crushed dried oregano
⅛ teaspoon crushed dried thyme
2 cloves garlic, peeled and minced
½ tablespoon tamari soy sauce or ¼ teaspoon salt

For the Topping

3 tablespoons vegetable stock (see page 108)
1 medium onion, sliced in thin rounds
1 or 2 green peppers, sliced thin
½ cup low-fat mozzarella cheese
¼ pound mushrooms, rinsed, dried, and sliced
1 small zucchini, sliced or grated
¼ cup grated Parmesan cheese

1. Pour the water into a large bowl. Sprinkle the yeast on the water, cover the bowl with a tea towel, and let it sit for 5 minutes.

2. Add 1 cup of the flour, the apple juice concentrate, and the milk. Mix well. Cover again with the towel and let a "sponge" form—about 10 minutes.

3. Add the optional tamari and the remaining flour, ½ cup at a time, until dough is light and just slightly sticky. Knead on a floured board, adding a little flour to dough if necessary, for 5 minutes. Return the dough to the bowl, cover the bowl with the towel, and let it sit for 1 hour, until dough is doubled in bulk. Meanwhile, prepare the sauce and topping.

4. Combine all the ingredients in a saucepan, bring to a boil, and simmer, uncovered, for 15 minutes. Set aside.

5. In a large skillet, simmer all the vegetables in the stock until tender. Set aside.

6. Preheat the oven to 350° F.

7. Punch down the dough and knead it for a few minutes. Roll it out on a floured board to form a 10-inch circle. Place the dough on a non-stick pan. Spoon the sauce evenly over the dough, except for the edge, and bake for 15 minutes. Remove the pizza from the oven, add the topping, and bake for another 15 minutes.

Makes 4 servings; 420 mg. sodium per serving.

Salad Nicoise

For the Salad

½ head bibb lettuce, rinsed, patted dry, and torn into bite-size pieces
1 6½-ounce can low-salt tuna fish, packed in water, rinsed
½ 8-ounce can pitted black olives, rinsed well
¼ pound whole green beans, lightly steamed

½ tablespoon capers, rinsed well
4 new potatoes, cooked and halved
1 tomato, cut into 8 wedges
1 hard-cooked egg, peeled and sliced thin

For the Dressing

Juice of 1½ lemons
½ teaspoon Dijon mustard
1 medium clove garlic, peeled
1 tablespoon fresh basil
2 tablespoons chopped scallions (bulb plus 1 inch of green stem)
1 tablespoon pure olive oil
1 tablespoon water

1. Lay half the lettuce attractively on each of two dinner-size plates. Arrange the rest of the ingredients on each plate in the following manner: spread the tuna on the lettuce, then the olives, green beans, and capers. Put the potato halves and tomato wedges in a ring around the tuna, and lay the egg slices over all.

2. Place the dressing ingredients in a blender or food processor and blend until smooth. Pour over the salad and serve.

Makes 4 medium salads; 381 mg. sodium per salad.

Side Dishes

Marinated Vegetables Vinaigrette

For the Vegetables

1 cup carrots, peeled and cut in strips
1 cup broccoli flowerets
1 cup cauliflower flowerets
1 cup green beans, trimmed
1 cup red bell pepper strips, seeded
1 red onion, sliced in rounds
1 cup rinsed, dried, sliced whole mushrooms
10 green olives, pitted and rinsed

Place the vegetables, except mushrooms and olives, in a large saucepan, adding just enough water to cover. Bring to a boil and cook for 3 minutes, then rinse under cold water to stop the cooking process. Set aside with the other vegetables.

For the Vinaigrette

¼ cup white wine
¼ cup safflower oil
¼ cup balsamic vinegar*
½ cup frozen apple juice concentrate
¼ cup apple cider vinegar
4 cups water
2 shallots
2 tablespoons chopped celery
2 cloves peeled garlic
2 tablespoons mixed dried herbs such as marjoram, basil, dill, rosemary

1. Simmer the wine for 2 to 3 minutes to burn off the alcohol. Set aside.

2. When the wine has cooled, combine all the vinaigrette ingredients except the herbs in a blender and purée. Add the herbs.

3. Combine the vegetables and the vinaigrette, stirring well. Refrigerate for 6 to 8 hours or overnight.

4. To serve, arrange leaves of romaine or red leaf lettuce on 6 plates. Center a portion of the marinated vegetables on each leaf.

— *Available at food specialty shops.

Makes 6 servings; 147 mg. sodium per serving.

Caponata

2 medium eggplants, cut in large cubes
2 tablespoons oil
2 medium onions, chopped
2 stalks celery, chopped
1½ pounds fresh plum tomatoes, peeled, or one 28-ounce can Italian plum tomatoes, drained (salt-free if possible)
2 tablespoons unsalted tomato paste
1 tablespoon capers, rinsed well and drained
½ cup frozen apple juice concentrate
⅓ cup red wine vinegar
Few drops Tabasco
½ cup halved pitted black olives, rinsed well and drained

1. In a large non-stick pot, sauté the eggplant in 1 tablespoon oil over low heat for 5 to 10 minutes, until eggplant is semi-soft. Remove the eggplant from the pot and set aside.

2. Add the remaining oil to the pot and sauté the onions in it until they are translucent. Add the celery, tomatoes, tomato paste, and capers, and cook for 5 minutes. Add the eggplant, apple juice concentrate, vinegar, and Tabasco to taste, and cook for 10 minutes or until the vinegar cooks off.

3. Add the olives and cook for 5 minutes more. Serve hot or cold, as an hors d'oeuvre, main dish or a side dish.

Makes 5 cups; 191 mg. sodium per cup.

Carrots with Orange Glaze

4 large carrots, peeled and sliced thin
½ cup fresh-squeezed orange juice
1 teaspoon grated orange rind
2 teaspoons arrowroot or cornstarch, mixed in ½ cup cold
  water

1. In a medium-size saucepan, steam carrots over low heat for 10 minutes or until tender. Drain and set carrots aside.

2. In the same saucepan heat the orange juice and orange rind until simmering. Add the arrowroot or cornstarch mixture to the orange juice and stir until thickened. Add the carrots and cook over low heat for 2 minutes.

Makes 4 servings; 34 mg. sodium per serving.

Parsleyed Carrots

1 pound baby carrots, peeled and trimmed, or regular-size
  carrots peeled and cut in 2-inch diagonal slices
1 tablespoon unsalted butter
1 clove garlic, minced
3 tablespoons finely chopped fresh parsley or spinach
1 teaspoon tamari soy sauce

1. Boil or steam the carrots for 8 to 10 minutes until tender. Drain thoroughly in a colander and set aside.

2. In a saucepan, melt the butter over medium-low heat. Add the garlic, parsley or spinach, and tamari. Cook for 1 minute, then add the carrots, toss together for 2 minutes, and serve.

Makes four 4-ounce servings; 127 mg. sodium per serving.

Mexican Corn

2 cups (6 medium cobs) fresh corn kernels or frozen kernels
Vegetable stock or chicken stock (see page 108), or skim milk
1 to 2 scallions, chopped (bulb plus 1 inch of green stem)
1 red bell pepper, cored, seeded, and chopped fine
1 to 2 tablespoons chopped fresh coriander leaves (cilantro)
   or 1 to 2 teaspoons ground
¼ teaspoon ground cumin (optional)
Freshly ground black pepper to taste

1. In a skillet, cook the corn in a few tablespoons of stock or skim milk for a few minutes.
2. Add the scallions, bell pepper, coriander, and cumin (if you are using it), and cook for 2 minutes more.
3. Season the mixture with pepper and serve.
Makes six ½-cup servings; 32 mg. sodium per serving.

Ratatouille

1 tablespoon olive oil
1 tablespoon peeled minced garlic
1 medium onion, chopped
1 medium eggplant, cut in ½-inch cubes
1 large green pepper, cored, seeded, and cut into 1-inch pieces
4 tomatoes, peeled, seeded, and chopped
2 zucchini, cut into ¼-inch rounds
2 teaspoons chopped fresh basil or 1 teaspoon dry
¾ tablespoon tamari soy sauce or ¼ teaspoon salt
15 whole mushrooms, rinsed, dried, and cut in half
3 tablespoons chopped fresh parsley
¼ cup shredded Swiss cheese

1. Warm the oil in a large skillet or sauté pan over medium heat. Add the garlic and onion and sauté until the onion is translucent.
2. Add the eggplant and green pepper, and cook for a few minutes, stirring occasionally.
3. Add the tomatoes, zucchini, basil, and tamari or salt. Lower the heat and cook, covered, for 25 minutes.
4. Add the mushrooms and parsley and cook for 5 minutes more. Serve warm or cold, topped with the Swiss cheese.
Makes 4 to 5 servings; 205 mg. sodium per ¼ recipe.

Asparagus with Orange-Lemon Sauce

1½ pounds asparagus
Juice of ½ orange or 2 tablespoons frozen orange juice
  concentrate
Juice of ½ lemon (1½ tablespoons)
1 teaspoon arrowroot powder or cornstarch
¼ cup cold water

1. Discard the white lower stalk of the asparagus. Rinse the asparagus well and leave whole or cut diagonally into large pieces.
2. Place the asparagus in a vegetable steamer over 1 inch of boiling water (or in a special asparagus steamer), and cook until tender. Set aside and keep warm.
3. In a small saucepan, mix the orange and lemon juices and heat until warm. Blend the cornstarch or arrowroot and water thoroughly in a small bowl, and add the mixture to the juices. Stir until thick and pour over the asparagus.
Makes 4 servings; 4 mg. sodium per serving.

Peas and Onions

1 tablespoon unsalted butter
½ large onion, sliced
1 10-ounce package frozen peas
Freshly ground black pepper to taste

1. Melt the butter in a medium-size saucepan over medium heat, and add the onion. Sauté for 4 minutes, until translucent.
2. Add the peas, cover, and cook for 8 minutes.
3. When the peas are ready, add freshly ground pepper.
Makes 4 servings; 70 mg. sodium per serving.

Parsley Potatoes

12 to 15 new red potatoes, unpeeled
1 tablespoon unsalted butter
2 tablespoons finely chopped fresh parsley

1. In a large pot bring 1½ quarts water to a boil. Add the potatoes and return the water to a boil. Lower the heat, and simmer, partially covered, for 25 minutes.
2. When cooked, rinse the potatoes under cold water to stop

the cooking process and drain well. Peel the potatoes, removing any green spots.

3. Melt the butter in a large skillet or sauté pan over medium heat. Add the parsley, then the potatoes. Toss to mix well, and serve.

Makes 4 servings; 5 mg. sodium per serving.

Spanish Rice

1 cup brown rice
2 cups beef stock or vegetable stock or chicken stock (see pages 108 and 109) or water
½ cup chopped onion
½ cup chopped green pepper
2 cloves garlic, peeled and minced
1 tablespoon tamari soy sauce
2 tomatoes, peeled and chopped
2 scallions, chopped (bulb plus 1 inch of green stem)
2 teaspoons chili powder

1. Rinse the rice under cold water, picking out any pebbles or odd-looking kernels.

2. In a medium-size saucepan, add a few tablespoons of the stock, reserving the rest, and all the rest of the ingredients, except the rice. Cook over medium heat for a few minutes, then add the rice, stir, and cook an additional 2 minutes.

3. Add the rest of the stock, bring it to a boil, reduce the heat to low, and simmer, covered, for 40 minutes.

Makes six ½-cup servings; 164 mg. sodium per serving.

Barley Almond Pilaf

1 cup barley
1 tablespoon safflower oil
4 tablespoons chopped scallions (bulb and 1 inch of green stem)
1 tablespoon tamari soy sauce or ⅓ teaspoon salt
2 tablespoons slivered almonds (optional)
3 cups vegetable stock or chicken stock (see page 108), or water

1. Place the barley in a colander and rinse it under cold water to remove the outside starch.

2. In a medium-size saucepan, heat the oil, then add the scallions, and sauté over medium heat for 1 minute. Add the barley and sauté for 3 to 4 minutes, stirring constantly. Add the tamari or salt, the almonds (if you are using them), and the stock or water. Bring the mixture to a boil, then reduce the heat and simmer, covered, for 1 hour, or until liquid evaporates.

3. After an hour, turn off the heat and let the barley sit for 5 minutes, covered.

Makes eight ½-cup servings; 114 mg. sodium per serving.

Brown and Wild Rice

A nice change from the standard rice recipes.

¾ cup brown rice (long or short grain)
¾ cup wild rice
1 tablespoon safflower oil
3 cups chicken stock or fish stock or beef stock (see pages 107 to 109) or water
1 teaspoon crushed dried basil, rosemary, thyme, or other seasoning of your choice, or a pinch of saffron
2 teaspoons tamari soy sauce

1. Rinse the brown and wild rice under cold water.

2. In a 1-quart saucepan heat the oil, then add the rice and sauté until toasted, about 5 minutes. Stir the rice as it toasts to coat the individual grains with oil.

3. Add the stock or water, seasoning, and tamari, and bring to a boil, covered. Lower the heat and cook undisturbed for 45 minutes or until all the water has been absorbed. Remove the rice from the heat, and keeping the lid on, allow it to steam for 5 to 10 minutes.

Makes eight ½-cup servings; 81 mg. sodium per serving.

Bulgur-Wheat Pilaf

2 cups water
1 cup bulgur wheat
2 tablespoons vegetable stock (see page 108)
1 cup chopped onion
Juice of 1 lemon
¼ cup chopped scallion
2 tablespoons chopped fresh dill or 1 tablespoon dried

1 clove garlic, minced
¼ cup chopped parsley

1. In a small saucepan, boil the water and add the bulgur. Remove from the heat, and allow to sit for 1 hour, until the bulgur has absorbed all the water.

2. In a large skillet, heat 2 tablespoons vegetable stock and the lemon juice and cook the remaining ingredients in the liquid until the onion is translucent.

3. Add the vegetable mixture to the bulgur and toss thoroughly. Serve hot or cool. (To reheat, sauté pilaf briefly in a skillet with a little more vegetable stock.)

Makes eight ½-cup servings; 4 mg. sodium per serving.

Barley Rice

1 tablespoon oil or 2 tablespoons water or stock
1 cup brown rice, short or long grain, rinsed
1 cup barley, rinsed
4 cups water or stock
2 scallions, chopped fine

1. Heat the oil (or water or stock) in a 4-quart saucepan. Add the rice and barley and sauté over medium heat for 2 minutes.

2. Add the 4 cups of water or stock and bring to a boil. Reduce the heat, cover, and simmer for 1 hour, until all the liquid is absorbed.

3. Serve with a garnish of chopped scallions.

Makes four ¾-cup servings; 22 mg. sodium per serving.

Carrot-Raisin Salad

3 cups grated peeled carrots (approximately 7 carrots)
½ cup raisins
1 cup nonfat plain yogurt
¼ cup orange juice
1 teaspoon lemon juice

1. Set aside ½ cup of the grated carrots. Combine the remaining carrots and the raisins in a bowl.

2. Place the ½ cup of carrots in a food processor or blender. Add the yogurt, and orange and lemon juices, and purée.

3. Stir the yogurt mixture into the remaining carrot-raisin mixture, combining it well.

4. Chill and serve cold.

Makes eight ½-cup servings; 53 mg. sodium per serving.

Polenta

5 cups water
1 cup cornmeal, preferably undegerminated (whole-grain)*
1 tablespoon tamari soy sauce
2 tablespoons grated Parmesan cheese for garnish

1. Bring 2½ cups of the water to a boil in a large saucepan. Meanwhile mix the cornmeal with the remaining 2½ cups water. Mix well to remove any lumps.

2. Add the cornmeal mixture and tamari to the boiling water, return to a boil, then lower the heat and simmer, uncovered, stirring constantly, for 30 to 40 minutes. The mixture will thicken. Pour the mixture into a 9- x 9-inch cake pan and allow it to cool.

Polenta may be served warm or cool. To warm, heat it in a 300° F oven for 30 minutes. Serve with warm Tomato Basil Sauce (see page 100) and grated Parmesan cheese.

Makes 8 servings; 128 mg. sodium per serving.

*Available in health-food stores.

Corn Bread

This corn bread bakes up dense and grainy, not light and fluffy like some. It goes really well with any of the chicken dishes.

1¾ cups cornmeal, preferably undegerminated (whole-grain)*
¾ cup whole-wheat flour, preferably pastry or graham
1½ teaspoons low sodium baking powder
1 small carrot, peeled and grated
3 tablespoons safflower oil or melted unsalted butter
3 tablespoons honey or sugar
1 egg
1½ cups skim milk

1. Preheat the oven to 425° F (if you are using a glass baking pan, preheat to 400° F). In a large bowl, combine the first four ingredients.

2. In a second bowl, combine the oil or butter with the honey or other sweetener, then add the egg, and finally the milk. Beat well after each addition.

3. Add the flour mixture to the oil mixture and stir until the ingredients are well combined.

4. Pour the batter into an oiled 8- x 8-inch baking pan. Bake for 20 minutes or until a knife inserted in the center comes out clean.

*Available at health-food shops.

Makes nine 2½- x 2½-inch squares; 122 mg. sodium per square.

Desserts

Mom's Baked Fruit Compote

1 pound pitted dried prunes
1 pound pitted dried apricots
1 16-ounce can pineapple chunks or 1 pound fresh pineapple
  cut in chunks
1 8-ounce can sliced peaches (without sugar) or ½ pound
  fresh peaches, peeled and sliced
1 8-ounce bag frozen pitted cherries (without sugar)
½ cup red or white wine

1. Soak the dried fruit in water to cover for 1 hour. Preheat the oven to 350° F.

2. Drain the dried fruit and the canned fruit.

3. Place layers of fruit in a large baking dish: first the prunes, then the apricots, the pineapple, the peaches, and finally the cherries.

4. Pour the wine over the fruit and bake, covered, 1 hour.

Makes 8 to 10 servings; about 10 mg. sodium per serving.

Fruity Frozen Yogurt

2 very ripe medium-size bananas
2 cups nonfat plain yogurt
1 6-ounce can frozen orange juice concentrate
1 6-ounce can frozen pineapple juice concentrate
1 teaspoon vanilla extract

1. Purée the bananas, yogurt, frozen juices, and vanilla in a blender or food processor. (Do half the ingredients at a time if the bowl is not big enough to hold it all.)

2. If you are using an ice cream freezer, follow the machine

instructions. If you are using your refrigerator freezer, pour the mixture into a large glass or ceramic bowl or a plastic container, cover, and place it in the freezer. Stir every 2 hours. The frozen yogurt will be ready in 8 hours. If you wish, smooth the frozen yogurt in the blender or processor before serving.

Makes 1 quart; 84 mg. sodium per cup.

Poached Bananas

1 cup frozen apple juice concentrate
½ cup port wine or burgundy
1 teaspoon vanilla extract
1 cup water
6 to 8 medium bananas
2 cups plain yogurt (optional)

1. Place all the ingredients, except the bananas and yogurt, in a saucepan large enough to hold the bananas, and bring to a boil over medium heat.

2. Peel the bananas, then add them to the saucepan and cook, uncovered, for 15 minutes.

3. Remove the saucepan from the heat and allow the bananas to cool in the poaching liquid. Serve them either whole, halved lengthwise, or in slices. Spoon the poaching liquid over the bananas and pass the yogurt (if you are serving it), allowing guests to help themselves.

Makes 6 to 8 servings; about 58 mg. sodium per serving.

Apple Mocha Kanten (Fruit Gel)

1 Roastaroma tea bag* or 1 tablespoon Postum beverage
*   powder or 1 tablespoon instant coffee*
1 tablespoon unflavored gelatin or 2 tablespoons agar flakes
1 cup boiling water
1 cup frozen apple juice concentrate
1 cup sugar-free apple sauce
Ground cinnamon to taste

1. Brew the Roastaroma tea, Postum, or coffee in the boiling water. Add the gelatin or agar and soak for 20 minutes.

2. Put the brewed liquid in a small saucepan and bring it to a boil. Add the apple juice concentrate, apple sauce, and cinnamon to taste. Lower the heat and simmer for 5 minutes.

3. Pour the mixture into a glass or ceramic mold and refrigerate until set, an hour or two.

Makes four ¾-cup servings; 6 mg. sodium per serving.

*Celestial Seasonings makes these—available in supermarkets and health-food stores.

Prune and Apricot Soufflé

A deliciously light but rich-tasting dessert.

2 ounces dried, pitted prunes
2 ounces dried apricots
1⅛ cups water
⅓ cup honey
1 tablespoon lemon juice
Unsalted butter
3 large egg whites

1. Place the prunes and apricots in the water in a medium-size saucepan and soak for 1 hour.

2. After an hour, bring the mixture to a boil over medium heat, then lower the heat and simmer, uncovered, for 15 minutes. Remove the pan from the heat and allow the fruit to cool for a few minutes.

3. When cooled, place the fruit in a processor or blender and purée the mixture for 30 seconds. Add the honey and lemon juice, process for a few seconds, then transfer the mixture to a bowl, and refrigerate for 30 minutes.

4. Lightly butter 5 individual soufflé molds and chill them for 20 minutes.

5. Preheat the oven to 375° F.

6. Whip the egg whites with a beater or wire whisk until they form stiff peaks. Gently fold half the egg whites into the fruit mixture. When it is well incorporated, fold in the second half of the egg whites. Spoon the soufflé mixture into the chilled molds.

7. Place the soufflé molds in a baking dish. Add enough hot water to reach halfway up the molds. Put the baking dish in the oven and bake for 30 minutes.

Makes 5 servings; 167 mg. sodium per serving.

Brown Rice Pudding

1 cup brown rice, rinsed and picked over
1 quart skim milk

1 can evaporated skim milk
1 to 2 tablespoons unsalted butter
1 teaspoon vanilla extract
1 teaspoon ground cinnamon, plus extra for serving
2 eggs
⅓ cup honey
2 to 3 tablespoons rum or praline liqueur or Kahlúa
¾ cup raisins
¼ teaspoon ground nutmeg
¼ cup slivered almonds, lightly toasted

1. Put the rice, milk, evaporated milk, butter, vanilla, and cinnamon in a 2-quart saucepan. Bring to a boil over medium heat, stirring occasionally. Reduce the heat and simmer, uncovered, for 1 hour, until the rice cooks and the milk reduces in volume.

2. In a bowl, mix the eggs with the honey and liquor. Add this to the rice mixture, then add the raisins, nutmeg, and almonds. Continue cooking. Stir as the mixture thickens. When it is of custard thickness, after 20 minutes or so, spoon the mixture into custard cups, sprinkle with additional cinnamon, and refrigerate.

Makes seven to eight ½-cup servings; about 102 mg. sodium per serving.

Oatmeal Raisin Cookies

1 cup rolled oats
1 cup whole-wheat pastry flour
¼ cup unbleached white flour
1 cup raisins
1 teaspoon low sodium baking powder
¾ cup honey or barley malt extract syrup
½ teaspoon vanilla extract
½ cup evaporated skim milk
2 egg whites
Cooking parchment or vegetable oil

1. Preheat the oven to 350° F. In a large bowl mix together the first five ingredients. Set aside. In another bowl, mix the honey or malt syrup, vanilla, and skim milk. Set aside.

2. Beat the egg whites until stiff peaks form. Mix the liquid ingredients into the dry ingredients, and then gently fold in the egg whites.

3. Line 2 large cookie sheets with parchment paper or rub them thoroughly with oil. Spoon approximately 1 tablespoon of batter per cookie onto the sheets. Bake for 15 to 18 minutes.

4. Remove the cookies from the sheets and set them on a wire rack to cool.

Makes about 30 cookies; 10 mg. sodium per cookie.

Whole-Wheat Chocolate or Carob Chip Cookies

1½ cups whole-wheat pastry flour
1½ teaspoon low sodium baking powder
6 ounces chocolate
¼ cup chopped walnuts
½ cup unsalted butter, softened
⅔ cup honey
1 egg
⅓ cup skim milk

1. Preheat the oven to 375° F. In a bowl, mix together the first four ingredients.

2. In a large bowl, blend the butter with the honey, then add the egg and the skim milk.

3. Add the flour mixture to the butter mixture and stir to mix well.

4. Drop the batter by the tablespoonful onto two non-stick cookie sheets and bake for 12 to 15 minutes.

5. Remove the cookies from the sheets and cool on a wire cooling rack.

Makes about 24 cookies; 8 mg. sodium per cookie.

PART THREE
THE SODIUM LISTINGS

The Brand Name Listings

We believe that the following brand name listing contains the most accurate, complete, and current information available on the sodium content of the more than 5,000 foods and drugs included. We have divided the foods into about 30 categories, such as "Baby Food," "Cereal," and "Soups." Many of these categories are then subdivided by another relevant characteristic. For example, "Fruits" breaks down into "Fruits: Canned," "Fruits: Dried," "Fruits: Fresh," and "Fruits: Frozen." Under each of these subdivisions come products listed by food manufacturer (or under "USDA" if there is no brand name). The 30 categories of food appear in alphabetical order, as do the names of the companies and foods within each subdivision. Take a moment to glance through the listing and become familiar with the way it is organized.

You may be wondering how we gathered so much information about sodium. This listing was pieced together using several sources. Chief among them was information we solicited from food companies first in late 1982 and again, in late 1984. Unfortunately many other companies have not made public the sodium content of their foods. And still others, especially small regional firms, have not even analyzed their products. In a few cases, we obtained data from food labels or by telephone conversations with company spokespersons. For fresh or dried nonbranded foods, such as dried beans, peas, milk, or spinach, we used sodium data from the U.S. Department of Agriculture (*Home and Garden Bulletin No. 233* and *Handbook No. 456*).

At this time, it is a precarious task to list the sodium contents of brand name foods. Many companies are responding to public pressure for less salt by reformulating their products. Some of these companies will list the new sodium level on their nutrition label. We suggest you regard available label information as the more current.

You should know, however, that the sodium levels printed on labels are often higher than the levels companies have sent us (or any inquiring consumer). That's because the companies often send out averages, and the average sodium content of, let's say, 10 cans of chicken noodle soup, may understate the sodium content of the one can of soup you take off the shelf. FDA and USDA allow small understatements, but not large ones. Therefore, companies often raise the label levels above the average. That means that your can of soup may contain somewhat less sodium than the label says, but it may also contain as much or slightly more.

We must add just one word of caution. When comparing our figures to food labels or comparing one of our brand listings to another, pay close attention to serving sizes. Two slices of Brand X bread may appear to be lower in sodium than Brand Y simply because Brand X's serving sizes are smaller. For example, Brand X bread may be lower in sodium because it is sliced thinner. In preparing the brand name listing, we tried to choose serving sizes that represent what most people eat. For some foods, these estimates came from the U.S. Department of Agriculture (*Home Economics Research Report No. 44*). In other cases, we used serving sizes recommended by individual companies.

The best way to use this listing is to identify foods that you normally consume. If a food is high in sodium, check to see whether a comparable brand name food contains less sodium than your regular product. If no lower-sodium brand exists, consider substituting another type of food in your diet. When you compare foods, make sure the difference in sodium levels is significant. A 10- or 20-milligram difference is not enough to worry about; a 100-milligram difference is.

Use the brand name listing in good health.

ABBREVIATIONS

The brand name listing includes a few abbreviations with which you should become familiar.

fl. oz. ....*fluid ounce(s)*
env. ....*envelope(s)*
oz. ....*ounce(s)*
recon. ....*reconstituted*
tbl. ....*tablespoon(s)*
tsp. ....*teaspoon(s)*
w/ ....*with*
w/o ....*without*

Baby Food

INFANT FOODS

Baked Goods   PORTION   SODIUM (MG.)

Gerber

Animal Shaped Cookies............0.4 oz. (2 cookies) 22
Arrowroot Cookies ....0.4 oz. (2 cookies) 43
Biscuits...................................0.4 oz. (1 biscuit) 31
Pretzels ....0.4 oz. (1 pretzel) 29
Zwieback Toast ....0.5 oz. (2 toasts)....................27

Dry and Strained Cereal

Beech-Nut

Barley Cereal ....0.5 oz. dry (4 tbl.) 10
Hi-Protein Cereal ....0.5 oz. dry (4 tbl.) 10
Mixed Cereal ....0.5 oz. dry (4 tbl.) 10
Mixed Cereal w/Applesauce &
  Bananas.................................4.5 oz. (1 jar) 10
Oatmeal ....0.5 oz. dry (4 tbl.) 10
Oatmeal w/Applesauce &
  Bananas.................................4.5 oz. (1 jar) 10
Rice Cereal ....0.5 oz. dry (4 tbl.) 10
Rice w/Applesauce &
  Bananas.................................4.5 oz. (1 jar) 30

Gerber

Barley Cereal ....0.5 oz. dry (4 tbl.) 4
High Protein Cereal....................0.5 oz. dry (4 tbl.) 3
High Protein Cereal w/Apple
  & Orange.................................0.5 oz. dry (4 tbl.) 15
Mixed Cereal ....0.5 oz. dry (4 tbl.) 4
Mixed Cereal w/Applesauce
  & Bananas ....4.5 oz. (1 jar) 3
Mixed Cereal w/Banana............0.5 oz. dry (4 tbl.) 13
Oatmeal.....................................0.5 oz. dry (4 tbl.) 4
Oatmeal w/Applesauce &
  Bananas.................................4.5 oz. (1 jar) 3
Oatmeal w/Banana ....0.5 oz. dry (4 tbl.) 14
Rice Cereal................................0.5 oz. dry (4 tbl.) 4

PORTION   SODIUM (MG.)

Rice Cereal w/Applesauce
   & Bananas ....4.8 oz. (1 jar) 8
Rice Cereal w/Banana ....0.5 oz. dry (4 tbl.) 11

Heinz

Barley Instant Cereal ....8 oz. prepared 52
Cereal & Eggs ....4.5 oz. (1 jar) 26
Hi-Protein Instant Cereal ....8 oz. prepared 222
Instant Mixed Cereal................8 oz. prepared 141
Instant Oatmeal ....8 oz. prepared 32
Instant Rice Cereal ....8 oz. prepared 45
Mixed Cereal w/Apples
   & Bananas ....4.8 oz. (1 jar) 11
Oatmeal w/Apples &
   Bananas....................................4.8 oz. (1 jar) 5
Rice Cereal w/Apples &
   Bananas....................................4.8 oz. (1 jar) 4

Strained Desserts

Beech-Nut

Apple Betty ....4.5 oz. (1 jar) 15
Apple Custard Pudding ....4.5 oz. (1 jar) 40
Banana Custard Pudding..........4.5 oz. (1 jar) 35
Chocolate Custard Pudding......4.5 oz. (1 jar) 35
Cottage Cheese w/Pineapple
   Juice.....................................4.5 oz. (1 jar) 30
Fruit Dessert.............................4.5 oz. (1 jar) 25
Orange Pineapple Dessert ......4.5 oz. (1 jar) 15
Pineapple Dessert....................4.5 oz. (1 jar) 25
Rice Pudding ....4.5 oz. (1 jar) 40
Vanilla Custard Pudding............4.5 oz. (1 jar) 45

Gerber

Banana-Apple Dessert ....4.8 oz. (1 jar) 3
Cherry-Vanilla Pudding ....4.5 oz. (1 jar) 9
Chocolate Custard Pudding......4.5 oz. (1 jar) 31
Dutch Apple Dessert..................4.8 oz. (1 jar) 27
Fruit Dessert..............................4.8 oz. (1 jar) 14
Hawaiian Delight ....4.5 oz. (1 jar) 23

PORTION   SODIUM (MG.)

Orange Pudding	4.8 oz. (1 jar)	28
Peach Cobbler	4.8 oz. (1 jar)	9
Raspberry Dessert w/Nonfat Yogurt	4.5 oz. (1 jar)	29
Vanilla Custard Pudding	4.5 oz. (1 jar)	31

Heinz

Banana Pudding	4.8 oz. (1 jar)	12
Cottage Cheese w/Bananas	4.5 oz. (1 jar)	43
Custard Pudding	4.5 oz. (1 jar)	34
Dutch Apple Dessert	7.8 oz. (1 jar)	20
Fruit Dessert	4.8 oz. (1 jar)	19
Peach Cobbler	4.5 oz. (1 jar)	8
Pineapple Orange Dessert	4.5 oz. (1 jar)	19
Tutti Frutti	4.5 oz. (1 jar)	32

Strained Fruit, Fruit and Yogurt

Beech-Nut

Apples & Apricots	4.5 oz. (1 jar)	10
Apples & Grapes	4.5 oz. (1 jar)	10
Apples & Strawberries	4.5 oz. (1 jar)	10
Apples, Mandarin Oranges & Bananas	4.5 oz. (1 jar)	10
Apples, Peaches & Strawberries	4.5 oz. (1 jar)	10
Apples, Pears & Bananas	4.5 oz. (1 jar)	10
Apples, Pears & Pineapples	4.5 oz. (1 jar)	10
Applesauce	4.5 oz. (1 jar)	10
Applesauce & Bananas	4.5 oz. (1 jar)	10
Applesauce & Cherries	4.5 oz. (1 jar)	10
Applesauce & Raspberries	4.5 oz. (1 jar)	10
Apricots w/Tapioca	4.5 oz. (1 jar)	25
Bananas & Pineapple w/Tapioca	4.5 oz. (1 jar)	15
Bananas w/Tapioca	4.5 oz. (1 jar)	15
Guava w/Tapioca	4.5 oz. (1 jar)	10
Island Fruits	4.5 oz. (1 jar)	10
Mango w/Tapioca	4.5 oz. (1 jar)	10
Mixed Fruit w/Yogurt	4.5 oz. (1 jar)	25

PORTION   SODIUM (MG.)

Papaya w/Tapioca	4.5 oz. (1 jar)	10
Peach Apple w/Yogurt	4.5 oz. (1 jar)	30
Peaches	4.5 oz. (1 jar)	10
Pears	4.5 oz. (1 jar)	10
Pears & Pineapples	4.5 oz. (1 jar)	10
Plums w/Tapioca	4.5 oz. (1 jar)	10
Prunes w/Tapioca	4.5 oz. (1 jar)	10

Gerber

Apple & Nonfat Yogurt	4.5 oz. (1 jar)	22
Apple Blueberry	4.5 oz. (1 jar)	4
Applesauce	4.5 oz. (1 jar)	3
Applesauce & Apricots	4.5 oz. (1 jar)	3
Applesauce w/Pineapple	4.5 oz. (1 jar)	3
Apricots w/Tapioca	4.8 oz. (1 jar)	5
Bananas & Nonfat Yogurt	4.5 oz. (1 jar)	22
Bananas w/Pineapple & Tapioca	4.5 oz. (1 jar)	6
Bananas w/Tapioca	4.5 oz. (1 jar)	17
Guava	4.5 oz. (1 jar)	5
Guava & Papaya	4.5 oz. (1 jar)	6
Mango	4.8 oz. (1 jar)	5
Mixed Fruit & Nonfat Yogurt	4.5 oz. (1 jar)	23
Papaya & Applesauce	4.5 oz. (1 jar)	6
Peaches	4.8 oz. (1 jar)	5
Pears	4.5 oz. (1 jar)	4
Pears & Pineapple	4.5 oz. (1 jar)	3
Plums w/Tapioca	4.8 oz. (1 jar)	5
Prunes w/Tapioca	4.8 oz. (1 jar)	13

Heinz

Apples & Apricots	4.5 oz. (1 jar)	1
Apples & Cranberries w/Tapioca	4.8 oz. (1 jar)	9
Apples & Pears	4.5 oz. (1 jar)	4
Applesauce	4.5 oz. (1 jar)	4
Apricots w/Tapioca	4.8 oz. (1 jar)	17
Bananas & Pineapple w/Tapioca	4.8 oz. (1 jar)	20
Bananas w/Tapioca	4.8 oz. (1 jar)	10

PORTION   SODIUM (MG.)

Peaches	4.5 oz. (1 jar)	8
Pears	4.5 oz. (1 jar)	0
Pears & Pineapple	4.5 oz. (1 jar)	1
Plums w/Tapioca	4.5 oz. (1 jar)	9
Prunes w/Tapioca	4.8 oz. (1 jar)	12

Strained Juices

Beech-Nut

Apple Cherry Juice	4.2 oz. (1 jar)	10
Apple Cranberry	4.2 oz. (1 jar)	10
Apple Grape Juice	4.2 oz. (1 jar)	10
Apple Juice	4.2 oz. (1 jar)	10
Apple Peach Juice	4.2 oz. (1 jar)	10
Apple Prune Juice	4.2 oz. (1 jar)	10
Mixed Juice	4.2 oz. (1 jar)	10
Orange Banana Juice	4.2 oz. (1 jar)	10
Orange Juice	4.2 oz. (1 jar)	10
Orange Pineapple Juice	4.2 oz. (1 jar)	10

Gerber

Apple Juice	4.2 oz. (1 jar)	2
Apple-Banana Juice	4.2 oz. (1 jar)	3
Apple-Cherry Juice	4.2 oz. (1 jar)	0
Apple-Grape Juice	4.2 oz. (1 jar)	2
Apple-Peach Juice	4.2 oz. (1 jar)	4
Apple-Plum Juice	4.2 oz. (1 jar)	4
Apple-Prune Juice	4.2 oz. (1 jar)	5
Mixed Fruit Juice	4.2 oz. (1 jar)	2
Orange Juice	4.2 oz. (1 jar)	4
Orange-Apple Juice	4.2 oz. (1 jar)	5
Orange-Apricot Juice	4.2 oz. (1 jar)	7
Orange-Pineapple Juice	4.2 oz. (1 jar)	0

Heinz

Apple Juice	4.2 oz. (1 jar)	5
Apple-Apricot	4.2 oz. (1 jar)	5
Apple-Cherry	4.2 oz. (1 jar)	6
Apple-Grape	4.2 oz. (1 jar)	7
Apple-Peach	4.2 oz. (1 jar)	5

PORTION   SODIUM (MG.)

	PORTION	SODIUM (MG.)
Apple-Pineapple	4.2 oz. (1 jar)	7
Apple-Prune	4.2 oz. (1 jar)	8
Mixed Fruit	4.2 oz. (1 jar)	14
Orange	4.2 oz. (1 jar)	7
Orange-Apple-Banana	4.2 oz. (1 jar)	14

Strained Main Dishes

Beech-Nut

	PORTION	SODIUM (MG.)
Beef & Egg Noodles	4.5 oz. (1 jar)	35
Beef w/Vegetables & Cereal (High Meat)	4.5 oz. (1 jar)	40
Cereal, Egg Yolks & Bacon	4.5 oz. (1 jar)	90
Chicken Noodle	4.5 oz. (1 jar)	40
Chicken Rice	4.5 oz. (1 jar)	15
Chicken w/Vegetables & Cereal (High Meat)	4.5 oz. (1 jar)	35
Ham w/Vegetables & Cereal (High Meat)	4.5 oz. (1 jar)	40
Macaroni, Tomato & Beef	4.5 oz. (1 jar)	35
Turkey Rice	4.5 oz. (1 jar)	40
Vegetable Bacon	4.5 oz. (1 jar)	140
Vegetable Beef	4.5 oz. (1 jar)	35
Vegetable Chicken	4.5 oz. (1 jar)	35
Vegetable Ham	4.5 oz. (1 jar)	30
Vegetable Lamb w/Rice & Barley	4.5 oz. (1 jar)	25
Vegetable Liver w/Rice & Barley	4.5 oz. (1 jar)	35
Vegetable Turkey	4.5 oz. (1 jar)	25

Gerber

	PORTION	SODIUM (MG.)
Beef & Egg Noodles w/Vegetables	4.5 oz. (1 jar)	19
Beef w/Vegetables (High Meat)	4.5 oz. (1 jar)	31
Cereal & Egg Yolk	4.5 oz. (1 jar)	17
Chicken & Noodles	4.5 oz. (1 jar)	20
Chicken w/Vegetables (High Meat)	4.5 oz. (1 jar)	33

PORTION   SODIUM (MG.)

Cottage Cheese w/Pineapple	4.5 oz. (1 jar)	194
Cream of Chicken Soup	4.5 oz. (1 jar)	29
Ham w/Vegetables		
(High Meat)	4.5 oz. (1 jar)	22
Macaroni & Cheese	4.5 oz. (1 jar)	102
Macaroni-Tomato w/Beef	4.5 oz. (1 jar)	18
Turkey & Rice w/Vegetables	4.5 oz. (1 jar)	23
Turkey w/Vegetables		
(High Meat)	4.5 oz. (1 jar)	37
Veal w/Vegetables		
(High Meat)	4.5 oz. (1 jar)	27
Vegetables & Bacon	4.5 oz. (1 jar)	79
Vegetables & Beef	4.5 oz. (1 jar)	17
Vegetables & Chicken	4.5 oz. (1 jar)	14
Vegetables & Ham	4.5 oz. (1 jar)	15
Vegetables & Lamb	4.5 oz. (1 jar)	15
Vegetables & Liver	4.5 oz. (1 jar)	19
Vegetables & Turkey	4.5 oz. (1 jar)	20

Heinz

Beef & Egg Noodles	4.5 oz. (1 jar)	24
Beef w/Vegetables		
(High Meat Dinner)	4.5 oz. (1 jar)	37
Chicken Noodle Dinner	4.5 oz. (1 jar)	31
Chicken Soup	4.5 oz. (1 jar)	27
Chicken w/Vegetables (High		
Meat Dinner)	4.5 oz. (1 jar)	34
Macaroni, Tomatoes & Beef	4.5 oz. (1 jar)	24
Turkey Rice Dinner		
w/Vegetables	4.5 oz. (1 jar)	23
Turkey w/Vegetables (High		
Meat Dinner)	4.5 oz. (1 jar)	36
Vegetables & Bacon	4.5 oz. (1 jar)	56
Vegetables & Beef	4.5 oz. (1 jar)	27
Vegetables & Ham	4.5 oz. (1 jar)	20
Vegetables & Lamb	4.5 oz. (1 jar)	33
Vegetables, Dumplings &		
Beef	4.5 oz. (1 jar)	19
Vegetables, Egg Noodles &		
Turkey	4.5 oz. (1 jar)	33

PORTION   SODIUM (MG.)

Vegetables, Egg Noodles &
  Chicken ....4.5 oz. (1 jar)........................36

Strained Meats and Eggs

Beech-Nut

Beef & Beef Broth..................3.5 oz. (1 jar) 75
Chicken & Chicken Broth.........3.5 oz. (1 jar) 70
Ham & Ham Broth..................3.5 oz. (1 jar) 60
Lamb & Lamb Broth ....3.5 oz. (1 jar) 75
Turkey & Turkey Broth ....3.5 oz. (1 jar) 60
Veal & Veal Broth ....3.5 oz. (1 jar)........................70

Gerber

Beef ....3.5 oz. (1 jar) 52
Beef Liver ....3.5 oz. (1 jar) 47
Beef w/Beef Heart..................3.5 oz. (1 jar) 58
Chicken ....3.5 oz. (1 jar) 39
Egg Yolks ....3.3 oz. (1 jar) 45
Ham ....3.5 oz. (1 jar) 40
Lamb ....3.5 oz. (1 jar) 51
Pork ....3.5 oz. (1 jar) 38
Turkey....................................3.5 oz. (1 jar) 56
Veal..3.5 oz. (1 jar) 52

Heinz

Beef & Beef Broth..................3.5 oz. (1 jar) 46
Chicken & Chicken Broth.........3.5 oz. (1 jar) 44
Lamb & Lamb Broth ....3.5 oz. (1 jar) 52
Liver & Liver Broth ....3.5 oz. (1 jar) 37
Turkey & Turkey Broth ....3.5 oz. (1 jar) 43
Veal & Veal Broth ....3.5 oz. (1 jar) 47

Strained Vegetables

Beech-Nut

Beets ....4.5 oz. (1 jar) 90
Carrots ....4.5 oz. (1 jar) 120
Creamed Corn ....4.5 oz. (1 jar) 20
Garden Vegetables ....4.5 oz. (1 jar) 50

PORTION   SODIUM (MG.)

Green Beans	4.5 oz. (1 jar)	10
Mixed Vegetables	4.5 oz. (1 jar)	35
Peas	4.5 oz. (1 jar)	10
Peas & Carrots	4.5 oz. (1 jar)	50
Squash	4.5 oz. (1 jar)	10
Sweet Potatoes	4.5 oz. (1 jar)	60

Gerber

Beets	4.5 oz. (1 jar)	119
Carrots	4.5 oz. (1 jar)	47
Creamed Corn	4.5 oz. (1 jar)	11
Creamed Spinach	4.5 oz. (1 jar)	49
Garden Vegetables	4.5 oz. (1 jar)	28
Green Beans	4.5 oz. (1 jar)	4
Mixed Vegetables	4.5 oz. (1 jar)	27
Peas	4.5 oz. (1 jar)	4
Squash	4.5 oz. (1 jar)	5
Sweet Potatoes	4.5 oz. (1 jar)	23

Heinz

Beets	4.5 oz. (1 jar)	128
Carrots	4.5 oz. (1 jar)	28
Creamed Corn	4.5 oz. (1 jar)	31
Creamed Peas	4.5 oz. (1 jar)	15
Green Beans	4.5 oz. (1 jar)	1
Mixed Vegetables	4.5 oz. (1 jar)	24
Squash	4.5 oz. (1 jar)	0
Sweet Potatoes	4.8 oz. (1 jar)	36

JUNIOR FOODS

Cereal

Heinz

Cereal & Eggs	7.5 oz. (1 jar)	68

Cereal and Fruit

Heinz

Mixed Cereal w/Applesauce & Bananas	7.5 oz. (1 jar)	9

PORTION   SODIUM (MG.)

Oatmeal w/Applesauce & Bananas	7.5 oz. (1 jar)	4
Rice Cereal w/Mixed Fruit	7.8 oz. (1 jar)	17

Desserts

Beech-Nut

Apple Betty	7.5 oz. (1 jar)	25
Apple Custard Pudding	7.5 oz. (1 jar)	60
Banana Custard Pudding	7.5 oz. (1 jar)	70
Banana Dessert	7.5 oz. (1 jar)	10
Cottage Cheese w/Pineapple	7.5 oz. (1 jar)	50
Fruit Dessert	7.5 oz. (1 jar)	45
Rice Pudding	7.5 oz. (1 jar)	65
Tropical Fruit Dessert	7.5 oz. (1 jar)	25
Vanilla Custard Pudding	7.5 oz. (1 jar)	70

Gerber

Banana-Apple Dessert	7.8 oz. (1 jar)	9
Cherry Vanilla Pudding	7.5 oz. (1 jar)	17
Dutch Apple Dessert	7.8 oz. (1 jar)	46
Fruit Dessert	7.8 oz. (1 jar)	64
Hawaiian Delight	7.8 oz. (1 jar)	40
Peach Cobbler	7.8 oz. (1 jar)	20
Raspberry Dessert w/Nonfat Yogurt	7.5 oz. (1 jar)	51
Vanilla Custard Pudding	7.8 oz. (1 jar)	51

Heinz

Cottage Cheese w/Bananas	7.8 oz. (1 jar)	72
Custard Pudding	7.8 oz. (1 jar)	59
Dutch Apple Dessert	7.8 oz. (1 jar)	20
Fruit Dessert	7.8 oz. (1 jar)	33
Peach Cobbler	7.8 oz. (1 jar)	13
Pineapple Orange Dessert	7.8 oz. (1 jar)	33
Tutti Frutti	7.8 oz. (1 jar)	55

PORTION   SODIUM (MG.)

Fruit, Fruit and Yogurt

Beech-Nut

Apples & Apricots	7.5 oz. (1 jar)	10
Apples & Grapes	7.5 oz. (1 jar)	10
Apples & Strawberries	7.5 oz. (1 jar)	15
Apples, Mandarin Oranges & Bananas	7.5 oz. (1 jar)	10
Apples, Peaches & Strawberries	7.5 oz. (1 jar)	15
Apples, Pears & Bananas	7.5 oz. (1 jar)	10
Apples, Pears & Pineapples	7.5 oz. (1 jar)	10
Applesauce	7.5 oz. (1 jar)	10
Applesauce & Bananas	7.5 oz. (1 jar)	10
Applesauce & Cherries	7.5 oz. (1 jar)	10
Applesauce & Raspberries	7.5 oz. (1 jar)	10
Apricots w/Tapioca	7.5 oz. (1 jar)	40
Banana Pineapple w/Tapioca	7.5 oz. (1 jar)	10
Bananas w/Tapioca	7.5 oz. (1 jar)	40
Guava w/Tapioca	7.5 oz. (1 jar)	15
Island Fruit	7.5 oz. (1 jar)	15
Mango w/Tapioca	7.5 oz. (1 jar)	15
Mixed Fruit w/Yogurt	7.5 oz. (1 jar)	45
Papaya w/Tapioca	7.5 oz. (1 jar)	15
Peach Apple w/Yogurt	7.5 oz. (1 jar)	50
Peaches	7.5 oz. (1 jar)	10
Pears	7.5 oz. (1 jar)	10
Pears & Pineapple	7.5 oz. (1 jar)	10
Plums w/Tapioca	7.5 oz. (1 jar)	20

Gerber

Apple Blueberry	7.5 oz. (1 jar)	6
Applesauce	7.5 oz. (1 jar)	2
Applesauce & Apricots	7.5 oz. (1 jar)	6
Apricots w/Tapioca	7.8 oz. (1 jar)	9
Bananas w/Pineapple & Tapioca	7.5 oz. (1 jar)	9
Bananas w/Tapioca	7.5 oz. (1 jar)	28
Peaches	7.8 oz. (1 jar)	7
Pears	7.5 oz. (1 jar)	6

PORTION   SODIUM (MG.)

Pears & Pineapple	7.5 oz. (1 jar)	4
Plums w/Tapioca	7.8 oz. (1 jar)	9
Prunes w/Tapioca	7.8 oz. (1 jar)	20

Heinz

Apples & Apricots	7.5 oz. (1 jar)	2
Apples & Cranberries w/Tapioca	7.8 oz. (1 jar)	15
Apples & Pears	7.8 oz. (1 jar)	4
Applesauce	7.5 oz. (1 jar)	6
Apricots w/Tapioca	7.8 oz. (1 jar)	29
Bananas & Pineapple w/Tapioca	7.8 oz. (1 jar)	33
Bananas w/Tapioca	7.8 oz. (1 jar)	18
Peaches	7.5 oz. (1 jar)	13
Pears	7.5 oz. (1 jar)	0
Pears & Pineapple	7.5 oz. (1 jar)	2

Main Dishes

Beech-Nut

Beef & Egg Noodle Dinner	7.5 oz. (1 jar)	60
Beef & Egg Noodles w/Vegetables	7.5 oz. (1 jar)	36
Beef w/Vegetables & Cereal (High Meat)	4.5 oz. (1 jar)	40
Chicken Noodle Dinner	7.5 oz. (1 jar)	70
Chicken w/Vegetables & Cereal (High Meat)	4.5 oz. (1 jar)	35
Ham w/Vegetables & Cereal (High Meat)	4.5 oz. (1 jar)	40
Macaroni, Tomato & Beef Dinner	7.5 oz. (1 jar)	70
Spaghetti, Tomato & Beef Dinner	7.5 oz. (1 jar)	75
Split Peas & Ham	7.5 oz. (1 jar)	80
Turkey Rice Dinner	7.5 oz. (1 jar)	110
Vegetable Bacon Dinner	7.5 oz. (1 jar)	160
Vegetable Beef Dinner	7.5 oz. (1 jar)	70
Vegetable Chicken Dinner	7.5 oz. (1 jar)	60

PORTION   SODIUM (MG.)

	PORTION	SODIUM (MG.)
Vegetable Lamb w/Rice & Barley Dinner	7.5 oz. (1 jar)	180
Vegetable Liver w/Rice & Barley Dinner	7.5 oz. (1 jar)	60

Gerber

	PORTION	SODIUM (MG.)
Beef w/Vegetables	4.5 oz. (1 jar)	36
Cereal & Egg Yolk	7.5 oz. (1 jar)	26
Chicken & Noodles	7.5 oz. (1 jar)	28
Chicken w/Vegetables	4.5 oz. (1 jar)	33
Ham w/Vegetables	4.5 oz. (1 jar)	24
Macaroni & Cheese	7.5 oz. (1 jar)	168
Macaroni-Tomato w/Beef	7.5 oz. (1 jar)	34
Spaghetti-Tomato Sauce & Beef	7.5 oz. (1 jar)	57
Split Peas & Ham	7.5 oz. (1 jar)	36
Turkey & Rice w/Vegetables	7.5 oz. (1 jar)	36
Turkey w/Vegetables	4.5 oz. (1 jar)	40
Veal w/Vegetables	4.5 oz. (1 jar)	31
Vegetables & Bacon	7.5 oz. (1 jar)	121
Vegetables & Beef	7.5 oz. (1 jar)	26
Vegetables & Chicken	7.5 oz. (1 jar)	21
Vegetables & Ham	7.5 oz. (1 jar)	30
Vegetables & Lamb	7.5 oz. (1 jar)	26
Vegetables & Liver	7.5 oz. (1 jar)	28
Vegetables & Turkey	7.5 oz. (1 jar)	32

Heinz

	PORTION	SODIUM (MG.)
Beef w/Vegetables & Cereal (High Meat Dinner)	4.5 oz. (1 jar)	29
Chicken Noodle Dinner	7.5 oz. (1 jar)	34
Chicken w/Vegetables (High Meat Dinner)	4.5 oz. (1 jar)	29
Egg Noodles & Beef	7.5 oz. (1 jar)	47
Macaroni, Tomatoes & Beef	7.5 oz. (1 jar)	30
Spaghetti, Tomato Sauce & Meat	7.5 oz. (1 jar)	38
Turkey Rice Dinner w/Vegetables	7.5 oz. (1 jar)	49

PORTION   SODIUM (MG.)

Turkey w/Vegetables (High
  Meat Dinner)..........................4.5 oz. (1 jar) 41
Vegetables & Bacon ....7.5 oz. (1 jar)....................115
Vegetables & Beef.....................7.5 oz. (1 jar)....................36
Vegetables & Ham ....7.5 oz. (1 jar)....................36
Vegetables, Dumpling &
  Beef ....7.5 oz. (1 jar) 32
Vegetables, Egg Noodles &
  Chicken ....7.5 oz. (1 jar) 49
Vegetables, Egg Noodles
  & Turkey ....7.5 oz. (1 jar) 51

Meats

Beech-Nut

Beef & Beef Broth.....................3.5 oz. (1 jar) 75
Chicken & Chicken Broth..........3.5 oz. (1 jar) 70
Lamb & Lamb Broth ....3.5 oz. (1 jar) 80

Gerber

Beef ....3.5 oz. (1 jar) 52
Chicken ....3.5 oz. (1 jar) 39
Chicken Sticks ....2.5 oz. (1 jar) 322
Ham ....3.5 oz. (1 jar) 39
Lamb ....3.5 oz. (1 jar) 53
Meat Sticks ....2.5 oz. (1 jar) 325
Turkey.....................................3.5 oz. (1 jar) 50
Turkey Sticks ....2.5 oz. (1 jar) 330
Veal...3.5 oz. (1 jar) 55

Heinz

Beef & Beef Broth.....................3.5 oz. (1 jar) 46
Chicken & Chicken Broth..........3.5 oz. (1 jar) 44
Lamb & Lamb Broth ....3.5 oz. (1 jar) 52
Veal & Veal Broth ....3.5 oz. (1 jar) 47

Vegetables

Beech-Nut

Carrots....................................7.5 oz. (1 jar)....................220

PORTION SODIUM (MG.)

Creamed Corn	7.5 oz. (1 jar)	30
Green Beans	7.5 oz. (1 jar)	10
Mixed Vegetables	7.5 oz. (1 jar)	60
Peas	7.5 oz. (1 jar)	10
Squash	7.5 oz. (1 jar)	10
Sweet Potatoes	7.5 oz. (1 jar)	100

Gerber

Carrots	7.5 oz. (1 jar)	111
Creamed Corn	7.5 oz. (1 jar)	21
Creamed Green Beans	7.5 oz. (1 jar)	19
Mixed Vegetables	7.5 oz. (1 jar)	77
Peas	7.5 oz. (1 jar)	11
Squash	7.5 oz. (1 jar)	4
Sweet Potatoes	7.8 oz. (1 jar)	51

Heinz

Carrots	7.5 oz. (1 jar)	47
Creamed Corn	7.8 oz. (1 jar)	26
Creamed Green Beans	7.5 oz. (1 jar)	38
Creamed Peas	7.5 oz. (1 jar)	38
Sweet Potatoes	7.5 oz. (1 jar)	58

TODDLER FOODS

Juices

Gerber

Apple	4.0 oz. (1 jar)	1
Apple-Cherry	4.0 oz. (1 jar)	2
Apple-Grape	4.0 oz. (1 jar)	2
Mixed Fruit	4.0 oz. (1 jar)	1

Meals

Beech-Nut

Beef Stew	1 jar	630
Chicken & Stars	1 jar	580
Hearty Vegetable Soup	1 jar	580
Pasta Squares w/Meat Sauce	1 jar	530

PORTION   SODIUM (MG.)

Spaghetti Rings w/Meat Sauce	1 jar	680
Vegetable Stew w/Chicken	1 jar	630

Gerber

Beef & Egg Noodles	1 jar	705
Noodles & Chicken	1 jar	665
Potatoes & Ham	1 jar	489
Spaghetti, Tomato Sauce & Beef	1 jar	870
Vegetables & Beef	1 jar	665
Vegetables & Chicken	1 jar	725
Vegetables & Ham	1 jar	670
Vegetables & Turkey	1 jar	805

Baking Ingredients

FLAVORING BARS AND CHIPS

Baker's

Flavored Chocolate Chips	1.5 oz. (¼ cup)	9
German's Sweet Chocolate	1 oz.	1
Semi-Sweet Chocolate	1 oz.	1
Unsweetened Chocolate	1 oz.	1

Borden

Semi-Sweet Chocolate Chips	1 oz.	5

Hershey

Baking Chocolate	1 oz.	1
Milk Chocolate Chips	1.5 oz. (¼ cup)	55
Semi-Sweet Chocolate Chips	1.5 oz. (¼ cup)	5
Semi-Sweet Chocolate Mini Chips	1.5 oz. (¼ cup)	5

Nestlé

Butterscotch Artificial Flavored Morsels	1 oz.	20
Choco-Bake	1 oz.	10

PORTION   SODIUM (MG.)

Milk Chocolate Morsels ....1 oz.................................25
Semi-Sweet Morsels ....1 oz.................................10

Reese's

Peanut Butter Flavor Chips ......1.5 oz. (¼ cup) 90

FLOUR

USDA

Corn ....1 cup 1
Gluten................................1 cup 3
Rye ....1 cup 1
Self-rising (sifted).................1 cup 1241
Self-rising (unsifted) ....1 cup 1349
Soy ....1 cup 1
White (sifted) ....1 cup 2
White (unsifted ....1 cup 3
White Cake ....1 cup 2
Whole Wheat ....1 cup 4

Ballard

Self-rising...........................1 cup 1290

Gold Medal

Self-rising...........................about 1 cup 1520

Pillsbury's Best

Self-rising...........................1 cup 1290

Red Band

Self-rising...........................about 1 cup 1520

Red Star ....about 1 cup 1520

YEAST

Fleischmann's

Active Dry Yeast ....0.3 oz. 10
Active Fresh Yeast.................0.6 oz. 5
Household Yeast ....0.5 oz. 5

PORTION   SODIUM (MG.)

OTHER

Argo/Kingsford

Cornstarch................................1 tbl. 5

Calumet

Baking Powder ....1 tsp. 405

Fearn Soya

Rice Baking Mix.........................½ cup dry 857

Featherweight

Low Sodium Baking Powder ....1 tsp. 2

USDA

Baking Soda...............................1 tsp. 821
Vanilla Extract............................1 tsp. 0

Beverages

ALCOHOLIC

USDA

Beer ....12 fl. oz. 25
Dessert Wines ....2 fl. oz 2
Gin ....1 fl. oz. 0
Rum ....1 fl. oz. 0
Table Wine ....3.5 fl. oz. 5
Vodka...1 fl. oz 0
Whiskey......................................1 fl. oz. 0

CLUB SODA, SELTZER, AND WATERS*

Club Soda

A & P..8 fl. oz. 50
Canada Dry.................................8 fl. oz. 50
Cragmont....................................12 fl. oz. 42

*©1983 by the New York Times Co. Reprinted by permission.

	PORTION	SODIUM (MG.)
Foodtown	8 fl. oz.	9
Giant, No Salt Added	12 fl. oz.	10
Gristede's	8 fl. oz.	46
Grand Union	8 fl. oz.	38
Hoffman	8 fl. oz.	38
Krasdale	8 fl. oz.	3-7
No Cal	8 fl. oz.	under 10
Pathmark	8 fl. oz.	6
Red Apple	8 fl. oz.	0
Safeway	12 fl. oz.	42
Schweppes	8 fl. oz.	40-51
Shopwell	8 fl. oz.	3-7
Waldbaum's	8 fl. oz.	38
White Rock	8 fl. oz.	46

Seltzer

	PORTION	SODIUM (MG.)
Ann Page	8 fl. oz.	under 3
Best Health's	8 fl. oz.	under 5
Canada Dry	8 fl. oz.	under 10
Foodtown	8 fl. oz.	1
Good Health	8 fl. oz.	under 5
Gristede's	8 fl. oz.	under 10
Grand Union	8 fl. oz.	under 5
Hoffman	8 fl. oz.	under 5
Krasdale	8 fl. oz.	under 10
Manischewitz	8 fl. oz.	under 10
Old Fashioned	8 fl. oz.	under 10
Pathmark	8 fl. oz.	under 10
Seagram's	8 fl. oz.	under 10
Shopwell	8 fl. oz.	under 10
Snapple Sparkling Water	8 fl. oz.	0
Syfo	8 fl. oz.	0
Vintage	8 fl. oz.	under 10
White Rock	8 fl. oz.	under 10

Waters

	PORTION	SODIUM (MG.)
Appolinaris Natural Mineral Water	8 fl. oz.	135

PORTION   SODIUM (MG.)

Artesia Pure Sparkling Texas
  Mineral Water ....8 fl. oz. 2
Badoit Naturally Sparkling
  Mineral Water ....8 fl. oz...................................49
Bru ....8 fl. oz. 2
Crystal Geyser Sparkling
  Mineral Water ....8 fl. oz. 30
Ferrarelle Naturally Sparkling
  Mineral Water ....8 fl. oz. 9
Gerolsteiner Sprudel Natural
  Mineral Water ....8 fl. oz. 32
Montclair Sparkling Natural
  Mineral Water ....8 fl. oz..........................under 3
Perrier Naturally Sparkling
  Mineral Water ....8 fl. oz. 3
Peters Val Naturally Sparkling
  Mineral Water ....8 fl. oz. 79
Poland Spring Natural
  Spring Water........................8 fl. oz..........................under 5
Ramlosa Sparkling Mineral
  Water ....8 fl. oz. 47
S. Pellegrino ....8 fl. oz. 11
Safeway...................................12 fl. oz. 1
Saratoga Original Vichy
  Water ....8 fl. oz. 46
Saratoga Naturally Sparkling
  Mineral Water ....8 fl. oz..........................under 5
Vichy Celestins........................8 fl. oz. 290
Vichy Saint Yorre Royale
  Mineral Water ....8 fl. oz. 397
Vittelloise ....8 fl. oz. 11
White Rock Sparkling
  Mineral Water ....8 fl. oz. 2

COCOA

Alba

Chocolate Marshmallow ....1 env.....................................180
Milk Chocolate ....1 env.....................................180

PORTION   SODIUM (MG.)

Mocha ....1 env......................................180
Superman ....1 env......................................210

Carnation

Hot Chocolate & Mini
   Marshmallows.........................4 tsp. (heaping) 114
Instant Hot Cocoa Mix,
   Marshal Mallow ....1 env.......................................135
Milk Chocolate Hot
   Cocoa Mix................................4 tsp. (heaping) 111
Rich Chocolate Hot
   Cocoa Mix................................4 tsp. (heaping) 94
70-Calorie Rich Hot Cocoa
   Mix ....3 tsp. (heaping) 139
70-Calorie Hot Chocolate
   w/Marshmallows......................3 tsp. (heaping) 137

Hershey

Cocoa..⅓ cup..5
Hot Cocoa Mix ....1 oz..145
Instant Cocoa (w/whole
   milk) ....8 oz. prepared......................160

Nestlé

Hot Cocoa Mix ....1 oz..30
Hot Cocoa Mix
   w/Marshmallows......................1 oz..110
Quik, Chocolate Flavor.............8 oz. prepared......................155

Ovaltine

50-Calorie Cocoa Mix................1 oz..183
Hot'n Rich Cocoa Mix...............0.5 oz. 88

Swiss Miss

Lite Cocoa Mix............................6 fl. oz. 109
Milk Chocolate Cocoa Mix........6 fl. oz. 110
Mini Marshmallow Cocoa
   Mix ....6 fl. oz. 125

PORTION   SODIUM (MG.)

COFFEE (all plain coffees have less than 10 mg.)

General Foods International Coffees

Café Amaretto	6 fl. oz.	30
Café Français	6 fl. oz.	30
Café Vienna	6 fl. oz.	95
Irish Mocha Mint	6 fl. oz.	25
Orange Cappuccino	6 fl. oz.	100
Suisse Mocha	6 fl. oz.	25

USDA

Coffee, black	1 cup	1
Regular Instant	1 tsp.	1
Regular Freeze-Dried	1 tsp.	1

DIET DRINKS (see also Soft Drinks page 174)

Canned

Carnation Slender

Banana	10 fl. oz. (1 can)	431
Chocolate	10 fl. oz. (1 can)	515
Chocolate Fudge	10 fl. oz. (1 can)	550
Chocolate Malt	10 fl. oz. (1 can)	530
Chocolate Marshmallow	10 fl. oz. (1 can)	550
Peach	10 fl. oz. (1 can)	431
Strawberry	10 fl. oz. (1 can)	431
Milk Chocolate	10 fl. oz. (1 can)	520
Vanilla	10 fl. oz. (1 can)	550

Sego Liquid Diet Food

Lite Chocolate	10 fl. oz. (1 can)	378
Lite Dutch Chocolate	10 fl. oz. (1 can)	423
Lite Vanilla	10 fl. oz. (1 can)	533
Milk Chocolate flavored	10 fl. oz. (1 can)	509
Very Banana	10 fl. oz. (1 can)	265
Very Butterscotch	10 fl. oz. (1 can)	265
Very Cherry Vanilla	10 fl. oz. (1 can)	265
Very Chocolate Coconut	10 fl. oz. (1 can)	509

PORTION   SODIUM (MG.)

	PORTION	SODIUM (MG.)
Very Chocolate Flavored	10 fl. oz. (1 can)	509
Very Chocolate Fudge	10 fl. oz. (1 can)	509
Very Chocolate Malt	10 fl. oz. (1 can)	509
Very Chocolate Marshmallow	10 fl. oz. (1 can)	509
Very Dutch Chocolate Flavored	10 fl. oz. (1 can)	509
Very French Vanilla	10 fl. oz. (1 can)	265
Very Strawberry	10 fl. oz. (1 can)	265
Very Vanilla	10 fl. oz (1 can)	265
WEIGHT WATCHERS Frosted Treat, all flavors	6 fl. oz.	90

Dry

Alba '77 Fit 'n Frosty

Chocolate	1 env.	170
Chocolate Marshmallow	1 env.	170
Strawberry	1 env.	170
Vanilla	1 env.	170

Cambridge

The Cambridge Diet	1.2 oz. (1 meal)	500

Carnation Slender

Chocolate	1 oz.	192
Chocolate Malt	1 oz.	192
Coffee	1 oz.	217
Dutch Chocolate	1 oz.	192
French Vanilla	1 oz.	192
Wild Strawberry	1 oz.	227

FRUIT DRINKS

Bottled and Canned

CapriSun

Apple	6.8 fl. oz.	2
Fruit Punch	6.8 fl. oz.	1
Grape	6.8 fl. oz.	19
Lemonade	6.8 fl. oz.	2
Orange	6.8 fl. oz.	2

PORTION   SODIUM (MG.)

Country Time

Lemonade ....6 fl. oz. 60

Del Monte

Pineapple Grapefruit Juice
  Drink...................................6 fl. oz. 50
Pineapple Orange Juice
  Drink...................................6 fl. oz. 20
Pineapple Pink Grapefruit
  Juice Drink ....6 fl. oz. 50

Hi-C

Apple ....6 oz..12
Apple-Cranberry.......................6 oz..0
Cherry ....6 oz..4
Citrus Cooler............................6 oz..4
Fruit Punch ....6 oz..0
Grape ....6 oz..0
Orange.....................................6 oz..58
Peach ....6 oz..0
Pineapple-Orange ....6 oz..1
Strawberry...............................6 oz..0
Tangerine.................................6 oz..0
Wild Berry ....6 oz..5

Land O Lakes

Fruit Drinks, all flavors ....8 fl. oz. 10

Ocean Spray

Cranapple ....6 fl. oz. 4
Cranapple, Low Calorie ....6 fl. oz. 8
Cranberry Juice Cocktail ....6 fl. oz. 3
Cranberry Juice Cocktail,
  Low Calorie...........................6 fl. oz. 6
Crangrape ....6 fl. oz. 5
Cranicot ....6 fl. oz. 4
Pink Grapefruit Juice
  Cocktail ....6 fl. oz. 15

PORTION   SODIUM (MG.)

Frozen Concentrate

Birds Eye

Awake	6 fl. oz.	15
Orange Plus	6 fl. oz.	10

FRUIT FLAVORED MIXES

Borden

Natural Orange Flavor Instant Breakfast Drink	4 fl. oz.	40

Country Time

Lemon-Lime, sugar-sweetened	8 fl. oz.	30
Pink Lemonade, sugar-sweetened	8 fl. oz.	30
Lemonade, sugar-sweetened	8 fl. oz.	30
Lemonade, sugar-free	8 fl. oz.	0

Cragmont

Cherry	8 fl. oz.	11
Grape	8 fl. oz.	11
Lemonade	8 fl. oz.	13
Orange	8 fl. oz.	36
Punch	8 fl. oz.	11
Strawberry	8 fl. oz.	11

Crystal Light

Fruit Punch	8 fl. oz.	0
Iced Tea	8 fl. oz.	20
Lemonade	8 fl. oz.	0
Lemon-Lime	8 fl. oz.	0
Orange	8 fl. oz.	0

Elam's

Lemonade mix, sweetened	8 fl. oz.	61
Lemonade mix, unsweetened	8 fl. oz.	61

PORTION   SODIUM (MG.)

Kool-Aid, Sugar-sweetened

	PORTION	SODIUM (MG.)
Apple	8 fl. oz.	0
Cherry	8 fl. oz.	5
Grape	8 fl. oz.	0
Lemonade	8 fl. oz.	0
Orange	8 fl. oz.	0
Pink Lemonade	8 fl. oz.	0
Rainbow Punch	8 fl. oz.	5
Raspberry	8 fl. oz.	0
Strawberry	8 fl. oz.	5
Sunshine Punch	8 fl. oz.	5
Tropical Punch	8 fl. oz.	0

Kool-Aid, Sugar Free

	PORTION	SODIUM (MG.)
Cherry	8 fl. oz.	5
Grape	8 fl. oz.	0
Lemonade	8 fl. oz.	0
Rainbow Punch	8 fl. oz.	5
Sunshine Punch	8 fl. oz.	5
Tropical Punch	8 fl. oz.	0

Kool-Aid, Unsweetened

	PORTION	SODIUM (MG.)
Apple	8 fl. oz.	0
Black Cherry	8 fl. oz.	0
Cherry	8 fl. oz.	0
Grape	8 fl. oz.	0
Lemon-Lime	8 fl. oz.	0
Lemonade	8 fl. oz.	0
Orange	8 fl. oz.	0
Pink Lemonade	8 fl. oz.	1
Rainbow Punch	8 fl. oz.	5
Raspberry	8 fl. oz.	0
Strawberry	8 fl. oz.	35
Sunshine Punch	8 fl. oz.	0
Tropical Punch	8 fl. oz.	5

Tang

	PORTION	SODIUM (MG.)
Grape	6 fl. oz.	5
Grapefruit	6 fl. oz.	0
Orange	6 fl. oz.	0

PORTION   SODIUM (MG.)

Wyler's

Artificial Crisp Apple Flavor Crystals	8 fl. oz.	25

Wyler's (Sugar Free)

Lemonade Flavor	8 fl. oz.	35
Tropical Punch Artificial Flavor	8 fl. oz.	30
Wild Cherry Artificial Flavor	8 fl. oz.	20
Wild Grape Artificial Flavor	8 fl. oz.	20

FRUIT JUICE
Bottled and Canned
Del Monte

Apricot Nectar	6 fl. oz.	under 10
Grapefruit Juice	6 fl. oz.	under 10
Orange Juice	6 fl. oz.	under 10
Pineapple Juice	6 fl. oz.	under 10
Prune Juice	6 fl. oz.	under 10

Kraft

Grapefruit Juice, Unsweetened	6 fl. oz.	0
Orange-Grapefruit Juice, Unsweetened	6 fl. oz.	0
Orange-Pineapple Juice, Unsweetened	6 fl. oz.	0
Pasteurized Orange Juice, Unsweetened	6 fl. oz.	0

Libby's

Apricot Nectar	6 fl. oz.	5
Grapefruit Juice, Unsweetened	6 fl. oz.	5
Orange Juice, Sweetened	6 fl. oz.	5
Orange Juice, Unsweetened	6 fl. oz.	5
Orange-Grapefruit Juice	6 fl. oz.	5
Peach Nectar	6 fl. oz.	5
Pear Nectar	6 fl. oz.	5

PORTION   SODIUM (MG.)

Ocean Spray

Grapefruit Juice	6 fl. oz.	7

Realemon

Lemon Juice from Concentrate	2 tbl	10
Lime Juice from Concentrate	2 tbl	10

USDA

Apple Juice	6 fl. oz.	6
Apricot Nectar	6 fl. oz.	6
Grape Juice	6 fl. oz.	6
Grapefruit Juice	6 fl. oz.	2
Orange Juice	6 fl. oz.	5
Orange-Grapefruit Juice	6 fl. oz.	6
Papaya Nectar	6 fl. oz.	11
Peach Nectar	6 fl. oz.	13
Pear Nectar	6 fl. oz.	7
Pineapple Juice	6 fl. oz.	2
Prune Juice	6 fl. oz.	8
Tangerine Juice	6 fl. oz.	2

Welch's

Grape Juice	6 fl. oz.	under 10
Red Grape Juice	6 fl. oz.	under 10
White Grape Juice	6 fl. oz.	under 10

Frozen Concentrate
Minute Maid

Grapefruit Juice	6 fl. oz.	2
Orange Juice	6 fl. oz.	2

USDA

Apple Juice	6 fl. oz.	12
Grape Juice	6 fl. oz.	4
Grapefruit Juice	6 fl. oz.	2
Orange Juice	6 fl. oz.	2
Pineapple Juice	6 fl. oz.	2
Tangerine Juice	6 fl. oz.	2

PORTION   SODIUM (MG.)

MILK (see Milk, Cream, and Sour Cream, page 350)

SOFT DRINKS

Bottled and Canned

Aspen

Aspen	12 fl. oz.	11

Canada Dry

Barrelhead Root Beer	12 fl. oz.	*26
Barrelhead Root Beer (diet)	12 fl. oz.	*39
Birch Beer	12 fl. oz.	*29
Bitter Lemon	12 fl. oz.	*26
Cactus Cooler	12 fl. oz.	*32
California Strawberry	12 fl. oz.	*32
Collins Mixer	12 fl. oz.	*26
Concord Grape	12 fl. oz.	*32
Diet Cola	12 fl. oz.	*39
Diet Orange	12 fl. oz.	*39
Diet Tonic Water	12 fl. oz.	*11
Ginger Ale	12 fl. oz.	*11
Ginger Ale (diet)	12 fl. oz.	*44
Golden Ginger Ale	12 fl. oz.	*36
Half & Half	12 fl. oz.	*26
Hi-Spot	12 fl. oz.	*38
Island Lime	12 fl. oz.	*29
Jamaica Cola	12 fl. oz.	*0
Pineapple	12 fl. oz.	*32
Purple Passion	12 fl. oz.	*29
Rooti	12 fl. oz.	*26
Sunripe Orange	12 fl. oz.	*32
Tahitian Treat	12 fl. oz.	*32
Tonic Water	12 fl. oz.	*11
Vanilla Cream	12 fl. oz.	*29
Whiskey Sour	12 fl. oz.	*26
Wild Cherry	12 fl. oz.	*32
Wink	12 fl. oz.	*29

*In addition to that present in the water source.

PORTION   SODIUM (MG.)

Coca-Cola

	PORTION	SODIUM (MG.)
Coca-Cola	12 fl. oz.	14
Tab	12 fl. oz.	30

Cragmont

	PORTION	SODIUM (MG.)
Black Cherry	12 fl. oz.	24
Black Cherry (diet	12 fl. oz.	101
Cherry Cola	12 fl. oz.	24
Cherry Cola (diet)	12 fl. oz.	74
Chocolate (diet)	12 fl. oz.	84
Cola	12 fl. oz.	0
Cola (diet)	12 fl. oz.	47
Collins	12 fl. oz.	25
Cream	12 fl. oz.	24
Cream (diet)	12 fl. oz.	58
Ginger Ale	12 fl. oz.	25
Ginger Ale (diet)	12 fl. oz.	36
Grape	12 fl. oz.	24
Grapefruit	12 fl. oz.	23
Grapefruit (diet)	12 fl. oz.	72
Lemon Lime	12 fl. oz.	62
Lemon Lime (diet)	12 fl. oz.	59
Orange	12 fl. oz.	24
Orange (diet)	12 fl. oz.	114
Punch	12 fl. oz.	24
Raspberry	12 fl. oz.	24
Raspberry (diet)	12 fl. oz.	24
Root Beer	12 fl. oz.	25
Root Beer (diet)	12 fl. oz.	40
Skipper	12 fl. oz.	61
Skipper (diet)	12 fl. oz.	61
Strawberry	12 fl. oz.	24
Tonic	12 fl. oz.	0

Fanta

	PORTION	SODIUM (MG.)
Ginger Ale	12 fl. oz.	27
Grape	12 fl. oz.	14
Orange	12 fl. oz.	14
Root Beer	12 fl. oz.	21

PORTION   SODIUM (MG.)

Fresca

Fresca ....12 fl. oz. 36

Health Valley

Cola ....11 fl. oz. 25
Ginger Ale ....11 fl. oz. 30
Lemon-Lime Soda ....11 fl. oz. 40
Root Beer, New Old
   Fashioned ....11 fl. oz. 30
Root Beer, New Sarsaparilla ....11 fl. oz. 25
Wild Berry Soda....................11 fl. oz. 25
Mandarin Lime ....11 fl. oz. 40

Hi-C

Grape ....12 fl. oz. 12
Lemonade ....12 fl. oz. 12
Orange..................................12 fl. oz. 15
Punch....................................12 fl. oz. 12

Mello Yello

Mello Yello............................12 fl. oz. 30

Mountain Dew

Mountain Dew ....12 fl. oz. 31

Mr. Pibb

Mr. Pibb ....12 fl. oz. 41
Mr. Pibb, Sugar-free ....12 fl. oz. 45

On Tap

Root Beer ....12 fl. oz. 18

Pepsi-Cola

Grape....................................12 fl. oz. 42
Orange..................................12 fl. oz. 42
Pepsi (diet) ....12 fl. oz. 62
Pepsi Light ....12 fl. oz. 42
Pepsi-Cola.............................12 fl. oz. 10
Root Beer ....12 fl. oz. 4

PORTION   SODIUM (MG.)

Safeway

	PORTION	SODIUM (MG.)
Black Cherry	12 fl. oz.	24
Black Cherry (diet)	12 fl. oz.	101
Cherry	12 fl. oz.	0
Cherry Cola	12 fl. oz.	24
Cherry Cola (diet)	12 fl. oz.	74
Chocolate Mint (diet)	12 fl. oz.	73
Citrus Cooler	12 fl. oz.	0
Cola	12 fl. oz.	0
Cola (diet)	12 fl. oz.	47
Collins	12 fl. oz.	25
Cream	12 fl. oz.	24
Cream (diet)	12 fl. oz.	58
Ginger Ale	12 fl. oz.	25
Ginger Ale (diet)	12 fl. oz.	36
Grape	12 fl. oz.	24
Grape (noncarbonated)	12 fl. oz.	50
Grapefruit	12 fl. oz.	23
Grapefruit (diet)	12 fl. oz.	72
Lemon Lime	12 fl. oz.	62
Lemon Lime (diet)	12 fl. oz.	59
Lemonade	12 fl. oz.	23
Mineral Water	12 fl. oz.	1
Orange	12 fl. oz.	24
Orange (diet)	12 fl. oz.	114
Orange (noncarbonated, 46 oz. can)	12 fl. oz.	34
Pineapple-Orange (noncarbonated, 46 oz. can)	12 fl. oz.	50
Punch	12 fl. oz.	24
Punch (noncarbonated, 46 oz. can)	12 fl. oz.	19
Raspberry	12 fl. oz.	24
Raspberry (diet)	12 fl. oz.	114
Root Beer	12 fl. oz.	25
Root Beer (diet)	12 fl. oz.	40
Skipper	12 fl. oz.	23
Strawberry	12 fl. oz.	24
Tonic	12 fl. oz.	13

PORTION   SODIUM (MG.)

	PORTION	SODIUM (MG.)
Tonic (diet)	12 fl. oz.	34
Wild Berry (noncarbonated, 46 oz. can)	12 fl. oz.	0

7-Up

7-Up	12 fl. oz.	60

Shasta

Black Cherry	12 fl. oz.	37
Black Cherry (diet)	12 fl. oz.	48
Cherry Cola	12 fl. oz.	29
Cherry Cola (diet)	12 fl. oz.	53
Cola	8 fl. oz.	1
Cola	12 fl. oz.	2
Cola (diet)	8 fl. oz.	30
Cola (diet)	12 fl. oz.	44
Creme	12 fl. oz.	24
Creme (diet)	12 fl. oz.	50
Fruit Punch	12 fl. oz.	30
Ginger Ale	8 fl. oz.	14
Ginger Ale	12 fl. oz.	21
Ginger Ale (diet)	8 fl. oz.	24
Ginger Ale (diet)	12 fl. oz.	36
Grape	12 fl. oz.	27
Grape (diet)	12 fl. oz.	60
Grapefruit	12 fl. oz.	21
Grapefruit (diet)	12 fl. oz.	50
Lemon Lime	8 fl. oz.	37
Lemon Lime	12 fl. oz.	55
Lemon Lime (diet)	8 fl. oz.	33
Lemon Lime (diet)	12 fl. oz.	49
Orange	12 fl. oz.	30
Orange (diet)	12 fl. oz.	52
Root Beer	8 fl. oz.	17
Root Beer	12 fl. oz.	25
Root Beer (diet)	12 fl. oz.	54
Strawberry	12 fl. oz.	46
Strawberry (diet)	12 fl. oz.	50

PORTION   SODIUM (MG.)

Sprite

Sprite	12 fl. oz.	47
Sprite, Sugar-free	12 fl. oz.	45

Teem

Teem	12 fl. oz.	31

Vess

Cola	12 fl. oz.	33
Fruit Punch	12 fl. oz.	48
Ginger Ale	12 fl. oz.	41
Grape	12 fl. oz.	42
Lemon Lime	12 fl. oz.	28
Orange	12 fl. oz.	37
Strawberry	12 fl. oz.	48

Weight Watchers

All Sodas	12 fl. oz.	100

TEA (all plain teas have less than 10 mg.)

Bags

Celestial Seasonings

Almond Sunset	8 fl. oz.	2
Chamomile Flowers	8 fl. oz.	5
Cinnamon Rose	8 fl. oz.	2
Comfrey Leaves	8 fl. oz.	3
Country Apple	8 fl. oz.	5
Diet Partner	8 fl. oz.	2
Emperor's Choice	8 fl. oz.	2
Foenugreek Seeds	8 fl. oz.	3
Ginseng Plus	8 fl. oz.	4
Grandma's Tummy Mint	8 fl. oz.	7
Lemon Iced Delight	8 fl. oz.	2
Lemon Mist	8 fl. oz.	3
Lemon Verbena Leaves	8 fl. oz.	2
Mandarin Orange Spice	8 fl. oz.	2
Mellow Mint	8 fl. oz.	4
Mint Iced Delight	8 fl. oz.	6

PORTION   SODIUM (MG.)

	PORTION	SODIUM (MG.)
Morning Thunder	8 fl. oz.	1
Mo's 24	8 fl. oz.	5
Orange Iced Delight	8 fl. oz.	2
Pelican Punch	8 fl. oz.	6
Peppermint Leaves	8 fl. oz.	9
Red Zinger	8 fl. oz.	3
Roastaroma Grain & Spice	8 fl. oz.	4
Rosehips	8 fl. oz.	3
Sleepytime	8 fl. oz.	5
Spearmint Leaves	8 fl. oz.	6

Kaffree Herb Tea

Apple Delight	8 fl. oz.	0
Hush-A-Bye	8 fl. oz.	3
Mint Blend	8 fl. oz.	2
Original	8 fl. oz.	1
Spicy Orange	8 fl. oz.	0

USDA

Tea	8 fl. oz.	2

Canned

Safeway

Iced Tea (diet)	12 fl. oz.	68
Iced Tea (sweetened w/sugar)	12 fl. oz.	59

Loose

Celestial Seasonings

Chamomile Flowers	8 fl. oz.	8
Comfrey Leaves	8 fl. oz.	4
Emperor's Choice	8 fl. oz.	2
Lemon Mist	8 fl. oz.	3
Mellow Mint	8 fl. oz.	5
Mo's 24	8 fl. oz.	5
Morning Thunder	8 fl. oz.	0
Pelican Punch	8 fl. oz.	2
Peppermint Leaves	8 fl. oz.	2

	PORTION	SODIUM (MG.)
Raspberry Leaves	8 fl. oz.	0
Red Clover Tops	8 fl. oz.	1
Red Zinger	8 fl. oz.	2
Roastaroma Grain & Spice	8 fl. oz.	3
Rosehips	8 fl. oz.	1
Sleepytime	8 fl. oz.	7
Spearmint Leaves	8 fl. oz.	6
Winterberry	8 fl. oz.	1

Mix

Nestea

	PORTION	SODIUM (MG.)
Iced Tea Mix	0.02 oz. dry	under 10
Iced Tea Mix, Lemon Flavor	0.03 oz. dry	under 10
Iced Tea Mix, Sugar & Lemon	0.06 oz. dry	under 10

VEGETABLE JUICE

Bottled and Canned

Campbell's

	PORTION	SODIUM (MG.)
Tomato Juice	6 fl. oz.	625

Del Monte

	PORTION	SODIUM (MG.)
Snap-E-Tom	6 fl. oz.	980

Diet Delight

	PORTION	SODIUM (MG.)
Tomato Juice	6 fl. oz.	16

Featherweight

	PORTION	SODIUM (MG.)
Tomato Juice	6 fl. oz.	20

Hunt's

	PORTION	SODIUM (MG.)
Tomato Juice	6 fl. oz.	550
Tomato Juice, No Salt Added	6 fl. oz.	30

Libby's

	PORTION	SODIUM (MG.)
Tomato Juice	6 fl. oz.	455

PORTION   SODIUM (MG.)

Ocean Spray

Tomato-Vegetable Juice
   Cocktail ....6 fl. oz. 599

Stokely-Van Camp

Tomato Juice ....6 fl. oz. 495

V-8

Cocktail ....6 fl. oz. 625
Cocktail, Low Sodium ....6 fl. oz. 50
Spicy Hot Cocktail ....6 fl. oz. 625

Welch's

Tomato Juice ....6 fl. oz. 550

OTHER BEVERAGES

Carnation

Instant Malted Milk,
   Chocolate ....3 tsp. (heaping) 47
Instant Malted Milk,
   Natural ....3 tsp. (heaping) 98

Carnation Instant Breakfast

Chocolate ....1 env. 256
Chocolate Malt ....1 env. 307
Coffee ....1 env. 250
Egg Nog ....1 env. 316
Strawberry ....1 env. 314
Vanilla ....1 env. 265

Celestial Seasonings

Breakaway Carob Mint ....8 fl. oz. 4
Breakaway Cinnamon
   Splendor ....8 fl. oz. 4
Breakaway Orange
   Cappuccino ....8 fl. oz. 5

Frosted

Chocolate ....8 fl. oz. 205
Strawberry ....8 fl. oz. 190

PORTION   SODIUM (MG.)

Kraft

Instantized Malted Milk,
  Chocolate Flavor ....1 fl. cup.............................160
Instantized Malted Milk,
  Natural ....1 fl. cup.............................210

Land O Lakes

Egg Nog ....8 fl. oz. 142

Nestlé

Quik, Chocolate Flavor..............2 tsp. dry...............................35
Quik, Strawberry Flavor ....2 tsp. dry.....................under 10

Ovaltine

Chocolate Ovaltine ....0.8 oz. dry 145
Malt Ovaltine..............................0.8 oz. dry 92

PDQ

Chocolate ....¾ tsp. dry...............................36
Egg Nog ....2 tbl. dry...............................52
Strawberry.................................¾ tsp. dry...............................21

Pillsbury Instant Breakfast

Chocolate ....1 env...................................305
Chocolate Malt ....1 env...................................310
Strawberry.................................1 env...................................300
Vanilla...1 env...................................315

Postum

Instant Coffee Flavored
  Grain Beverage ....6 fl. oz. 5
Instant Grain Beverage ....6 fl. oz. 5

Sealtest

Egg Nog ....8 fl. oz. 172

Bread Products

BAGELS

Lenders

Egg	2 oz. (1 bagel)	420
Garlic	2 oz. (1 bagel)	411
Onion	2 oz. (1 bagel)	319
Plain	2 oz. (1 bagel)	352
Poppy Seeds	2 oz. (1 bagel)	344
Pumpernickel	2 oz. (bagel)	369
Raisin 'n Honey	2.5 oz. (1 bagel)	385
Rye	2 oz. (1 bagel)	456
Sesame Seeds	2 oz. (1 bagel)	301
Wheat 'n Honey	2.5 oz. (1 bagel)	465

BISCUITS

Mixes and Doughs

Arrowhead Mills

Whole Grain Mix	2 biscuits	340

Ballard

Oven Ready	1.5 oz. (2 biscuits)	355
Oven Ready Buttermilk	1.5 oz. (2 biscuits)	355

Bisquick

Bisquick Baking Mix	2 oz. (½ cup)	700
Baking Powder	2 biscuits	590
Buttermilk	2 biscuits	590
Butter Tastin'	2 biscuits	590

Hillbilly

Hillbilly	(2 biscuits)	730
Butter Tastin'	2.2 oz. (2 biscuits)	590
Buttermilk	2.2 oz. (2 biscuits)	590

Merrico

Butter-Me-Not (dough)	1.9 oz. (2 biscuits)	435

	PORTION	SODIUM (MG.)
Homestyle (dough)	1.6 oz. (2 biscuits)	450
Texas Style (dough)	2.4 oz (2 biscuits)	540
Texas Style Butter Flavor (dough)	2.4 oz. (2 biscuits)	460

Pillsbury

Big Country Buttermilk	2 biscuits	645
Big Premium Heat 'n Eat Buttermilk	2 biscuits	605
Butter	2 biscuits	355
Buttermilk	2 biscuits	355
Country Style	2 biscuits	355
Extra Lights Flaky Buttermilk	2 biscuits	340
Good 'n Buttery Big Country	2 biscuits	605
Heat 'n Eat Buttermilk	2 biscuits	530
Tenderflake Baking Powder Dinner	2 biscuits	355
Tenderflake Buttermilk	2 biscuits	355

Pillsbury Hungry Jack

Butter Tastin'	2 oz. (2 biscuits)	550
Buttermilk Flaky	2 oz. (2 biscuits)	590
Buttermilk Fluffy	2 oz. (2 biscuits)	560
Extra Rich Buttermilk	1.5 oz. (2 biscuits)	345
Flaky	2 oz. (2 biscuits)	585

Prepared

Merico

Butter Flavor Deluxe	2 biscuits	520
Butter Flavored	2 biscuits	460
Butter-Me-Not	2 biscuits	435
Buttermilk Deluxe	2 biscuits	515
Buttermilk Flaky Deluxe	2 biscuits	565
Homestyle	2 biscuits	450
Maple Flavor Deluxe	2 biscuits	520
Mountain Man Buttermilk	2 biscuits	516
Mountain Man Flaky	2 biscuits	555
Texas Style	2 biscuits	540

PORTION   SODIUM (MG.)

Wonder

Biscuits...2 oz. (2 biscuits) 280

BREAD CRUMBS & CRACKER MEAL

Contadina

Seasoned Bread Crumbs..........2 tbl..397

Kellogg's

Corn Flake Crumbs....................1 oz. (¼ cup)........................305

Nabisco

Cracker Meal ....½ cup 15

BREAD LOAVES

Mixes and Doughs

Arrowhead Mills

Multigrain Bread Mix.................2 oz. prepared....................227
Stoneground Whole Wheat
    Bread Mix ....2 oz. prepared....................227

Elam's

Rye Bread Mix ....2 oz................................500
Stone Ground Whole Wheat
    Bread Mix ....2 oz................................60
Unbleached White Bread Mix   2 oz.................................494

Pillsbury

Pipin' Hot Wheat Loaf...............1 in. slice 170
Pipin' Hot White Loaf ....1 in. slice 170

Pillsbury Poppin' Fresh

Rye ....½ in. slice...........................195
Wheat..½ in. slice...........................195
White ....½ in. slice...........................195

Rich's

Honey Wheat Bread Dough ....2 slices, prepared 370
White Bread Dough...................2 slices, prepared 300

PORTION   SODIUM (MG.)

Prepared
Arnold/Orowheat/Brownberry

	PORTION	SODIUM (MG.)
Bran'nola	2.4 oz. (2 slices)	355
Bran'nola Country Oat	2.7 oz. (2 slices)	405
Bran'nola Hearty Wheat	2.7 oz. (2 slices)	445
Bran'nola Old Style White	2.7 oz. (2 slices)	415
Brick Oven Small Family White	1.6 oz. (2 slices)	205
Brick Oven Small Family Whole Wheat	1.6 oz. (2 slices)	190
Brick Oven White (16 oz.)	1.6 oz. (2 slices)	205
Brick Oven White (32 oz.)	2.1 oz. (2 slices)	370
Brick Oven Whole Wheat (16 oz.)	1.6 oz. (2 slices)	190
Brick Oven Whole Wheat (32 oz.)	2.1 oz. (2 slices)	420
Country White	2.4 oz. (2 slices)	490
Dill Rye (seeded)	2.0 oz. (2 slices)	405
Hearthstone White	2.1 oz. (2 slices)	450
Honey Wheat Berry	2.4 oz. (2 slices)	410
Jewish Rye (seeded)	2.3 oz. (2 slices)	425
Jewish Rye (unseeded)	2.3 oz. (2 slices)	415
Measure Up White	1.1 oz. (2 slices)	135
Measure Up Whole Wheat	1.1 oz. (2 slices)	140
Melba Thin Rye	1.5 oz. (2 slices)	270
Pumpernickel	2.3 oz. (2 slices)	460
Raisin Tea	1.8 oz. (2 slices)	225
Sprouted Wheat	1.8 oz. (2 slices)	265
Stone Ground 100% Whole Wheat	1.6 oz. (2 slices)	220

B & M

	PORTION	SODIUM (MG.)
Brown Bread (Plain)	1.6 oz.	220
Brown Bread (Raisin)	1.6 oz.	220

Colonial

	PORTION	SODIUM (MG.)
Country Meal Bread	2 oz. (2 slices)	225
Honey Grain Bread	2 oz. (2 slices)	225
Wheat Bread	2 oz. (2 slices)	245
White Enriched Bread	2 oz. (2 slices)	245

PORTION   SODIUM (MG.)

Country Hearth

	PORTION	SODIUM (MG.)
Bran 'N Honey Bread	3 oz (2 slices)	870
Butter Split Wheat Bread	3 oz. (2 slices)	710
Butter Split White Bread	3 oz. (2 slices)	1090
Deli Rye	3 oz. (2 slices)	1290
Grainola Bread	3 oz. (2 slices)	710
Old Fashioned Buttermilk Bread	3 oz. (2 slices)	980
Old Fashioned White Bread	3 oz. (2 slices)	980
Sesame French Bread	3 oz. (2 slices)	750
Sesame Vienna Bread	3 oz. (2 slices)	750
7 Whole Grain	3 oz. (2 slices)	560
Wheat Berry Bread	3 oz. (2 slices)	1340
Whole Grain Bread	3 oz. (2 slices)	560
Whole Wheat Bread	3 oz. (2 slices)	780

DiCarlo

	PORTION	SODIUM (MG.)
Parisian French	2 oz. (2 slices)	360
Sour Dough	2 oz. (2 slices)	280

Earth Grains

	PORTION	SODIUM (MG.)
Cracked Wheat Bread	3 oz. (2 slices)	370
Earth Bread	2 oz. (2 slices)	480
Gold 'N Bran Bread	3 oz. (2 slices)	430
Honey Wheat Berry Bread	3 oz. (2 slices)	390
Light Rye Bread	2 oz. (2 slices)	295
100% Whole Wheat Bread	2 oz. (2 slices)	480
Very Thin Wheat Bread	1 oz. (2 slices)	100
Very Thin White Bread	1 oz. (2 slices)	210

Francisco

	PORTION	SODIUM (MG.)
French Style	2 oz. (2 slices)	325
Vienna French	2 oz. (2 slices)	340

Fresh Horizons

	PORTION	SODIUM (MG.)
Wheat	2 oz. (2 slices)	280
White	2 oz. (2 slices)	280

Friend's

	PORTION	SODIUM (MG.)
Brown Bread (plain)	1.6 oz.	220
Brown Bread (raisins)	1.6 oz.	220

PORTION   SODIUM (MG.)

Giant

White Bread, no salt added ....2 oz. (2 slices).......................10

Hillbilly

Hillbilly ....2 oz. (2 slices)....................280

Hollywood

Dark ....2 oz. (2 slices)....................320
Light ....~....2 oz. (2 slices)....................300

Home Pride

Butter Top Wheat...................2 oz. (2 slices)....................280
Butter Top White ....2 oz. (2 slices)....................320
7 Grain................................2 oz. (2 slices)....................280
Wheatberry ....2 oz. (2 slices)....................320

Kilpatrick's

Country Meal Bread ....2 oz. (2 slices)....................225
Honey Grain Bread ....2 oz. (2 slices)....................225
Wheat Bread.........................2 oz. (2 slices)....................245
White Enriched Bread..............2 oz. (2 slices)....................245

Levy

Real Jewish Rye (seeded)........2.3 oz. (2 slices) 370
Real Jewish Rye (unseeded)   2.3 oz. (2 slices) 370
Real Pumpernickel..................2.3 oz. (2 slices) 385

Manor

Country Meal Bread ....2 oz. (2 slices)....................225
Honey Grain Bread ....2 oz. (2 slices)....................225
Wheat Bread.........................2 oz. (2 slices)....................245
White Enriched Bread..............2 oz. (2 slices)....................245

Mrs. Wright's

Raisin Bread, Low-Sodium ......1.4 oz. (2 slices) 10
Wheat Bread, Low-Sodium ......1.4 oz. (2 slices) 10
White Bread, Low-Sodium ....1.4 oz. (2 slices) 10

Natural Hearth

Butter Split White Bread ....3 oz. (2 slices) 1090

	PORTION	SODIUM (MG.)
Deli Rye	3 oz. (2 slices)	1290
Enriched Bread Made w/Buttermilk	3 oz. (2 slices)	980
Homestyle Bread	3 oz. (2 slices)	980
Honey Bran Bread	3 oz. (2 slices)	870
Honey Nugget Wheat Bread	3 oz. (2 slices)	1340
Natural Whole Grain	3 oz. (2 slices)	560
100% Stone Ground Whole Wheat Bread	3 oz. (2 slices)	780
7 Whole Grain Bread	3 oz. (2 slices)	560
Split Butter Top Wheat Bread	3 oz. (2 slices)	710
Pepperidge Farm		
Apple Bread w/Cinnamon	1.8 oz. (2 slices)	210
Bran with Diced Raisins	1.8 oz. (2 slices)	225
Cinnamon	1.9 oz. (2 slices)	195
Cinnamon Apple & Walnut	1.8 oz. (2 slices)	195
Corn & Molasses, thin sliced	1.8 oz. (2 slices)	280
Cracked Wheat, thin sliced	1.8 oz. (2 slices)	290
Date Walnut	1.8 oz. (2 slices)	215
Family Rye, 1½ lb.	2.4 oz. (2 slices)	485
Family Wheat, 2 lb.	1.9 oz. (2 slices)	270
French, Brown & Serve	2 oz.	360
French Style, Fully Baked	2 oz.	315
French Style, Fully Baked, (1 lb. loaf)	2 oz.	430
French Twin	2 oz.	250
Honey Bran, 1½ lb.	2.4 oz. (2 slices)	345
Honey Wheatberry	1.8 oz. (2 slices)	315
Honey Wheatberry, 1½ lb.	2.4 oz. (2 slices)	325
Italian, Brown & Serve	2 oz.	315
Multi-Grain Very Thin Sliced	1.1 oz. (2 slices)	150
Oatmeal 1½ lb.	2.4 oz. (2 slices)	340
Oatmeal, thin sliced	1.8 oz. (2 slices)	370
Onion Party, Slices	0.8 oz. (4 slices)	100
Orange & Raisin	1.8 oz. (2 slices)	160
Pumpernickel, Family	2.3 oz. (2 slices)	610
Pumpernickel, Family, 1½ lb.	2.4 oz. (2 slices)	610
Pumpernickel, Party, slices	0.8 oz. (4 slices)	195
Raisin with Cinnamon	1.8 oz. (2 slices)	185
Rye, Dijon Mustard	1.6 oz. (2 slices)	305

	PORTION	SODIUM (MG.)
Rye, Family	2.3 oz. (2 slices)	485
Rye, Jewish	2.3 oz. (2 slices)	555
Rye, Party, slices	0.8 oz. (4 slices)	325
Rye, Sandwich, 1½ lb.	2.4 oz. (2 slices)	445
Rye, Seedless	2.3 oz. (2 slices)	500
Rye, very thin sliced	1.1 oz. (2 slices)	220
Sprouted Wheat, sliced	1.8 oz. (2 slices)	215
Vienna, thick sliced	1.8 oz. (2 slices)	350
Wheat Germ, thin sliced	1.8 oz. (2 slices)	280
Wheat, 1½ lb.	2.4 oz. (2 slices)	385
Wheat, Sandwich	1.6 oz. (2 slices)	230
White, Lg. Family, thin sliced	1.9 oz. (2 slices)	350
White, Sandwich	1.6 oz. (2 slices)	270
White Sandwich Pockets	2.5 oz. (1 pocket)	361
White, thin sliced	1.8 oz. (2 slices)	270
White, Toasting, sliced	2.3 oz. (2 slices)	455
White, very thin	1.1 oz. (2 slices)	165
White with Cracked Wheat, 1½ lb.	2.4 oz. (2 slices)	235
White, 1½ lb.	2.4 oz. (2 slices)	350
Whole Wheat Sandwich Pockets	2.5 oz. (1 pocket)	347
Whole Wheat, thin sliced	1.8 oz. (2 slices)	245
Whole Wheat, very thin sliced	1.1 oz. (2 slices)	155

Profile

Dark	2 oz. (2 slices)	310
Light	2 oz. (2 slices)	340

Rainbo

Country Meal Bread	2 oz. (2 slices)	225
Honey Grain Bread	2 oz. (2 slices)	225
Wheat Bread	2 oz. (2 slices)	245
White Enriched Bread	2 oz. (2 slices)	245

Roman Meal

Roman Meal	2 oz. (2 slices)	280

Stop & Shop

Cracked Wheat, no salt added	2 slices	35

PORTION   SODIUM (MG.)

Daisy Loaf White, no salt
  added ....2 slices 40
Oatmeal, no salt added ....2 slices 25
100% Whole Wheat, no
  salt added ....2 slices 50
Premium White, no salt
  added ....2 slices 9

Thomas

Protein Bread (Fresh) ....1.4 oz. (2 slices) 188
Protein Bread (Frozen) ....1.7 oz. (2 slices) 230
Sahara Bread, Regular............1 oz.............................145
Sahara Bread, Wheat...............1 oz.............................187

Weight Watchers

Thin Sliced White Bread ....1 oz. (2 slices).....................200

Wonder

Cinnamon Raisin ....2 oz. (2 slices).....................280
Cracked Wheat ....2 oz. (2 slices).....................360
Family Italian ....2 oz. (2 slices).....................320
French ....2 oz. (2 slices).....................360
Fresh & Natural Wheat............2 oz. (2 slices).....................550
100% Whole Wheat ....2 oz. (2 slices).....................320
Rye, Hearty............................2 oz. (2 slices).....................360
Rye, Mild................................2 oz. (2 slices).....................360
Soft 100% Whole Wheat ....2 oz. (2 slices).....................280
Wheat.....................................2 oz. (2 slices).....................280
White ....2 oz. (2 slices).....................280
White w/Buttermilk...................2 oz. (2 slices).....................320

BREAD STICKS

Pillsbury

Pipin' Hot Soft Bread
  Sticks ....1 bread stick 240

Stella D'Oro

Plain Bread Sticks (dietetic) ....0.9 oz. (3 sticks) ....under 10
Sesame Bread Sticks
  (dietetic).............................1.2 oz. (3 sticks) ....under 10

PORTION   SODIUM (MG.)

CRACKERS

Devonsheer

Plain Unsalted Melba Rounds	½ oz. (5 crackers)	2.5

Earth Grains

Cheddar Cheese Safari Crackers	½ oz.	225
Garlic Butter 'N Cheese Safari Crackers	½ oz.	345
Onion 'N Cheese Safari Crackers	½ oz.	375
Pizza Safari Crackers	½ oz.	240

Featherweight

Unsalted Crackers	4 crackers	under 10

Health Valley

Amaranth Crackers	1 cracker	15
Cheese Wheels	0.5 oz.	85
Herb Crackers	0.5 oz.	122
Herb Crackers, unsalted	0.5 oz.	55
No Salt 7-Grain Crackers	0.5 oz.	23
Sesame Crackers	0.5 oz.	95
Sesame Crackers, unsalted	0.5 oz.	8
7-Grain Crackers	0.5 oz.	68
Stoned Wheat Crackers, regular	0.5 oz.	95
Stoned Wheat Crackers, unsalted	0.5 oz.	2
Wheat and Sweet Rye Crackers	0.5 oz.	33
Wheat and Sweet Rye Crackers, unsalted	0.5 oz.	15
Yogurt and Green Onion Crackers	0.5 oz.	68
Yogurt and Green Onion Crackers, unsalted	0.5 oz.	30

Ideal

Flatbread, Bran	0.5 oz. (3 pieces)	132

PORTION   SODIUM (MG.)

Flatbread, Extra-Thin ....0.5 oz. (5 pieces) 117
Flatbread, Whole Grain ....0.5 oz. (3 pieces) 132

Interbake Foods Limited

Stoned Wheat Thins ....0.5 oz. (2 crackers) 222

Keebler

Cinnamon Crisp Grahams ....0.5 oz. (4 crackers) 90
Club Crackers...........................0.4 oz. (4 crackers) 155
Toasted Rye Crackers ....0.6 oz. (5 crackers) 140
Toasted Sesame Crackers........0.6 oz. (5 crackers) 135
Toasted Wheat Crackers ....0.6 oz. (5 crackers) 145
Town House Crackers................0.6 oz. (5 crackers) 145
TUC Snack Crackers ....0.5 oz. (3 crackers) 85
Zesta Saltines...........................0.5 oz. (5 crackers) 165

Manischewitz

Matzo Crackers ....0.5 oz. 1
Matzo Thins ....0.5 oz. 0
Thin Tea Matzos ....0.5 oz. 1
Unsalted Matzos ....0.5 oz. 1
Wheat Crackers ....0.5 oz. 1
Whole Wheat Matzos w/Bran 0.5 oz. 1

Nabisco

Bacon Flavored Thins................0.5 oz. (7 pieces) 205
Buttery Flavor Sesame
  Snack.....................................0.5 oz. (4½ pieces) 135
Cheddar Triangles ....0.5 oz. (8½ pieces) 130
Cheese Nips.............................0.5 oz. (13 pieces) 180
Cheese Peanut Butter
  Sandwich................................0.5 oz. (2 pieces) 165
Cheese Sandwich ....0.5 oz. (2 pieces) 200
Cheese Tid-Bit ....0.5 oz. (16 pieces) 215
Chicken in a Biskit ....0.5 oz. (7 pieces) 130
Cinnamon Treats ....0.5 oz. (2 pieces) 120
Country Crackers......................0.5 oz. (5 crackers) 145
Crown Pilot ....0.5 oz. (1 cracker) 80
Dandy Soup & Oyster
  Crackers ....0.5 oz. (20 crackers) 220

PORTION   SODIUM (MG.)

Dip in a Chip Cheese 'n
  Chives ....0.5 oz. (7½ pieces) 135
Dixies Drumstick ....0.5 oz. (9 pieces) 185
Escort...0.5 oz (3½ crackers) 120
French Onion Crackers ....0.5 oz. (6 crackers) 140
Gitana Sodas ....0.5 oz. (4 crackers) 120
Graham Crackers ....0.5 oz. (2 crackers) 105
Holland Rusk ....0.5 oz. (1½ crackers)...........40
Honey Maid Graham
  Crackers ....0.5 oz. (2 crackers) 105
Malted Milk Peanut Butter
  Sandwich ....0.5 oz. (2 pieces) 165
Meal Mates Sesame Bread
  Wafer ....0.5 oz. (3 wafers) 180
Oysterettes Soup & Oyster
  Crackers ....0.5 oz. (18 crackers) 220
Premium Saltines.......................0.5 oz. (5 crackers) 230
Premium Unsalted Tops ....0.5 oz. (5 crackers) 120
Ritz Crackers ....0.5 oz. (4½ crackers)..........135
Royal Lunch Milk Crackers ......0.5 oz. (1¼ crackers) 95
Sea Rounds ....0.5 oz. (1¼ crackers) 155
Sesame Wheats.........................0.5 oz. (4½ crackers) 125
Snack Shapes ....0.5 oz. (7½ crackers)..........200
Sociables...................................0.5 oz. (6½ crackers)..........165
Swiss Cheese Natural
  Flavor Snack Crackers ....0.5 oz. (7½ crackers)..........165
Triscuit Wafers ....0.5 oz. (3 crackers) 90
Twigs Sesame/Cheese
  Sticks ....0.5 oz. (5 sticks) 200
Uneeda Biscuits, Unsalted
  Top ....0.5 oz. (3 pieces) 115
Vegetable Thins ....0.5 oz. (6½ crackers)..........150
Waverly Wafers ....0.5 oz. (4 crackers) 155
Wheat Thins ....0.5 oz. (8 crackers) 120
Wheatsworth Stone Ground
  Wheat Crackers ....0.5 oz. (4½ crackers)..........165
Zwieback ....0.5 oz. (2 crackers) 40

Penn-Dutch

Cheddar Cheese Safari
  Crackers ....0.5 oz. 225

PORTION   SODIUM (MG.)

Garlic Butter & Cheese Safari
   Crackers ....0.5 oz. 345
Onion 'n Cheese Crackers ......0.5 oz. 375
Pizza Safari Crackers..............0.5 oz. 240
Salt-free Safari Crackers ....0.5 oz. 20

Pepperidge Farm Distinctive Crackers

Cracked Wheat ....0.5 oz. (2½ crackers)..........130
English Water Biscuit ....0.5 oz. (3½ crackers)............85
Hearty Wheat ....0.5 oz. (2½ crackers)..........123
Sesame ....0.5 oz. (3 crackers) 135

Pepperidge Farm Goldfish Thins

Butter Flavored...........................0.5 oz. (4 crackers) 48
Cheese.......................................0.5 oz. (4 crackers) 60
Salted...0.5 oz. (4 crackers) 60
Wheat...0.5 oz. (4 crackers) 63

Pepperidge Farm Tiny Goldfish Crackers

Cheddar Cheese ....0.5 oz. (22½ crackers)..........88
Pizza Flavored ....0.5 oz. (22½ crackers)..........90
Pretzel ....0.5 oz. (20 crackers) 80
Salted...0.5 oz. (22½ crackers)..........80

Pepperidge Farm Vending Pack Crackers

Butter Pretzels Nugget Style....0.5 oz. (10½ pretzels) 327
Butter Pretzels Thin Sticks ......0.5 oz. (12½ pretzels) 323
Butter Pretzels Tiny Twists ......0.5 oz. (7½ pretzels) 408
Cheddar Cheese Goldfish ....0.5 oz. (22½ crackers)..........88
Parmesan Cheese Goldfish......0.5 oz. (22½ crackers) 125
Salted Goldfish..........................0.5 oz. (22½ crackers)..........90

Ralston Purina

Animal Crackers.........................0.5 oz. (7.5 crackers) 65
Cheddar Snacks ....0.5 oz. (9 crackers) 130
Cheese Snacks ....0.5 oz. (12.5 crackers)........165
Cheese & Chive Snacks ....0.5 oz. (9 crackers) 130
Crackers (unsalted tops) ....0.5 oz. (5 crackers) 103
Oyster Crackers ....0.5 oz. (33 crackers) 190
Rich & Crisp ....0.5 oz. (4.5 crackers) 100
Rye Snacks................................0.5 oz. (7.5 crackers) 105

PORTION   SODIUM (MG.)

Rykrisp, Natural	0.4 oz. (2 crackers)	110
Rykrisp, Seasoned	0.5 oz. (2 crackers)	220
Rykrisp, Sesame	0.5 oz. (2 crackers)	220
Saltines	0.5 oz. (5 crackers)	200
Sesame & Wheat Snacks	0.5 oz. (9 crackers)	130
Snackers	0.5 oz. (4 crackers)	105
Wheat Snacks	0.5 oz. (7.5 crackers)	100

Venus

Bite-Size Wheat Wafers, No Salt Added	0.4 oz. (4 wafers)	2

CROUTONS

Devonsheer

Seasoned Croutons, unsalted	0.2 oz. (10 croutons)	1

Kellogg's

Croutettes	0.7 oz. (⅔ cup)	260

Pepperidge Farm

Bacon & Cheese	0.5 oz.	130
Bleu Cheese	0.5 oz.	170
Cheddar & Romano Cheese	0.5 oz.	190
Cheese & Garlic	0.5 oz.	190
Dijon Mustard Rye & Cheese	0.5 oz.	150
Onion & Garlic	0.5 oz.	160
Seasoned	0.5 oz.	210
Sour Cream & Chive	0.5 oz.	215

ENGLISH MUFFINS

Arnold/Orowheat/Brownberry

Bran'nola	2.3 oz. (1 muffin)	260
Extra Crisp 6s (and 12s)	2.3 oz. (1 muffin)	310
Raisin	2.5 oz. (1 muffin)	350

Earth Grains

Plain	2.3 oz. (1 muffin)	415

PORTION   SODIUM (MG.)

Raisin	2.3 oz. (1 muffin)	460
Whole Wheat	2.3 oz. (1 muffin)	495

Merico

English Muffin	2 oz. (1 muffin)	205

Pepperidge Farm

Cinnamon Apple	2 oz. (1 muffin)	200
Cinnamon Raisin	2 oz. (1 muffin)	185
Plain English Muffins	2 oz. (1 muffin)	220
6 Bacon & Cheese	2 oz. (1 muffin)	365
6 Cinnamon Chip	2 oz. (1 muffin)	245
Sourdough	2 oz. (1 muffin)	255
Stone Ground Wheat	2 oz. (1 muffin)	210

Thomas

English Muffins	2 oz. (1 muffin)	208
Honey Wheat	2 oz. (1 muffin)	228
Sourdough	2 oz. (1 muffin)	208

Wonder

English Muffins	2 oz. (1 muffin)	280

MUFFINS AND QUICKBREADS
(for sweet rolls see also Cakes, Pastries, and Pies, page 203)

Frozen

Pepperidge Farm Old Fashioned Muffins

Blueberry	1.9 oz. (1 muffin)	250
Bran w/Raisin	2.0 oz. (1 muffin)	295
Carrot Walnut	2.1 oz. (1 muffin)	220
Cinnamon Swirl	1.8 oz. (1 muffin)	170
Corn	1.9 oz. (1 muffin)	260
Orange-Cranberry	2.1 oz. (1 muffin)	200

Mixes

Arrowhead Mills

Multigrain Corn Bread Mix	two 2-in. squares	220

PORTION   SODIUM (MG.)

Aunt Jemima

Easy Mix Corn Bread ....1.7 oz. (⅙ cornbread) 516

Ballard

Corn Bread Mix ....⅛ recipe 570

Betty Crocker

Apple Cinnamon ....1/12 package 135
Corn Muffin Mix ....1 muffin............................315
Gingerbread Mix ....⅑ package 325
Tart Cherry ....1/12 package 120
Wild Blueberry Muffin Mix ....1 muffin............................150

Dromedary

Corn Bread Mix ....two 2-in. squares 475
Corn Muffin Mix ....1 muffin............................470
Gingerbread Mix ....2-in. square 185

Duncan Hines

Banana Nut................................1 muffin............................175
Bran Muffin Mix ....1 muffin............................165
Spicy Apple................................1 muffin............................185
Wild Blueberry Muffin Mix ....1 muffin............................155

Elam's

Corn Bread Mix ....1.0 oz. dry 340

Fearn Soya

Bran Muffin Mix ....1.5 oz. dry 117
Corn Bread Mix ....1.0 oz. dry 124

Flako

Corn Muffin mix ....1 oz. dry (1 muffin)............313

Pillsbury

Applesauce Spice Bread ....1/12 loaf................................155
Apricot Nut Bread ....1/12 loaf................................150
Banana Bread ....1/12 loaf................................150
Blueberry Nut Bread ....1/12 loaf................................155
Carrot Nut Bread ....1/12 loaf................................185

PORTION   SODIUM (MG.)

Cherry Nut Bread	1/12 loaf	150
Cranberry Bread	1/12 loaf	155
Date Bread	1/12 loaf	155
Gingerbread	3-in. square	310
Nut Bread	1/12 loaf	180

Prepared

Tastykake

Blueberry Muffin	1 oz. (1 muffin)	220
Corn Muffin	1 oz. (1 muffin)	225

Thomas Toast-r-Cakes

Blueberry	1.3 oz. (1 piece)	275
Bran	1.3 oz. (1 piece)	310
Corn	1.3 oz. (1 piece)	325

Wonder

Raisin Rounds	2 oz. (1 muffin)	280
Sour Dough	2 oz. (1 muffin)	250

ROLLS AND BUNS

Arnold/Orowheat/Brownberry

Dinner Party Rounds	1.3 oz. (2 rolls)	280
Dutch Egg Sandwich Buns	1.6 oz. (1 roll)	241
Hamburger Buns (8s)	1.4 oz. (1 roll)	285
Hot Dog Buns (6s)	1.3 oz. (1 roll)	290
Soft Sandwich, Plain	1.3 oz. (1 roll)	260
Soft Sandwich, Poppy Seeds	1.3 oz. (1 roll)	255
Soft Sandwich, Sesame Seeds	1.3 oz. (1 roll)	260

Colonial

Hamburger/Hot Dog Buns	2 oz. (1 bun)	250

Francisco

French Style, Enriched	2 oz. (1 roll)	325
Sandwich	2 oz. (1 roll)	325
Sourdough French	2.2 oz. (2 rolls)	320

PORTION   SODIUM (MG.)

Home Pride
Dinner Rolls ....2 oz. (2 rolls) 340

Kilpatrick's
Hamburger/Hot Dog Buns ....2 oz. (1 bun) 250

Manor
Hamburger/Hot Dog Buns ....2 oz. (1 bun) 250

Merico
Crescent Rolls ....2 oz. (2 rolls) 550
Super Crescent Rolls ....3 oz. (2 rolls) 825

Pepperidge Farm
Butter Crescent, Heat &
  Serve ....1.0 oz. (1 roll) 155
Club, Brown 'n Serve ....1.3 oz (1 roll) 220
Dinner Rolls ....0.7 oz. (1 roll) 90
Finger Rolls with Poppy
  Seeds...................................0.6 oz. (1 roll) 95
Finger Rolls with Sesame
  Seeds...................................0.6 oz. (1 roll) 80
Frankfurter Rolls, side sliced....1.7 oz. (1 roll) 315
Frankfurter Rolls, top sliced......1.7 oz. (1 roll) 315
French, Brown 'n Serve (2) ......2.5 oz. (½ roll)......................420
French, Brown 'n Serve (3) ......1.7 oz. (½ roll)......................295
French Style Rolls (4)................1.5 oz. (½ roll)......................285
French Style Rolls (9)................1.3 oz. (1 roll) 250
Golden Twist, Heat 'n Serve ....1.0 oz. (1 roll) 160
Hearth, Brown 'n Serve ....0.7 oz. (1 roll) 100
Old Fashioned Rolls ....0.6 oz. (1 roll) 90
Onion Sandwich Buns with
  Poppy Seeds, Sliced.............1.9 oz. (1 roll) 235
Parker House Rolls ....0.6 oz. (1 roll) 90
Party Rolls.................................0.7 oz. (2 rolls) 100
Party Rolls with Poppy
  Seeds...................................0.7 oz. (2 rolls) 110
Sandwich Buns with Poppy
  Seeds, Sliced ....1.6 oz. (1 roll) 200
Sandwich Buns with Sesame
  Seeds, sliced ....1.6 oz. (1 roll) 210

PORTION   SODIUM (MG.)

Sliced Hamburger Rolls	1.5 oz. (1 roll)	255
Soft Family Rolls	1.2 oz. (1 roll)	200
Sour French Rolls	1.3 oz. (1 roll)	235

Pillsbury

Bakery Style Country White Rolls	1.4 oz. (1 roll)	350
Butterflake Rolls	1.3 oz. (1 roll)	410
Crescent Rolls	2 oz. (2 rolls)	460
Hot Roll Mix	2 rolls, prepared	250
Parkerhouse Rolls	2 rolls	595

Rainbo

Hamburger/Hot Dog Buns	2 oz. (1 bun)	250

Sara Lee

Parkerhouse Rolls	1.6 oz. (2 rolls)	260
Party Rolls	1.8 oz. (3 rolls)	294
Sesame Seed Rolls	1.8 oz. (3 rolls)	294

Wonder

Buttermilk (Brown & Serve)	2 oz. (2 rolls)	280
Dinner Rolls	2 oz. (2 rolls)	280
Gem Style (Brown 'n Serve)	2 oz. (2 rolls)	280
Half & Half (Brown & Serve)	2 oz. (2 rolls)	280
Hamburger Buns	2 oz. (2 rolls)	300
Home Bake (Brown & Serve)	2 oz. (2 rolls)	260
Hot Dog Rolls	2 oz. (2 rolls)	300
Hoagie Rolls	5 oz. (1 roll)	840
Pan Rolls	2 oz. (2 rolls)	280

STUFFING

Pepperidge Farm

Chicken Pan Style	1 oz.	395
Corn Bread (8 oz.)	1 oz.	530
Corn Bread (16 oz.)	1 oz.	530
Cube (7 oz.)	1 oz.	440
Cube (14 oz.)	1 oz.	440
Herb Seasoned (8 oz.)	1 oz.	435

	PORTION	SODIUM (MG.)
Herb Seasoned (16 oz.)	1 oz.	435
Seasoned Pan Style	1 oz.	435

Stove Top

Americana New England, prepared w/butter	½ cup prepared	635
Americana San Francisco, prepared w/butter	½ cup prepared	635
Beef, prepared w/butter	½ cup prepared	585
Chicken Flavor, prepared w/butter	½ cup prepared	640
Cornbread, prepared w/butter	½ cup prepared	665
Pork, prepared w/butter	½ cup prepared	620
Stuffing w/Rice, prepared w/butter	½ cup prepared	505

TORTILLAS

El Charrito

Corn Tortillas	1.4 oz. (2 tortillas)	20

Old El Paso

Taco Shells	0.4 oz. (1 shell)	47
Tostada Shells	0.4 oz. (1 shell)	66

USDA

Corn Tortillas	1.2 oz. (1 tortilla)	26
Flour Tortillas	2 oz. (1 tortilla)	473

Cakes, Pastries, and Pies

CAKES AND PASTRIES
(for muffins see also Bread Products, page 184)

Frozen and Doughs	PORTION	SODIUM (MG.)

Health Valley

Apple Spice Cake	2 oz.	125
Carrot Nut Cake	2 oz.	125

PORTION   SODIUM (MG.)

Merico

Cinnamon Raisin Rolls	1.8 oz. (1 roll)	375
Fruit Danish	2 oz. (1 danish)	425

Morton

Bavarian Creme Donuts (12 oz.)	2 oz. (1 donut)	75
Blueberry Muffin Rounds (9 oz)	1.5 oz. (1 muffin)	180
Blueberry Muffins (9.5 oz.)	1.6 oz. (1 muffin)	130
Boston Creme Donuts (14 oz.)	2.3 oz. (1 donut)	90
Chocolate Iced Donuts (9 oz.)	1.5 oz. (1 donut)	75
Chocolate Iced Morning Light Donuts (12 oz.)	2 oz. (1 donut)	177
Corn Muffins (10 oz.)	1.7 oz. (1 muffin)	280
Devils Food Donut Holes	1.6 oz. (1 piece)	105
Glazed Donuts (9 oz.)	1.5 oz. (1 donut)	75
Glazed Morning Light Donuts (12 oz.)	2 oz. (1 donut)	176
Honey Wheat Donut Holes	1.6 oz. (1 piece)	140
Jelly Donuts (11 oz.)	1.8 oz. (1 donut)	75
Jelly Morning Light Donuts (15.5 oz.)	2.6 oz. (1 donut)	193
Mini Donuts (10 oz.)	1 oz. (1 donut)	200
Mini Honey Buns (16 oz.)	1 oz. (1 bun)	90
Regular Honey Buns (9 oz.)	2 oz. (1 bun)	150
Vanilla Donut Holes	1.6 oz. (1 piece)	123

Morton Great Little Desserts (cakes)

Cherry Cheesecake	6 oz. (1 cake)	350
Cream Cheesecake	6 oz. (1 cake)	350
Pineapple Cheesecake	6 oz. (1 cake)	355
Strawberry Cheesecake	6 oz. (1 cake)	350

Pepperidge Farm

All Butter Croissants	2 oz. (1 roll)	310
Almond Croissants	2 oz. (1 roll)	260
Cinnamon Croissants	2 oz. (1 roll)	280
Raisin Croissants	2 oz. (1 roll)	265

PORTION   SODIUM (MG.)

Pepperidge Farm Cakes Supreme

Boston Creme	2.9 oz. (¼ cake)	190
Carrot Supreme	2.1 oz.	184
Chocolate	2.9 oz. (¼ cake)	140
Lemon Coconut	3.1 oz. (¼ cake)	220
Pineapple Cream	2 oz. (¹⁄₁₂ cake)	145
Strawberry Cream	2 oz. (¹⁄₁₂ cake)	145
Walnut	2.5 oz. (¼ cake)	200

Pepperidge Farm Danish

Almond Deep Dish Danish	2.2 oz. (1 danish)	260
Apple Deep Dish Danish	2 oz. (1 danish)	235
Blueberry Deep Dish Danish	2.1 oz. (1 danish)	245
Cheese Deep Dish Danish	2.6 oz. (1 danish)	310
Cherry Deep Dish Danish	2.1 oz. (1 danish)	250

Pepperidge Farm Layer Cakes

Chocolate Fudge	1.7 oz. (¹⁄₁₀ cake)	140
Coconut	1.7 oz. (¹⁄₁₀ cake)	120
Devil's Food	1.7 oz. (¹⁄₁₀ cake)	135
German Chocolate	1.7 oz. (¹⁄₁₀ cake)	170
Golden	1.7 oz. (¹⁄₁₀ cake)	115
Orange	1.9 oz. (¹⁄₁₀ cake)	162
Vanilla	1.7 oz. (¹⁄₁₀ cake)	120

Pepperidge Farm Rich 'n Moist Cakes

Chocolate w/Chocolate Frosting	1.8 oz. (⅛ cake)	160
Chocolate w/Vanilla Frosting	1.8 oz. (⅛ cake)	170
Vanilla w/Chocolate Frosting	1.6 oz. (⅛ cake)	130
Yellow w/Chocolate Frosting	1.6 oz. (⅛ cake)	120

Pepperidge Farm Old Fashioned Cakes

Apple Walnut w/Cream Cheese Icing	1.5 oz. (⅛ cake)	140
Butter Pound	1.1 oz (¹⁄₁₀ cake)	145
Carrot w/Cream Cheese Icing	1.5 oz. (⅛ cake)	145

Pepperidge Farm Pastries

Apple Criss Cross	2 oz. (1 pastry)	140
Apple Dumplings	3 oz. (1 dumpling)	235

PORTION   SODIUM (MG.)

	PORTION	SODIUM (MG.)
Apple Strudel	3 oz. (⅕ package)	215
Apple Turnovers	3.1 oz. (1 turnover)	215
Blueberry Turnovers	3.1 oz. (1 turnover)	235
Cherry Turnovers	3.1 oz. (1 turnover)	285
Coconut Custard	1 turnover	292
Iced Brownies	1 brownie	164
Patty Shells	1.7 oz. (1 shell)	180
Peach Turnovers	3.1 oz. (1 turnover)	260
Puff Pastry Sheets	4.3 oz. (1 sheet)	580
Raspberry Turnovers	3.1 oz. (1 turnover)	265

Pillsbury

	PORTION	SODIUM (MG.)
Apple Turnovers	2 oz. (1 turnover)	305
Blueberry Turnovers	2 oz. (1 turnover)	305
Caramel Danish w/Nuts	2.7 oz. (2 rolls)	490
Cherry Turnovers	2 oz. (1 turnover)	310
Cinnamon Raisin Danish w/Icing	2 rolls	450
Cinnamon w/Icing	2.4 oz. (2 rolls)	520
Hungry Jack Butter Tastin' Cinnamon w/Icing	2.8 oz. (2 rolls)	570
Orange Danish	2.8 oz. (2 rolls)	485
Orange Danish w/Icing	2 rolls	460
Pipin' Hot Apple Danish	1 roll	250
Pipin' Hot Cinnamon Rolls	1 roll	360

Sara Lee

	PORTION	SODIUM (MG.)
Almond Coffee Cake	1.5 oz. (⅛ cake)	167
Almond Coffee Ring	1.2 oz. (⅛ cake)	151
Almond Light Coffee Round	1 oz. (⅛ cake)	124
Apple Crunch Rolls	1 oz. (1 roll)	105
Apple Individual Danish	1.3 oz. (1 Danish)	110
Apple Walnut Cake	1.6 oz. (⅛ cake)	140
Banana Cake	1.7 oz. (⅛ cake)	154
Banana Nut Layer Cake	2.5 oz. (⅛ cake)	166
Banana Nut Pound Cake	1.1 oz. (¹⁄₁₀ cake)	104
Black Forest Cake	2.6 oz. (⅛ cake)	167
Blueberry Coffee Ring	1.2 oz. (⅛ cake)	135
Blueberry Crumb Cake	1.7 oz. (1 cake)	201

PORTION   SODIUM (MG.)

Blueberry Light Coffee Round	1 oz. (1/8 cake)	116
Butter Streusel Coffee Cake, Large	1.4 oz. (1/8 cake)	197
Caramel Pecan Rolls	1.3 oz. (1 roll)	162
Caramel Sticky Buns	1 oz. (1 bun)	118
Carrot Cake	1.5 oz (1/8 cake)	125
Cheese Individual Danish	1.3 oz. (1 Danish)	125
Cherry Cream Cheese Cake	3.2 oz. (1/6 cake)	127
Chocolate Bavarian Cake	2.8 oz. (1/8 cake)	78
Chocolate Cake	1.7 oz. (1/8 cake)	168
Chocolate Cupcake	1.7 oz. (1 cupcake)	159
Chocolate 'n Cream Layer Cake	2.3 oz. (1/8 cake)	136
Chocolate Pound Cake	1.1 oz. (1/10 cake)	143
Chocolate Swirl Pound Cake	1.2 oz. (1/10 cake)	116
Cinnamon Raisin Individual Danish	1.3 oz. (1 Danish)	130
Cinnamon Rolls	0.9 oz. (1 roll)	96
Cinnamon Streusel Coffee Cake	1.4 oz. (1/8 cake)	173
Coconut Cake	1.2 oz. (1/8 cake)	82
Cream Cheese Cake, Large	2.8 oz. (1/6 cake)	161
Cream Cheese Cake, Small	3.3 oz. (1/3 cake)	193
Croissants	0.9 oz. (1 croissant)	140
Double Chocolate Cupcake	1.7 oz. (1 cupcake)	161
Double Chocolate Layer Cake	2.3 oz. (1/8 cake)	138
French Cream Cheese Cake	2.9 oz. (1/8 cake)	136
French Crumb Cake	1.7 oz. (1 cake)	200
German Chocolate Cake	1.5 oz. (1/8 cake)	133
Homestyle Pound Cake	1 oz. (1/10 cake)	105
Honey Rolls	1 oz. (1 roll)	119
Maple Crunch Coffee Ring	1.2 oz. (1/8 cake)	131
Maple Light Coffee Round	1 oz. (1/8 cake)	121
Orange Cake	1.7 oz (1/8 cake)	170
Pecan Coffee Cake, Large	1.4 oz. (1/8 cake)	172
Pecan Coffee Cake, Small	1.6 oz. (1/4 cake)	199

PORTION   SODIUM (MG.)

Pound Cake	1.1 oz (1/10 cake)	104
Pound Cake, Family Size	1.1 oz. (1/15 cake)	107
Raisin Pound Cake	1.3 oz. (1/10 cake)	120
Raspberry Coffee Ring	1.2 oz. (1/8 cake)	136
Raspberry Light Coffee Round	1 oz. (1/8 cake)	113
Strawberries 'n Cream Layer Cake	2.6 oz. (1/8 cake)	150
Strawberry Cream Cheese Cake	3.2 oz. (1/6 cake)	129
Strawberry French Cream Cheese Cake	3.2 oz. (1/8 cake)	128
Strawberry Shortcake	2.6 oz. (1/8 cake)	136
Walnut Layer Cake	2.3 oz. (1/8 cake)	153
Yellow Cupcake	1.7 oz. (1 cupcake)	176

Mixes—All figures for food as prepared

Arrowhead Mills

Whole Grain Carob Cake Mix	1/12 cake	170

Aunt Jemima

Easy Mix Coffee Cake	1/3 oz. (1/8 cake)	226

Betty Crocker Angel Food Cake Mixes

Chocolate	1/12 package	300
Confetti	1/12 package	255
Lemon Chiffon	1/12 package	190
Lemon Custard	1/12 package	260
Strawberry	1/12 package	260
Traditional White	1/12 package	165
White	1/12 package	260

Betty Crocker Classics Dessert Mixes

Chocolate Flavored Eclair	1/6 package	455
Chocolate Pudding Cake	1/6 package	255
German Black Forest Cake	1/12 package	325
Golden Pound Cake	1/12 package	155
Lemon Pudding Cake	1/6 package	270
Pineapple Upside Down Cake	1/6 package	215

PORTION   SODIUM (MG.)

Betty Crocker Snackin' Cake Mixes

Applesauce Raisin	⅑ package	250
Banana Walnut	⅑ package	260
Butter Pecan	⅑ package	250
Carrot Nut	⅑ package	240
Chocolate Fudge Chip	⅑ package	205
Coconut Pecan	⅑ package	255
Fudge Peanut Butter Chip	⅑ package	250
German Chocolate Coconut Pecan	⅑ package	255
Golden Chocolate Chip	⅑ package	255
Mint Fudge Chip	⅑ package	210

Betty Crocker Stir 'n Streusel

Cinnamon	⅙ package	230
German Chocolate	⅙ package	245

Betty Crocker Stir n' Frost Cake Mixes with Frosting

Carrot Cake w/Cream Cheese Frosting	⅙ package	215
Chocolate Chip Cake w/Chocolate Frosting	⅙ package	205
Chocolate-Chocolate Chip Cake w/Chocolate-Chocolate-Chip Frosting	⅙ package	260
Chocolate Devils Food Cake w/Chocolate Frosting	⅙ package	260
Chocolate Fudge Cake w/Vanilla Frosting	⅙ package	275
Spice Cake w/Vanilla Frosting	⅙ package	315
Yellow Cake w/Chocolate Frosting	⅙ package	215

Betty Crocker Supermoist Cake Mixes

Apple Cinnamon	1/12 package	275
Banana	1/12 cake	255
Butter Brickle	1/12 cake	275
Butter Pecan	1/12 cake	250
Butter Recipe, Yellow	1/12 cake	350
Carrot	1/12 cake	255

	PORTION	SODIUM (MG.)
Cherry Chip	1/12 cake	265
Chocolate Chip	1/12 package	230
Chocolate Chocolate Chip	1/12 package	245
Chocolate Fudge	1/12 cake	435
Devils Food	1/12 cake	410
German Chocolate	1/12 cake	420
Lemon	1/12 cake	260
Marble	1/12 cake	255
Milk Chocolate	1/12 cake	290
Orange	1/12 cake	280
Sour Cream Chocolate	1/12 cake	430
Sour Cream White	1/12 cake	260
Spice	1/12 cake	275
Strawberry	1/12 cake	260
White	1/12 cake	245
Yellow	1/12 cake	270

Bundt

Boston Cream	1/16 cake	305
Chocolate Macaroon	1/16 cake	305
Fudge Nut Crown	1/16 cake	290
Lemon Blueberry	1/16 cake	270
Marble Supreme	1/16 cake	265
Pound	1/16 cake	260
Tunnel of Fudge	1/16 cake	315
Tunnel of Lemon	1/16 cake	295

Dia-Mel Cakes

Chocolate	1/10 cake	15
Lemon	1/10 cake	15
Pound	1/2-in. slice	15
Spice	1/10 cake	20

Dromedary

Pound Cake	3/4-in. slice	335

Duncan Hines

Angel Food	1/12 cake	130
Apple	1/12 cake	265
Banana Supreme	1/12 cake	285
Butter Recipe, Fudge	1/12 cake	270

PORTION   SODIUM (MG.)

Butter Recipe, Golden	¹⁄₁₂ cake	270
Carrot	¹⁄₁₂ cake	250
Cherry Supreme	¹⁄₁₂ cake	260
Chocolate Chip	¹⁄₁₂ cake	265
Deep Chocolate	¹⁄₁₂ cake	280
Devil's Food	¹⁄₁₂ cake	280
Fudge Marble	¹⁄₁₂ cake	260
Golden Vanilla	¹⁄₁₂ cake	285
Lemon Supreme	¹⁄₁₂ cake	285
Orange Supreme	¹⁄₁₂ cake	285
Pineapple Supreme	¹⁄₁₂ cake	285
Sour Cream Chocolate	¹⁄₁₂ cake	375
Spice	¹⁄₁₂ cake	285
Strawberry Supreme	¹⁄₁₂ cake	285
Swiss Chocolate	¹⁄₁₂ cake	375
Yellow	¹⁄₁₂ cake	285
White	¹⁄₁₂ cake	260

Estee Cakes

Chocolate	¹⁄₁₀ cake	15
Lemon	¹⁄₁₀ cake	15
White	¹⁄₁₀ cake	15

Fearn Soya

Carob Cake	⅙ cake	188
Carrot Cake	⅙ cake	211
Natural Banana Flavor Cake	⅙ cake	173
Spice Cake	⅙ cake	220

Featherweight

Cake & Cookie Mix	1-in. slice	5

Jell-O

Cheesecake	⅛ cake	365

Pillsbury

Apple Cinnamon Coffee Cake	⅛ cake	155
Butter Pecan Coffee Cake	⅛ cake	335
Cinnamon Streusel Coffee Cake	⅛ cake	225

PORTION   SODIUM (MG.)

Raspberry Angel Food	1/12 cake	300
Sour Cream Coffee Cake	1/8 cake	235
White Angel Food	1/12 cake	345

Pillsbury Plus

Applesauce Spice	1/12 cake	300
Banana	1/12 cake	290
Butter Recipe	1/12 cake	345
Carrot 'n Spice	1/12 cake	330
Chocolate Mint	1/12 cake	340
Dark Chocolate	1/12 cake	440
Devil's Food	1/12 cake	405
Fudge Marble	1/12 cake	300
German Chocolate	1/12 cake	340
Lemon	1/12 cake	310
Oats 'n Brown Sugar	1/12 cake	305
Strawberry	1/12 cake	300
White	1/12 cake	295
Yellow	1/12 cake	300

Pillsbury Streusel Swirl

Banana	1/16 cake	200
Cinnamon	1/16 cake	200
Dutch Apple	1/16 cake	200
Fudge Marble	1/16 cake	200
German Chocolate	1/16 cake	290
Lemon	1/16 cake	335
Pecan Brown Sugar	1/16 cake	200
Rich Butter	1/16 cake	235

Pillsbury Wiener Wrap

Cheese	1 wrap	395
Plain	1 wrap	430

Royal

Cheesecake	1/8 cake	450

Prepared

Drake's
Coffee Cake, Small	1 piece	220
Devil Dogs	1 piece	165

	PORTION	SODIUM (MG.)
Funny Bones	1 piece	130
Ring Ding Jr.	1 piece	120
Yankee Doodles	1 piece	130

Earth Grains

	PORTION	SODIUM (MG.)
Cinnamon Apple Doughnuts	3.0 oz. (1 doughnut)	515
Devil's Food Doughnuts	2.7 oz. (1 doughnut)	505
Old Fashioned Glazed Doughnuts	2.7 oz. (1 doughnut)	485
Old Fashioned Plain Doughnuts	2.3 oz. (1 doughnut)	345
Old Fashioned Powdered Doughnuts	2.5 oz. (1 doughnut)	400

Featherweight

	PORTION	SODIUM (MG.)
Low-Sodium Cake	½-in. slice	6

Hostess

	PORTION	SODIUM (MG.)
Big Wheels	1.4 oz. (1 piece)	130
Chip Flips	2.75 oz. (1)	165
Choco-Dile	2.2 oz. (1 piece)	280
Crumb Cakes	1.3 oz. (1 piece)	95
Cupcakes, Chocolate	1.8 oz. (1 cupcake)	250
Cupcakes, Orange	1.5 oz. (1 cupcake)	175
Danish, Apple	3.5 oz. (1)	410
Danish, Butterhorn	2.9 oz. (1)	520
Danish, Raspberry	2.9 oz. (1)	360
Dessert Cups	0.75 oz. (1)	120
Ding Dongs	1.4 oz. (1 piece)	130
Donettes, Chocolate Coated	0.5 oz. (1)	50
Donettes, Powdered	0.3 oz. (1)	40
Donuts, Chocolate Coated	1 oz. (1)	150
Donuts, Cinnamon	1 oz. (1 donut)	140
Donuts, Krunch	1 oz. (1 donut)	130
Donuts, Old Fashioned	1.5 oz. (1 donut)	220
Donuts, Old Fashioned Glazed	2 oz. (1)	200
Donuts, Plain	1 oz. (1 donut)	135
Donuts, Powdered Sugar	1 oz. (1 donut)	140
Ho Ho's	1 oz. (1 piece)	90
Honey Bun	4.8 oz. (1 piece)	650

                                          PORTION   SODIUM (MG.)

Hostess O's ....2.3 oz. (1)..........................265
Lil' Angels ....1 oz. (1) 95
Peanut Putters, Filled ....3 oz. (1) 240
Peanut Putters, Unfilled ....3 oz. (1) 240
Sno Ball....................................1.5 oz. (1 piece) 170
Suzy Q's, Banana ....2.3 oz. (1 piece) 195
Suzy Q's, Chocolate ....2.2 oz. (1 piece) 300
Tiger Tail ....2.2 oz. (1 piece) 240
Twinkies ....1.5 oz. (1 piece) 150

Kellogg's Pop Tarts

Blueberry ....1.8 oz. (1 pastry) 220
Blueberry, Frosted ....1.8 oz. (1 pastry) 220
Brown Sugar-Cinnamon ....1.8 oz. (1 pastry) 215
Brown Sugar-Cinnamon,
  Frosted...................................1.8 oz. (1 pastry) 205
Cherry ....1.8 oz. (1 pastry) 230
Cherry, Frosted.........................1.8 oz. (1 pastry) 230
Chocolate Chip ....1.8 oz. (1 pastry) 255
Chocolate Fudge, Frosted ....1.8 oz. (1 pastry) 255
Chocolate Vanilla Creme,
  Frosted...................................1.8 oz. (1 pastry) 285
Concord Grape, Frosted............1.8 oz. (1 pastry) 215
Dutch Apple, Frosted ....1.8 oz. (1 pastry) 215
Raspberry, Frosted ....1.8 oz. (1 pastry) 215
Strawberry ....1.8 oz. (1 pastry) 225
Strawberry, Frosted ....1.8 oz. (1 pastry) 215

Merico

Cinnamon Danish Roll ....1 iced roll 330
Cinnamon Raisin Roll................1 iced roll 375
Fruit Danish ....1 iced roll 425
Orange Danish..........................1 iced roll 585
Toaster Pastry..........................1.9 oz (1 piece)...................40

Nabisco Toastettes

Apple ....2 oz. (1 piece) 240
Blueberry ....2 oz. (1 piece) 250
Brown Sugar Cinnamon............2 oz. (1 piece) 280
Cherry ....2 oz. (1 piece) 240
Strawberry ....2 oz. (1 piece) 250

PORTION   SODIUM (MG.)

Pillsbury Toaster Strudel

Blueberry	1 pastry	205
Cinnamon	1 pastry	200
Raspberry	1 pastry	205
Strawberry	1 pastry	205

R&R

Plum Pudding	3.6 oz.	150

Rich's

Apple Cinnamon Cream Cake	2.8 oz. (⅛ cake)	150
Bavarian Cream Puffs	1.5 oz. (1 puff)	70
Chocolate Cream Puffs	1.4 oz. (1 puff)	85
Chocolate Eclairs	2 oz. (1 eclair)	110
Devil's Food Cream Cake	2.8 oz. (⅛ cake)	275

Tastykake

Apple Danish	2 oz. (1 piece)	360
Apple Filled Krumb Kake	1 oz. (1 piece)	150
Apple Spice Krimpet	1 oz. (1 piece)	89
Banana Treat	1 oz. (1 piece)	117
Blueberry Danish	2 oz. (1 piece)	170
Butterscotch Krimpet	1 oz. (1 piece)	95
Cheese Danish	2 oz. (1 piece)	182
Cherry Danish	2 oz. (1 piece)	178
Chocolate Creamie	1 oz. (1 piece)	116
Chocolate Cup	1 oz. (1 piece)	172
Chocolate Junior	3 oz. (1 piece)	292
Chocolate Kandy Kake	1 oz. (1 piece)	64
Chocolate Teen	2 oz. (1 piece)	310
Chocolate Tempty	1 oz. (1 piece)	100
Coconut Junior	3 oz. (1 piece)	327
Coconut Kandy Kake	1 oz. (1 piece)	44
Cream Filled Buttercream Cup	1 oz. (1 piece)	135
Cream Filled Chocolate Cup	1 oz. (1 piece)	161
Cream Filled Chocolate Krimpet	1 oz. (1 piece)	89
Cream Filled Koffee Kake	1 oz. (1 piece)	135
Cream Filled Vanilla Krimpet	1 oz. (1 piece)	96

	PORTION	SODIUM (MG.)
Delicious Fruit Cake	½-in. slice	103
Holiday Fruit Cake, vacuum pack	½ cake	118
Jelly Krimpet	1 oz. (1 piece)	104
Koffee Kake	2.5 oz. (1 piece)	314
Lemon Danish	2 oz. (1 piece)	321
Mint Kandy Kake	1 oz. (1 piece)	59
Peanut Butter Kandy Kake	1 oz. (1 piece)	48
Vanilla Creamie	1 oz. (1 piece)	115

FROSTINGS
Mixes—All figures for foods as prepared

Betty Crocker

Butter Brickle Creamy	¹⁄₁₂ cake	115
Butter Pecan Creamy	¹⁄₁₂ cake	100
Chiquita Banana Creamy	¹⁄₁₂ cake	80
Chocolate Almond Fudge	¹⁄₁₂ package	75
Chocolate Fudge Creamy	¹⁄₁₂ cake	75
Coconut Almond Creamy	¹⁄₁₂ cake	90
Coconut Pecan Creamy	¹⁄₁₂ cake	100
Cream Cheese and Nuts	¹⁄₁₂ package	110
Creamy Cherry Creamy	¹⁄₁₂ cake	100
Creamy White Creamy	¹⁄₁₂ cake	100
Dark Chocolate Fudge Creamy	¹⁄₁₂ cake	90
Milk Chocolate Creamy	¹⁄₁₂ cake	80
Sour Cream Chocolate Fudge Creamy	¹⁄₁₂ cake	75
Sour Cream White Creamy	¹⁄₁₂ cake	100
Lemon Creamy	¹⁄₁₂ cake	100
White Fluffy	¹⁄₁₂ cake	40

Pillsbury

Coconut Almond	¹⁄₁₂ cake	85
Coconut Pecan	¹⁄₁₂ cake	105
Fluffy White	¹⁄₁₂ cake	65
Rich 'N Easy Caramel	¹⁄₁₂ cake	35
Rich 'N Easy Chocolate Fudge	¹⁄₁₂ cake	70
Rich 'N Easy Double Dutch	¹⁄₁₂ cake	80

PORTION   SODIUM (MG.)

Rich 'N Easy Lemon	1/12 cake	15
Rich 'N Easy Milk Chocolate	1/12 cake	55
Rich 'N Easy Strawberry	1/12 cake	55
Rich 'N Easy Vanilla	1/12 cake	30

Ready to spread

Betty Crocker Creamy Deluxe

Butter Pecan	1.4 oz. (1/12 tub)	90
Cherry	1.4 oz. (1/12 tub)	100
Chocolate	1.4 oz. (1/12 tub)	100
Chocolate Chip	(1/12 tub)	90
Chocolate Chocolate Chip	(1/12 tub)	95
Chocolate Nut	1.4 oz. (1/12 tub)	100
Coconut Pecan	1/12 tub	80
Cream Cheese	1.4 oz. (1/12 tub)	100
Dark Dutch Fudge	1.4 oz. (1/12 tub)	100
Milk Chocolate	1.4 oz. (1/12 tub)	100
Orange	1.4 oz. (1/12 tub)	90
Sour Cream Chocolate	1.4 oz. (1/12 tub)	100
Sour Cream White	1.4 oz. (1/12 tub)	100
Lemon	1.4 oz. (1/12 tub)	100
Vanilla	1.4 oz. (1/12 tub)	100

Duncan Hines Creamy

Chocolate	1/12 tub	90
Dark Dutch Fudge	1/12 tub	95
Milk Chocolate	1/12 tub	85
Vanilla	1/12 tub	80

Durkee-Mower

Marshmallow Fluff	4 oz.	30

Pillsbury

Cake and Cookie Decorator (all colors)	1 tbl.	5

Pillsbury Supreme

Caramel Pecan	1/12 cake	70
Chocolate Fudge	1/12 cake	80
Chocolate Mint	1/12 cake	80

PORTION   SODIUM (MG.)

Coconut Almond	1/12 cake	60
Coconut Pecan	1/12 cake	60
Cream Cheese	1/12 cake	115
Double Dutch	1/12 cake	45
Lemon	1/12 cake	80
Milk Chocolate	1/12 cake	60
Sour Cream Vanilla	1/12 cake	80
Strawberry	1/12 cake	75
Vanilla	1/12 cake	75

PIES
Crusts

General Mills

Pie Crust Mix	1/16 package	140
Pie Crust Sticks	1/8 stick	140

Nabisco

Graham Cracker Crust	1/8 crust	110

Pillsbury

All Ready Pie Crust (frozen)	1/8 crust	275
Pie Crust Mix	1/6 pie	425
Pie Crust Sticks	1/6 pie	425

Fillings

Del Monte

Pumpkin	1/2 cup	under 10

Libby's

Pumpkin (Solid Pack)	1 cup	10
Pumpkin Pie Mix	1/6 pie, prepared	460

None Such

Condensed Mince Meat	1/4 package	330
Mince Meat w/Brandy and Rum	1/3 cup	265

Solo

Almond	2 oz.	40

	PORTION	SODIUM (MG.)
Apricot	2 oz.	14
Blueberry	2 oz.	14
Cherry	2 oz.	14
Date	2 oz.	50
Nut	2 oz.	38
Pecan	2 oz.	36
Pineapple	2 oz.	14
Poppy	2 oz.	26
Prune	2 oz.	22
Raspberry	2 oz.	14
Strawberry	2 oz.	14

Solo Glazes

Blueberry	2 oz.	66
Peach	2 oz.	66
Strawberry	2 oz.	66

Stokely-Van Camp

Pumpkin	4 oz.	13
Pumpkin Pie Filling	4 oz.	420

Frozen

Banquet Cream Pies

Banana	2.33 oz.	146
Chocolate	2.33 oz.	106
Coconut	2.33 oz.	113
Lemon	2.33 oz.	111
Strawberry	2.33 oz.	112

Banquet Family Size Fruit Pies

Apple	3.33 oz.	282
Blackberry	3.38 oz.	342
Blueberry	3.33 oz.	342
Cherry	3.33 oz.	258
Mincemeat	3.33 oz.	364
Peach	3.33 oz.	275
Pumpkin	3.33 oz.	341

Banquet Fruit Pies

Apple	8 oz.	598

	PORTION	SODIUM (MG.)
Cherry	8 oz.	534
Peach	8 oz.	579

Morton Great Little Desserts (pies)

Apple	8 oz.	510
Banana Cream	3.5 oz.	180
Blueberry	8 oz.	525
Cherry	8 oz.	520
Chocolate Cream	3.5 oz.	190
Coconut Cream	3.5 oz.	180
Coconut Custard	6.5 oz.	495
Dutch Apple	7.8 oz.	470
Lemon Cream	3.5 oz.	180
Peach	8 oz.	520

Morton Pies

Apple	4 oz. (1/6 pie)	240
Banana	2.3 oz. (1/6 pie)	130
Blueberry	4 oz. (1/6 pie)	250
Cherry	4 oz. (1/6 pie)	250
Chocolate	2.3 oz. (1/6 pie)	140
Coconut	2.3 oz. (1/6 pie)	130
Lemon	2.3 oz. (1/6 pie)	130
Mince	4 oz. (1/6 pie)	355
Peach	4 oz. (1/6 pie)	260
Pumpkin	4 oz. (1/6 pie)	310

Sara Lee Pies

Apple	5 oz. (1/6 pie)	555
Blueberry	5 oz. (1/6 pie)	220
Cherry	5 oz. (1/6 pie)	233
Dutch Apple	5 oz. (1/6 pie)	584
Peach	5 oz. (1/6 pie)	253
Pumpkin	5.6 oz. (1/8 pie)	403

Mixes

Betty Crocker Pies

Boston Cream	1/8 pie	405

PORTION   SODIUM (MG.)

Prepared

Drake's Pies

Apple ....1 pie..........................230

Frito-Lay Pies

Marshmallow.....................2 oz. (1 pie) 60
Pecan.................................3 oz. (1 pie) 455

Hostess Fruit Pies

Apple ....4.5 oz. (1 pie) 540
Berry..................................4.5 oz. (1 pie) 490
Blueberry............................4.5 oz. (1 pie) 450
Cherry ....4.5 oz. (1 pie) 530
Lemon ....4.5 oz. (1 pie) 470
Peach ....4.5 oz. (1 pie) 445
Strawberry ....4.5 oz. (1 pie) 400

Tastykake Pies

Apple ....4 oz. (1 pie) 458
Blackberry ....4 oz. (1 pie) 374
Blueberry............................4 oz. (1 pie) 406
Cherry ....4 oz. (1 pie) 389
Coconut Creme ....4 oz. (1 pie) 285
French Apple ....4 oz. (1 pie) 442
Lemon ....4 oz. (1 pie) 320
Peach ....4 oz. (1 pie) 391
Pineapple............................4 oz. (1 pie) 393
Pumpkin..............................4 oz. (1 pie) 339
Strawberry-Rhubarb..............4 oz. (1 pie) 362
Tasty Klair ....4 oz. (1 pie) 240

Cereals

COLD CEREALS

Arrowhead Mills

Arrowhead Crunch.................1 oz...........................40
Maple Nut Granola ....1 oz...........................13

PORTION   SODIUM (MG.)

Elam's

Miller's Bran	2 tbl.	2
Wheat Germ	2 tbl.	3

Fearn Soya

Naturfresh Corn Germ	¼ cup	1
Naturfresh Raw Wheat Germ	¼ cup	1

Featherweight

Corn Flakes	1 oz.	under 10
Crisp Rice Cereal	1 oz.	under 10

General Mills

Boo Berry	1 oz. (1 cup)	210
Brown Sugar & Honey Body Buddies	1 oz. (1 cup)	290
Buc Wheats	1 oz. (¾ cup)	235
Cheerios	1 oz. (1¼ cup)	330
Chocolate Crazy Cow	1 oz. (1 cup)	205
Cinnamon Toast Crunch	1 oz. (⅔ cup)	225
Cocoa Puffs	1 oz. (1 cup)	205
Corn Total	1 oz. (1 cup)	310
Count Chocula	1 oz. (1 cup)	205
Country Corn Flakes	1 oz. (1 cup)	310
Crispy Wheats 'N Raisins	1 oz. (¾ cup)	185
E.T.	1 oz. (¾ cup)	170
Franken Berry	1 oz. (1 cup)	205
Fruit Brute	1 oz. (1 cup)	215
Golden Grahams	1 oz. (¾ cup)	285
Honey Nut Cheerios	1 oz. (¾ cup)	255
Kaboom	1 oz. (1 cup)	370
Kix	1 oz. (1½ cup)	315
Lucky Charms	1 oz. (1 cup)	185
Natural Fruit Flavor Body Buddies	1 oz. (¾ cup)	285
Pac-Man	1 oz. (1 cup)	195
Powdered Donutz	1 oz. (1 cup)	185
Strawberry Shortcake	1 oz. (1 cup)	190
Total	1 oz. (1 cup)	375
Trix	1 oz. (1 cup)	170

PORTION   SODIUM (MG.)

Wheaties ....1 oz. (1 cup) 370

Health Valley

Real Almond Crunch....................1 oz..................................10
Amaranth w/Bananas ....1 oz..................................10

Bran Cereal w/Apples and
  Cinnamon ....1 oz..................................10
Bran Cereal w/Raisins ....1 oz....................................5
Real Hawaiian Fruit...................1 oz....................................5
Lites Puffed Cereals ....1 oz.........................under 3
Hearts O'Bran...........................1 oz..................................10
Orangeola w/Almonds &
  Dates ....1 oz....................................5
Orangeola w/Bananas &
  Coconut ....1 oz..................................10
Real Raisin Nut ....1 oz..................................10
Sprouted Seven Grain Cereal
  w/Bananas...............................1 oz....................................3
Sprouted Seven Grain Cereal
  w/Raisins ....1 oz....................................5
Wheat Germ w/Almonds &
  Dates ....1 oz....................................3

Health Valley

Brown Rice Baby Cereal ....1 oz..................................10
Hearts O'Bran w/Apples &
  Cinnamon ....1 oz..................................10
Hearts O'Bran w/Raisins &
  Spice ....1 oz..................................10
Raisin Bran Flakes ....1 oz....................................1
Real Apple Cinnamon................1 oz..................................10
Real Maple Nut ....1 oz..................................10
Sprouted Baby Cereal ....1 oz..................................10
Wheat Flakes ....1 oz..................................10
Wheat Germ w/Bananas ....1 oz..................................15

Heartland

Natural Cereal, coconut ....1 oz. (¼ cup)........................65
Natural Cereal, plain.................1 oz. (¼ cup)........................76
Natural Cereal, raisin ....1 oz. (¼ cup)........................68

PORTION   SODIUM (MG.)

Kellogg's

	PORTION	SODIUM (MG.)
All-Bran	1 oz. (⅓ cup)	320
Apple Jacks	1 oz. (1 cup)	125
Banana Frosted Flakes	1 oz. (⅔ cup)	180
Bran Buds	1 oz. (⅓ cup)	175
Bran Flakes	1 oz. (¾ cup)	220
C-3PO's	1 oz. (¾ cup)	160
Cocoa Krispies	1 oz. (¾ cup)	195
Corn Flakes	1 oz. (1¼ cup)	285
Cracklin' Bran	1 oz. (½ cup)	190
Crispix	1 oz. (¾ cup)	230
Froot Loops	1 oz. (1 cup)	135
Frosted Mini-Wheats, brown sugar	1 oz. (4 pieces)	5
Frosted Mini-Wheats, sugar frosted	1 oz. (4 pieces)	5
Frosted Rice	1 oz. (¾ cup)	200
Fruitful Bran	1.2 oz. (¾ cup)	240
Graham Crackos	1 oz. (¾ cup)	145
Honey & Nut Corn Flakes	1 oz. (¾ cup)	190
Honey Smacks	1 oz. (¾ cup)	70
Marshmallow Krispies	1.3 oz. (1¼ cups)	285
Most	1 oz. (½ cup)	30
Nutri-Grain Corn	1 oz. (½ cup)	185
Nutri-Grain Wheat	1 oz. (⅔ cup)	195
Nutri-Grain Wheat and Raisins	1.4 oz. (⅔ cup)	165
Product 19	1 oz. (¾ cup)	320
Raisin Bran	1 oz. (¾ cup)	205
Raisins, Rice & Rye	1 oz. (¾ cup)	235
Rice Krispies	1 oz. (1 cup)	285
Special K	1 oz. (1 cup)	220
Strawberry Krispies	1 oz. (¾ cup)	200
Sugar Corn Pops	1 oz. (1 cup)	95
Sugar Frosted Flakes	1 oz. (¾ cup)	190
Sugar Smacks	1 oz. (¾ cup)	60

Kretschmer

	PORTION	SODIUM (MG.)
Brown Sugar & Honey Wheat Germ	1 oz. (¼ cup)	3

PORTION   SODIUM (MG.)

Sun Country Granola w/Almonds	1 oz. (¼ cup)	13
Sun Country Granola w/Raisins	1 oz. (¼ cup)	13
Wheat Germ	1 oz. (¼ cup)	2

Nabisco

100% Bran Cereal	1 oz. (½ cup)	210
Shredded Wheat	1 oz. (1 biscuit)	10
Spoon Size Shredded Wheat	1 oz. (⅔ cup)	10
Team Flakes	1 oz. (1 cup)	185

Nature Valley

Cinnamon & Raisin Granola	1 oz. (⅓ cup)	35
Coconut & Honey Granola	1 oz. (⅓ cup)	35
Fruit & Nut Granola	1 oz. (⅓ cup)	35
Toasted Oat Mixture	1 oz. (⅓ cup)	35

Post

Alpha-Bits	1 oz.	195
Bran Flakes	1 oz.	225
C.W. Post Granola Cereal w/Raisins	1 oz.	50
C.W. Post Hearty Granola Cereal	1 oz.	55
Cocoa Pebbles	1 oz.	165
Fortified Oat Flakes	1 oz.	275
Frosted Rice Krinkles	1 oz.	185
Fruit & Fibre Apples & Cinnamon	1 oz.	195
Fruit & Fibre Dates, Raisin, Walnut	1 oz.	170
Fruity Pebbles	1 oz.	160
Grape-Nuts	1 oz.	195
Grape-Nuts Flakes	1 oz.	195
Honey Nut Crunch Raisin Bran	1 oz.	150
Honeycomb	1 oz.	195
Raisin Bran	1 oz.	170
Raisin Grape-Nuts	1 oz.	160

PORTION   SODIUM (MG.)

Smurf-berry Crunch	1 oz.	65
Strawberry Honeycomb	1 oz.	160
Super Sugar Crisp	1 oz.	25
Post Toasties Corn Flakes	1 oz.	305

Quaker

Bran, unprocessed	0.2 oz. (2 tbl.)	under 10
Cap'n Crunch	1 oz. (¾ cup)	193
Cap'n Crunch Peanut Butter	1 oz. (¾ cup)	218
Cap'n Crunch's Crunchberries	1 oz. (¾ cup)	165
Cinnamon Life	1 oz. (⅔ cup)	149
Corn Bran	1 oz. (⅔ cup)	245
King Vitaman	1 oz. (1¼ cup)	215
Life	1 oz. (⅔ cup)	160
Natural Cereal	1 oz. (¼ cup)	18
Natural Cereal w/Raisins, Dates	1 oz. (¼ cup)	15
Natural Cereal w/Apples, Cinnamon	1 oz. (¼ cup)	15
Puffed Rice	½ oz. (1 cup)	1
Puffed Wheat	1 oz. (1 cup)	1
Quisp	1 oz. (1.2 cup)	189
Shredded Wheat	1.3 oz. (2 pieces)	2

Ralston Purina

Bran Chex	1 oz. (⅔ cup)	300
Cookie Crisp, artificial chocolate chip	1 oz. (1 cup)	200
Cookie Crisp, artificial oatmeal	1 oz. (1 cup)	170
Cookie Crisp, artificial vanilla wafer	1 oz. (1 cup)	200
Corn Chex	1 oz. (1 cup)	310
Corn Flakes	1 oz. (1 cup)	270
Crispy Rice	1 oz. (1 cup)	205
Raisin Bran	1⅓ oz. (¾ cup)	315
Rice Chex	1 oz. (1⅛ cup)	280
Sugar Frosted Flakes	1 oz. (¾ cup)	180
Wheat Chex	1 oz. (⅔ cup)	200

PORTION   SODIUM (MG.)

HOT CEREALS

Elam's

Complete Cereal	1 oz. dry	5
Cracked Wheat Cereal	1 oz. dry	4
Scotch Style Oatmeal	1 oz. dry	3
Steel Cut Oatmeal	1 oz. dry	3
Wheat & Oatmeal Hot Cereal	1 oz. dry	1

Fearn Soya

Soy-o Wheat Cereal & Soya	¼ cup dry	38

H-O

Cream Enriched Farina	1.2 oz. dry (3 tbl.)	under 5
Instant Oatmeal (box)	1.2 oz. dry (½ cup)	under 5
Instant Oatmeal, Regular	1 oz. dry (1 packet)	230
Instant Oatmeal w/Apple & Brown Sugar	1.1 oz. dry (1 packet)	225
Instant Oatmeal w/Bran & Spices	1.5 oz. dry (1 packet)	300
Instant Oatmeal w/Cinnamon & Spice	1.6 oz. dry (1 packet)	305
Instant Oatmeal w/Maple & Brown Sugar	1.5 oz. dry (1 packet)	285
Instant Oatmeal, Sweet & Mellow	1.4 oz. dry (1 packet)	270
Old Fashioned Oats	1.3 oz. dry (½ cup)	under 5
Quick Oats	1.2 oz. dry (½ cup)	under 5

Maltex

Maltex	1 cup cooked	9

Malt-o-Meal

Malt-o-Meal	1 cup cooked	2

Maypo

Maypo	1 cup cooked	9

Nabisco Cream of Wheat

Instant	1 oz. dry (2½ tbl.)	10

PORTION   SODIUM (MG.)

Mix 'n Eat, Baked Apple
  Cinnamon Flavor ....1.2 oz. dry (1 packet) 240
Mix 'n Eat, Banana
  Artificial Flavor and
  Spice ....1.2 oz. dry (1 packet) 240
Mix 'n Eat Maple Artificial
  Flavor and Brown Sugar ......1.2 oz. dry (1 packet) 240
Mix 'n Eat, Regular ....1.2 oz. dry (1 packet) 240
Quick ....1 oz. dry (2½ tbl.)............130
Regular ....1 oz. dry (2½ tbl.)..................10

Pillsbury

Farina...⅔ cup prepared....................265

Quaker

Enriched White Hominy Grits,
  Regular & Quick ....1 oz. dry (3 tbl.).........................0
Hot 'n Creamy (Farina)..............1 oz. dry (⅙ cup) 1
Instant Grits Product ....0.8 oz. dry (1 packet) 379
Instant Grits w/Artificial
  Cheese Flavor ....1 oz. dry (1 packet) 497
Instant Grits w/Imitation
  Bacon Bits................................1 oz. dry (1 packet) 544
Instant Grits w/Imitation Ham
  Bits ....1 oz. dry (1 packet) 658
Instant Oatmeal, Regular..........1 oz. dry (1 packet) 252
Instant Oatmeal w/Apples &
  Cinnamon ....1.3 oz. dry (1 packet) 181
Instant Oatmeal w/Bran &
  Raisins......................................1.5 oz. dry (1 packet) 240
Instant Oatmeal w/Cinnamon
  & Spice ....1.6 oz. dry (1 packet) 258
Instant Oatmeal w/Maple &
  Brown Sugar............................1.5 oz. dry (1 packet) 228
Instant Oatmeal w/Raisins &
  Spice ....1.5 oz. dry (1 packet) 227
Quaker Oats, Old Fashioned 1 oz. dry (⅓ cup) 1
Quaker Oats, Quick ....1 oz. dry (⅓ cup) 1
Whole Wheat Hot Natural
  Cereal..1 oz. dry (⅓ cup) 1

PORTION   SODIUM (MG.)

Ralston-Purina

	PORTION	SODIUM (MG.)
Oats, Quick	1 oz. dry (⅓ cup)	5
Oats, Regular	1 oz. dry (⅓ cup)	5
Ralston, Instant	1 oz. dry (¼ cup)	5
Ralston, Regular	1 oz. dry (¼ cup)	5

Roman Meal

	PORTION	SODIUM (MG.)
Roman Meal, Plain	1 cup cooked	3
Roman Meal w/Oats	1 cup cooked	10

Van Camp's

	PORTION	SODIUM (MG.)
Golden Hominy	1 cup cooked	650
Golden Hominy w/Red & Green Peppers	8 oz.	635
White Hominy	1 cup cooked	635

Wheatena

	PORTION	SODIUM (MG.)
Wheatena	1 cup cooked	5

Cheese

NATURAL CHEESE
Hard

Bel

	PORTION	SODIUM (MG.)
Babybel, Mini Semisoft	1.5 oz.	340
Babybel, Semisoft, Part-Skim	1.5 oz.	340
Bonbel, Mini	1.5 oz.	340
Bonbel Reduced Calorie Semisoft Cheese Product	1.5 oz. (2 pieces)	340
Bonbel Semisoft	1.5 oz.	340
Bonbino Semisoft	1.5 oz.	340
Cheddar	1.5 oz.	340
Edam	1.5 oz.	340
Gouda	1.5 oz.	340

Colombo

	PORTION	SODIUM (MG.)
Braided	1.5 oz.	198

	PORTION	SODIUM (MG.)
Mozzarella	1.5 oz.	159
Syrian	1.5 oz.	474

Dorman's

Blue, Danish	1.5 oz.	594
Brick	1.5 oz.	239
Caraway	1.5 oz.	294
Cheddar (Colby)	1.5 oz	257
Edam	1.5 oz.	411
Gouda	1.5 oz.	348
Gruyere	1.5 oz.	143
Iceland Baby Swiss	1.5 oz.	435
Jarlsberg	1.5 oz.	215
Low Skim (Edam type)	1.5 oz.	425
Monterey Jack	1.5 oz.	228
Mozzarella, Part Skim	1.5 oz.	198
Muenster	1.5 oz.	200
Muenster, No Salt Added	1.5 oz.	15
Parmesan	1 oz.	454
Port Salut	1.5 oz.	227
Provolone	1.5 oz.	372
Romano	1.5 oz.	510
Roquefort	1.5 oz.	770
Skandor, Swedish	1.5 oz.	345
Swiss	1.5 oz.	111
Swiss Type, No Salt Added	1.5 oz.	15
Tilsiter	1.5 oz.	327

Featherweight

Cheddar	1.5 oz.	8
Colby	1.5 oz.	8

Health Valley

Cheddar	1.5 oz.	15
Jack	1.5 oz.	17
Muenster	1.5 oz.	19

Heluva Good Cheese

Cheddar	1.5 oz.	264
Cheddar, Unsalted	1.5 oz.	8
Gouda	1.5 oz.	349

PORTION   SODIUM (MG.)

Kolb-Lena
Brie	1.5 oz.	180
Camembert	1.5 oz.	180
Delico-Baby-Chalet, Swiss Type	1.5 oz.	150
Feta	1.5 oz.	435

Kraft
Blue	1.5 oz.	593
Brick	1.5 oz.	308
Caraway	1.5 oz.	293
CASINO Natural Romano	1 oz.	340
Cheddar, Mild	1.5 oz.	270
Cheddar, Sharp	1.5 oz.	270
Colby	1.5 oz.	263
Edam	1.5 oz.	413
Gouda	1.5 oz.	345
Grated Parmesan	1 oz.	425
Grated Romano	1.5 oz.	608
Monterey Jack	1.5 oz.	293
Mozzarella, Low Moisture, Part Skim	1.5 oz.	285
Muenster	1.5 oz.	255
Neufchatel	1.5 oz.	165
Parmesan, Natural	1 oz.	455
Provolone	1.5 oz.	375
Scamorze, Low Moisture, Part Skim	1.5 oz.	225
Swiss, aged	1.5 oz	113
Swiss, chunk	1.5 oz.	60
Swiss slices	1.5 oz.	158

Land O Lakes
Brick	1.5 oz.	240
Cheddar	1.5 oz.	263
Colby	1.5 oz.	255
Edam	1.5 oz.	413
Gouda	1.5 oz.	345
Monterey Jack	1.5 oz.	225
Mozzarella, Low Moisture, Part Skim	1.5 oz.	225

	PORTION	SODIUM (MG.)
Muenster	1.5 oz.	270
Provolone	1.5 oz.	375
Swiss	1.5 oz.	113

Mohawk Valley

Limburger, Little Gem Size	1.5 oz.	338

Pauly

Low Sodium Colby Cheese	1.5 oz.	8

Sargento

Baby Gouda	1.5 oz.	348
Caraway Gouda	1.5 oz.	348
Creamy Havarti	1.5 oz.	297
Danish Danko Brie	1.5 oz.	423
Danish Danko Camembert	1.5 oz.	335
Danish Feta Cups	1.5 oz.	474
Danish Samsoe	1.5 oz.	297
Edam	1.5 oz.	411
Edam Balls	1.5 oz.	411
85% Reduced Cholesterol Imitation Part-Skim Cheese	1.5 oz.	327
50% Reduced Sodium Cheddar	1.5 oz.	132
Kilo Norwegian Gjetost	1.5 oz.	255
Longhorn Cheddar, Midget	1.5 oz.	264
Longhorn Colby, Midget	1.5 oz.	257
Longhorn Monterey Jack, Midget	1.5 oz.	228
Mozzarella Bar Deluxe	1.5 oz.	225
Mozzarella Rounds	1.5 oz.	225
Mozzarella Squares	1.5 oz.	225
Natural Limburger	1.5 oz.	341
Parmesan Wedge	1 oz.	454
Romano Wedge	1 oz.	340
75% Reduced Sodium Swiss	1.5 oz.	26
Shredded Cheddar	1.5 oz.	264
Shredded Colby	1.5 oz.	257

	PORTION	SODIUM (MG.)
20% Reduced Calorie Cheddar	1.5 oz.	255
20% Reduced Calorie Longhorn Style	1.5 oz.	270

Sorrento

Mozzarella, part skim	1.5 oz.	298
Mozzarella, whole milk	1.5 oz.	298
Parmesan & Italian Style Grated	1 oz.	480
Provolone	1.5 oz.	298

Swiss Knight

Fondue	1.5 oz.	278
Plain Gruyere	1.5 oz.	540

Tillamook

Cheddar	1.5 oz.	299
Cheddar, Low Sodium	1.5 oz.	14

Treasure Cave

Blue Cheese, Crumbled	1.5 oz.	563

Weight Watchers

Natural Semisoft Skim Milk Cheese	1.5 oz.	285

Soft

Borden

Cottage Cheese, 4% milkfat minimum	½ cup	465

Breakstone

Cottage Cheese, Dry Curd w/added Skim Milk	½ cup	62
Cottage Cheese, Lowfat	½ cup	386
Cottage Cheese, Smooth & Creamy	½ cup	405

Colombo

Ricotta	4 oz.	96

PORTION   SODIUM (MG.)

Foremost

Cottage Cheese, Small Curd......4 oz..457
So-Lo Cottage Cheese,
  Lowfat.................................4 oz..459

Giant

Cottage Cheese & Pineapple
  (4% milkfat)........................½ cup 307
Cottage Cheese, Large
  Curd (4% milkfat)...............½ cup 380
Cottage Cheese, Small Curd
(4% milkfat).............................½ cup 380
Garden Salad Cottage
  Cheese (4% milkfat) ....½ cup 386
Lite One Cottage Cheese,
  Lowfat (1% milkfat) ....½ cup 283

Land O Lakes

Cottage Cheese..........................4 oz..460
Cottage Cheese, 2% milkfat ....4 oz..460

Lite-Line

Cottage Cheese, Lowfat............½ cup 375

Maggio

Whole Milk Ricotta ....4 oz..190

Philadelphia Cream Cheese (Kraft)

Cream Cheese............................1 oz..85
Cream Cheese w/Chives ....1 oz..115
Cream Cheese w/Pimentos......1 oz..115
Soft Cream Cheese ....1 oz..105
Soft Cream Cheese w/Chives
  & Onion ....1 oz..100
Soft Cream Cheese w/Olives
  & Pimento (sic) ....1 oz..165
Soft Cream Cheese
  w/Pineapple.........................1 oz..90
Soft Cream Cheese
  w/Strawberries ....1 oz..110

PORTION   SODIUM (MG.)

Soft Cream Cheese
   w/Toasted Onion ....1 oz..135
Whipped Cream Cheese ....1 oz..110
Whipped Cream Cheese
   w/Bacon & Horseradish ....1 oz..130
Whipped Cream Cheese
   w/Blue Cheese.....................1 oz..145
Whipped Cream Cheese
   w/Chives ....1 oz..145
Whipped Cream Cheese
   w/Onions ....1 oz..160
Whipped Cream Cheese
   w/Pimentos ....1 oz..140
Whipped Cream Cheese
   w/Smoked Salmon..................1 oz...90

Sargento

Farmer's Cheese ....1.5 oz. 198
Nibblin' Curds ....1.5 oz. 264
Pot Cheese, French Onion ......1.5 oz. 1
Pot Cheese, Garlic ....1.5 oz. 1
Pot Cheese, regular ....1.5 oz. 1
Ricotta, part skim......................4 oz. 141
Ricotta, whole milk ....4 oz. 96

Sealtest

Cottage Cheese, Large Curd......½ cup 455
Cottage Cheese, Small Curd......½ cup 456
Garden Salad Cottage
   Cheese.................................½ cup 481

Tuttle

Cottage Cheese, Dry Curd ......4 oz...14

Weight Watchers

Cottage Cheese, Lowfat............½ cup 262
Imitation Cream Cheese ....1 oz...67

PROCESS CHEESE

Borden

American Cheese Slices ....1.5 oz. 668
Swiss Cheese Slices.................1.5 oz. 533

PORTION   SODIUM (MG.)

Dorman's

American .... 1.5 oz. 609
Lo-Chol, Imitation .... 1.5 oz. 198
Swiss, pasteurized .... 1.5 oz. 582

Kraft

Deluxe American, loaf .... 1.5 oz. 638
Deluxe American, slices .... 1.5 oz. 698
Deluxe Pimento .... 1.5 oz. 480
Deluxe Swiss .... 1.5 oz. 728
Golden Image Imitation
  Colby .... 1.5 oz. 210
Golden Image Imitation
  Mild Cheddar .... 1.5 oz. 255
Old English Sharp American
  Cheese Loaf .... 1.5 oz. 608
Old English Sharp American
  Cheese Slices .... 1.5 oz. 600

Land O Lakes

American .... 1.5 oz. 608
American/Swiss .... 1.5 oz. 668

Sargento

American Topping, Shredded 1.5 oz. 269
Burgercheese, Sliced .... 1.5 oz. 609
Cracker Snacks, Sliced .... 1.5 oz. 609
Hot Pepper, Sliced .... 1.5 oz. 609
Parmesan Topping, Grated ...... 1.5 oz. 668
Pizza Topping, Shredded .... 1.5 oz. 459
Pizza Topping, Sliced .... 1.5 oz. 459

PROCESS CHEESE FOOD

Borden

American Cheese Food .... 1.5 oz. 735
CRACKER BARREL Sharp
  Cheddar Cold Pack .... 1.5 oz. 225
Port Wine Cheddar Cold
  Pack .... 1.5 oz. 248

PORTION SODIUM (MG.)

Smoke Flavored Cheddar
  Cold Pack ....1.5 oz. 225
Tangy Cheddar Cold Pack........1.5 oz. 240

Kraft

American Singles......................1.5 oz. 585
Cheez 'N Bacon Singles ....1.5 oz. 593
Golden Image American
  Flavor Imitation.......................1.5 oz. 585
Grated American ....1.5 oz. 1110
Jalapeño Singles ....1.5 oz. 450
Monterey Jack Singles ....1.5 oz. 608
Muenster Singles......................1.5 oz. 593
Pimento Singles........................1.5 oz. 608
Process Cheese Food
  w/Bacon..................................1.5 oz. 383
Process Cheese Food
  w/Garlic ....1.5 oz. 330
Process Cheese Food
  w/Jalapeño Peppers ....1.5 oz. 450
Sharp Singles............................1.5 oz. 593
Smokelle ....1.5 oz. 338
Swiss Singles ....1.5 oz. 420

Land O Lakes

Jalapeño ....1.5 oz. 540
La Cheddar...............................1.5 oz. 503
Onion ....1.5 oz. 495
Pepperoni ....1.5 oz. 263
Salami ....1.5 oz. 600

Laughing Cow

Wedges, Reduced Calorie ..1.5 oz. (2 pieces) 468
Wedges, Regular ....1.5 oz. (2 pieces) 468

Lite-Line

Low Cholesterol Cheese
  Food Substitute ....1.5 oz. 675

Sargento

Blue Cheese, cold-pack ....1.5 oz. 594

PORTION   SODIUM (MG.)

	PORTION	SODIUM (MG.)
Port Wine, cold-pack	1.5 oz.	411
Sharp Cheddar Cheese Sticks	1.5 oz.	264
Sharp Cheddar Crock	1.5 oz.	411

Wispride

	PORTION	SODIUM (MG.)
Blue Cheese, cold pack	1.5 oz.	398
Cheddar Flavor, cold pack	1.5 oz.	308
Port Wine, cold pack	1.5 oz.	375
Smoked, cold pack	1.5 oz.	353

PROCESS CHEESE PRODUCT
Harvest Moon

	PORTION	SODIUM (MG.)
American Flavor	1.5 oz.	638
Loaf	1.5 oz.	773

Light n' Lively

	PORTION	SODIUM (MG.)
American Flavor Singles	1.5 oz.	615
Sharp Cheddar Flavor Singles	1.5 oz.	570

Lite-Line

	PORTION	SODIUM (MG.)
American Flavor Slices	1.5 oz.	615
American Flavor Slices, Reduced Sodium	1.5 oz.	135
American Flavor Slices, Sodium Lite	1.5 oz.	300
Colby Flavor Slices	1.5 oz.	705
Monterey Jack Flavor Slices	1.5 oz.	705
Muenster Flavor Slices	1.5 oz.	705
Sharp Cheddar Flavor Slices	1.5 oz.	668
Swiss Flavor Slices	1.5 oz.	495

PROCESS CHEESE SPREAD
Dorman's

	PORTION	SODIUM (MG.)
Gruyere	1.5 oz.	360

Kraft

	PORTION	SODIUM (MG.)
American	1.5 oz.	533
Cheese Spread w/Bacon	1.5 oz.	540

	PORTION	SODIUM (MG.)
Cheese Spread w/Garlic	1.5 oz.	510
Cheez Whiz	1.5 oz.	555
Cheez Whiz Pimento	1.5 oz.	555
Cheez Whiz w/Jalapeño Peppers	1.5 oz.	495
Jalapeño	1.5 oz.	630
Jalapeño Peppers	1.5 oz.	143
Olives & Pimento	1.5 oz.	240
Pimento	1.5 oz.	180
Pineapple	1.5 oz.	113
Relish	1.5 oz.	135
Old English Sharp	1.5 oz.	495
Squeez-A-Snak Garlic Flavor	1.5 oz.	420
Squeez-A-Snak Hickory Smoke Flavor	1.5 oz.	413
Squeez-A-Snak Pimento	1.5 oz.	420
Squeez-A-Snak Sharp	1.5 oz.	413
Squeez-A-Snak w/Bacon	1.5 oz.	480
Squeez-A-Snak w/Jalapeño Peppers	1.5 oz.	893
Velveeta	1.5 oz.	645
Velveeta Pimento	1.5 oz.	675
Velveeta Slices	1.5 oz.	600

Land O Lakes

Golden Velvet	1.5 oz.	570

Mohawk Valley

Limburger	1.5 oz.	600

Nabisco Snack Mate

American	1.5 oz.	578
Cheddar	1.5 oz.	608
Cheese 'N Bacon	1.5 oz.	593
Chive 'N Green Onion Flavor	1.5 oz.	555
Sharp Cheddar	1.5 oz.	645

Price's

Pimento Spread	1.5 oz.	503

Roka

Blue Cheese Spread..................1.5 oz. 405

Condiments

(see also Gravies and Sauces, page 328 and Seasonings, page 373)

BARBECUE SAUCES

Cattleman's

Barbecue Sauce, regular	2 tbl	510
Barbecue Sauce, smoky	2 tbl	590

Chris' & Pitt's

Barbecue Sauce	2 tbl	282

Heinz

Hickory Smoke	2 tbl	270
Hot	2 tbl	240
Mushroom	2 tbl	260
Onion	2 tbl	260
Regular	2 tbl	280

Hellmann's

Big H Burger Sauce	2 tbl	290

Hunt's

Hickory BBQ Sauce	2 tbl	400
Hot & Zesty BBQ Sauce	2 tbl	400
Onion BBQ Sauce	2 tbl	400
Original BBQ Sauce	2 tbl	400

Kraft

Barbecue Sauce	2 tbl	495
Garlic Flavored Barbecue Sauce	2 tbl	525
Hickory Smoke Flavor Barbecue Sauce	2 tbl	485

PORTION   SODIUM (MG.)

Hickory Smoke Flavor Onion Bits Barbecue Sauce	2 tbl.	435
Hot Barbecue Sauce	2 tbl.	620
Hot Hickory Smoke Flavored Barbecue Sauce	2 tbl.	450
Onion Bits Barbecue Sauce	2 tbl.	430
Sweet 'N Sour Barbecue Sauce	2 tbl.	130

Open Pit

Hot 'N Spicy Barbecue Sauce	2 tbl.	320
Original Flavor Barbecue Sauce	2 tbl.	500

Western

Barbecue Sauce	2 tbl.	123
Barbecue Sauce, Smoked	2 tbl.	158

KETCHUP

Del Monte

Catsup	1 tbl.	169
Catsup, No Salt Added	1 tbl.	6

Dia-Mel

Catsup	1 tbl.	20

Featherweight

Catsup	1 tbl.	5

Hain

Natural Imitation Catsup, unsalted	1 tbl.	15

Health Valley

Catch-Up	1 tbl.	190
Catch-Up, No Salt	1 tbl.	70

Heinz

Hot Ketchup	1 tbl.	180

PORTION   SODIUM (MG.)

Lite Ketchup	1 tbl	110
Low Sodium Lite Ketchup	1 tbl	90
Tomato Ketchup	1 tbl	180

Hunt's

Ketchup	1 tbl	160
Ketchup, No Salt Added	1 tbl	5

Tillie Lewis

Imitation Tomato Catsup	1 tbl	under 10

Weight Watchers

Ketchup	1 tbl	176

MAYONNAISE

Bama

Mayonnaise	1 tbl	70

Dia-Mel

Mayonnaise	1 tbl	22

Diet-Delight

Mayolite	1 tbl	44

Featherweight

Sayomaise Imitation Mayonnaise	1 tbl	3

Hain

Safflower Mayonnaise	1 tbl	79
Soy Mayonnaise, unsalted	1 tbl	6

Hellmann's

Real Mayonnaise	1 tbl	80

Kraft

Real Mayonnaise	1 tbl	70

PORTION SODIUM (MG.)

Laura Scudder

Mayonnaise ....1 tbl............................146

Light n' Lively

Reduced Calorie Mayonnaise 1 tbl.................................85

NuMade

Real Mayonnaise ....1 tbl............................80

Scotch Buy

Imitation Mayonnaise ....1 tbl...........................115
Mayonnaise.................................1 tbl............................80

Weight Watchers

Reduced Calorie Mayonnaise....1 tbl............................79

Western

Mayonnaise ....1 tbl............................28

MUSTARD

Durkee

Mr. Mustard ....1 tbl............................270

Featherweight

Prepared Salad Mustard ....1 tbl. ....under 2

French's

Bold 'n Spicy Mustard ....1 tbl............................145
Medford Mustard ....1 tbl............................240
Mustard w/Horseradish..............1 tbl............................265
Mustard w/Onion ....1 tbl............................190
Prepared Yellow Mustard ....1 tbl............................180

Grey Poupon

Dijon Mustard ....1 tbl............................445

Gulden's

Creamy Mild Mustard ....1 tbl............................120
Diablo Mustard...........................1 tbl............................110

PORTION   SODIUM (MG.)

	PORTION	SODIUM (MG.)
Dijon Mustard	1 tbl.	120
Spicy Brown Mustard	1 tbl.	90

Hain

Stone Ground Mustard, unsalted	1 tbl.	10

Health Valley

Mustard with Herbs	0.5 oz.	84

Heinz

Brown Mustard	1 tsp.	58
Mild Mustard	1 tsp.	71
Pourable Mustard	1 tsp.	71

Kraft

Pure Prepared Mustard	1 tbl.	50

OLIVES

Black Olives, USDA (in size order)

Ascolano, Extra Large	2 oz. (10 olives)	385
Ascolano, Mammoth	2.3 oz. (10 olives)	454
Ascolano, Giant	2.8 oz. (10 olives)	559
Ascolano, Jumbo	3.4 oz. (10 olives)	664

Greek style, USDA

Medium	0.8 oz. (10 olives)	631
Extra Large	1.2 oz. (10 olives)	868

Green Olives USDA

Small	1.2 oz. (10 olives)	686
Large	1.6 oz. (10 olives)	926
Giant	2.8 oz. (10 olives)	1572

Manzanillo, USDA

Small	1.2 oz. (10 olives)	237
Medium	1.4 oz. (10 olives)	280
Large	1.6 oz. (10 olives)	322
Extra Large	1.9 oz. (10 olives)	385

PORTION   SODIUM (MG.)

Mission, USDA

Small	1.2 oz. (10 olives)	219
Medium	1.4 oz. (10 olives)	258
Large	1.6 oz. (10 olives)	297
Extra Large	1.9 oz. (10 olives)	355

Sevillano, USDA

Giant	2.8 oz. (10 olives)	570
Jumbo	3.4 oz. (10 olives)	676
Colossal	4.2 oz. (10 olives)	847
Supercolossal	5 oz. (10 olives)	1011

PICKLES AND RELISH

Featherweight

Kosher Dill Pickles	1 oz.	under 5
Sliced Cucumber Pickles	1 oz.	under 5
Whole Dill Pickles	1 oz.	under 5

Heinz

Bread 'n Butter Cucumber Slices	1 oz.	170
Hamburger Relish	1 oz.	325
Hot Dog Relish	1 oz.	200
India Relish	1 oz.	215
Picalilli	1 oz.	145
Sweet Cucumber Slices	1 oz.	195
Sweet Cucumber Stix	1 oz.	145
Sweet Gherkins	1 oz.	210
Sweet Midget Gherkins	1 oz.	205
Sweet Mixed Pickles	1 oz.	200
Sweet Pickles	1 oz.	210
Sweet Pickle Slices	1 oz.	205
Sweet Relish	1 oz.	205
Sweet Salad Cubes	1 oz.	270

Smucker's

Candied Dill Sticks	0.8 oz. (1 stick)	180
Candied Sweet Midgets	0.8 oz.	180
Candied Sweet Mix	1.2 oz. (4 pieces)	325

	PORTION	SODIUM (MG.)
Dill Spears	1.4 oz. (1 pickle)	515
Hamburger Dill Slices	0.4 oz. (3 slices)	140
Kosher Baby Dills	1.5 oz. (2 pickles)	565
Kosher Dill Spears	1.8 oz. (1 pickle)	645
Kosher Hamburger Dill Slices	0.4 oz. (3 slices)	140
Polish Dill Spears	1.8 oz. (1 pickle)	640
Sweet Gerkins	0.6 oz. (2 pickles)	180
Sweet Relish	1 tbl.	160
Sweet Slices	0.6 oz. (3 slices)	125
Sweet Stick	1.1 oz. (1 stick)	245
Whole Kosher Dills	1.8 oz. (1 pickle)	645
Whole Polish Dills	1.8 oz. (1 pickle)	645
Whole Sweet	0.8 oz. (2 pickles)	240

Vlasic

Hot & Spicy Garden Mix	1 oz.	380
Kosher Baby Dills	1 oz.	210
Kosher Crunchy Dills	1 oz.	210
Kosher Dill Gherkins	1 oz.	210
Kosher Dill Spears	1 oz.	175
Lightly Spiced Cocktail Onions	1 oz.	365
No Garlic Dills	1 oz.	210
Original Dills	1 oz.	375

Vlasic Bread & Butter

Bread & Butter Sweet Butter Chips	1 oz.	160
Bread & Butter Sweet Butter Stix	1 oz.	110
Old Fashioned Bread & Butter Chunks	1 oz.	120

Vlasic Refrigerated Pickles

Deli Bread & Butter Chunks	1 oz.	120
Kosher Deli Dills	1 oz.	290

Vlasic Half-The-Salt Pickles

Hamburger Dill Chips	1 oz.	175

	PORTION	SODIUM (MG.)
Kosher Crunchy Dills	1 oz	125
Kosher Dill Spears	1 oz	120
Sweet Butter Chips	1 oz	80

Vlasic Relish

Dill Relish	1 oz	415
Hamburg Relish	1 oz	255
Hot Dog Relish	1 oz	255
Sweet Relish	1 oz	220

OTHER
A.1.

Steak Sauce	2 tbl	550

Arrowhead Mills

Tamari	1 tbl	800

Bac*Os

Bac*Os	1 tbl	165

Bennett

Chili Sauce	1 tbl	125

Del Monte

Burrito Salsa	2 tbl	178
Chilies, Green, Whole Diced	¼ cup	345
Chilies, Jalapeño, Whole	¼ cup	845
Hot and Chunky Salsa Picante	2 tbl	203
Hot Salsa Picante	2 tbl	193
Mild Green Chile Salsa	2 tbl	295
Mild Salsa Roja	2 tbl	255
Taco Sauce, Hot	2 tbl	220
Taco Sauce, Mild	2 tbl	240

Durkee

Bacon Chips	1 tbl	35
Coconut (Shredded)	¼ cup	5
Famous Sauce	1 tbl	67
Frank's Hot Sauce	1 tbl	393

PORTION   SODIUM (MG.)

French Fried Onions	1 oz.	137
Imitation Bacon Bits	1 tbl.	687
Soup Greens	¼ jar	102

Featherweight

Chili Sauce	1 tbl.	10

French's

Bacon, Imitation Crumbles	1 tsp.	55
Worcestershire Sauce, Regular	1 tbl.	165
Worcestershire Sauce, Smoky	1 tbl.	165

Heinz

Chili Sauce	1 tbl.	191
57 Sauce	1 tbl.	265
Hot Banana Peppers	1 oz.	305
Hot Cherry Peppers	1 oz.	310
Hot Pepper Rings/Slices	1 oz.	300
Spiced Onions	1 oz.	600
Sweet Mild Cherry Peppers	1 oz.	315
Sweet Mild Pepper Rings/Slices	1 oz.	305
Sweet Onions	1 oz.	165
Sweet Pepper Mementos	1 oz.	320

Hellmann's

Sandwich Spread	1 tbl.	175
Tartar Sauce	1 tbl.	185

Kikkoman

Milder Soy Sauce	1 tsp.	182
Regular Soy Sauce	1 tsp.	320

Kraft

Cream Style Prepared Horseradish	1 tbl.	145
Horseradish Mustard	1 tbl.	45

PORTION   SODIUM (MG.)

Horseradish Sauce	1 tbl.	100
Prepared Horseradish	1 tbl.	50
Sandwich Spread	1 tbl.	95
Tartar Sauce	1 tbl.	160

La Choy

Chow Mein Noodles	1 oz. (½ cup)	205
Rice Noodles	1 oz. (½ cup)	360
Soy Sauce	1 tsp.	325

NuMade

Relish Sandwich Spread (East)	1 tbl.	120
Relish Sandwich Spread (West)	1 tbl.	176

Ocean Spray

Cran-orange Relish	2 oz.	18
Cran-raspberry Jellied Sauce	2 oz.	14
Cranberry Sauce, Jellied	2 oz.	17
Cranberry Sauce, Whole	2 oz.	16

Ortega

Green Chili Salsa	1 tbl.	175

Oscar Mayer

Bacon Bits	¼ oz.	189

Regina

Burgundy Cooking Wine	¼ cup	360
Sauterne Cooking Wine	¼ cup	360
Sherry Cooking Wine	¼ cup	370
Wine Vinegars, all	1 oz.	under 1

Soken

Gentle Soy Sauce	1 tbl.	420

Tostitos

Picante Sauce	3.1 oz.	490

PORTION   SODIUM (MG.)

USDA

Capers	1 tbl	203
Miso, Red	1 tbl	927
Miso, White	1 tbl	532
Tamari	1 tbl	857

Vlasic

Hot Banana Pepper Rings	1 oz.	465
Mexican Jalapeño Peppers	1 oz.	380
Mild Cherry Peppers	1 oz.	410
Mild Greek Pepperoncini	1 oz.	450

Western

Deluxe Tartar Sauce	1 tbl	112
Horseradish Sauce	1 tbl	97

Cookies, Brownies, and Bars

MIXES AND DOUGHS

Betty Crocker (as prepared)

Big Batch Chocolate Chip Cookies	2 cookies	95
Big Batch Oatmeal Cookies	2 cookies	100
Big Batch Sugar Cookies	2 cookies	95
Coconut Macaroon Mix	1/24 package	15
Date Bar	1/32 package	35
Fudge Brownie Mix, family size	1/24 package	95
Fudge Brownie Mix, regular size	1/16 package	100
Supreme Fudge Brownie Mix	1/24 package	90
Supreme Golden Brownie Mix, family size	1/24 package	105
Vienna Dream Bar	1/24 package prepared	65
Walnut Brownie Mix, family size	1/24 package	85

PORTION　SODIUM (MG.)

Walnut Brownie Mix, regular
size ....1/16 package 100

Duncan Hines (as prepared)

"Chewy" Brownie.......................1 brownie 90
Chocolate Chip...........................2 cookies...........................95
Double Chocolate Chip ....2 cookies...........................80
Golden Sugar ....2 cookies...........................70
Oatmeal Raisin..........................2 cookies...........................70
Peanut Butter ....2 cookies 120

Featherweight

Cake & Cookie Mix...................1-in. slice..............................5

Merico (refrigerated dough)

Chocolate Chip Cookies............0.9 oz. (2 cookies) 70
Peanut Butter Cookies ....0.9 oz. (2 cookies) 70
Sugar Cookies ....0.9 oz. (2 cookies) 70

Nestlé

Chocolate Chip Cookie Mix......2 cookies...............................65

Pillsbury (as prepared)

Brown Sugar Oatmeal Fudge
Jumbles ....1 bar 60
Chocolate Chip Oatmeal Fudge
Jumbles ....1 bar 60
Coconut Oatmeal Fudge
Jumbles ....1 bar 60
Fudge Brownie ....one 2-in square 95
Fudge Brownie, family size ......one 2-in. square 90
Peanut Butter Oatmeal Fudge
Jumbles ....1 bar 55
Walnut Brownie, family size......one 2-in. square 95

Pillsbury (refrigerated dough)

Chocolate Chip Cookies............3 cookies 125
Double Chocolate ....3 cookies 160
Fudge Brownies........................1 bar 115
Peanut Butter Cookies ....3 cookies 190
Sugar Cookies ....3 cookies 170

PORTION   SODIUM (MG.)

PREPARED
Amurol

Cookies	1 cookie	8
Filled Wafers	1 wafer	10
Wafer Bars	1 bar	36

Archway

Almond Shortbread	1 oz. (2 bars)	136
Apple Filled	0.7 oz. (1 cookie)	73
Apple 'N Raisin	1 oz. (1 cookie)	141
Applesauce	1 oz. (1 cookie)	121
Apricot Filled	0.9 oz. (1 cookie)	79
Blueberry Filled	1 oz. (1 cookie)	86
Cherry Filled	1 oz. (1 cookie)	74
Cherry Filled Fudge	1 oz. (1 cookie)	97
Chocolate Chip Drop	0.9 oz. (1 cookie)	102
Chocolate Chip Ice Box	0.8 oz. (1 cookie)	55
Chocolate Chip 'N Toffee	1.1 oz (1 cookie)	111
Chocolate Chip Supreme	1.1 oz. (1 cookie)	199
Cinnamon Apple	1.2 oz. (1 cookie)	183
Cocoa Chocolate Chip	1 oz. (2 cookies)	100
Coconut Chocolate Chip	1 oz. (2 cookies)	96
Coconut Macaroon	0.7 oz. (1 cookie)	38
Coconut Square	0.9 oz. (1 square)	92
Cookie Jar Hermit	1.8 oz. (1 cookie)	280
Creme Filled Fudge	0.9 oz. (1 cookie)	125
Date and Nut Bar	2.2 oz. (1 bar)	217
Date Filled Oatmeal	0.9 oz. (1 cookie)	99
Devil's Food	1.1 oz. (1 cookie)	82
Dutch Cocoa	0.9 oz. (1 cookie)	94
Frosty Lemon	1 oz. (1 cookie)	116
Frosty Orange	1 oz. (1 cookie)	134
Fruit & Honey	2 oz. (1 cookie)	246
Fruit Filled Oatmeal	1 oz. (1 cookie)	121
Fudge Nut Bar	1.7 oz. (1 bar)	198
Grandma's Molasses	1 oz. (1 cookie)	173
Honey 'N Nut	1 oz. (2 cookies)	118
Iced Molasses	1.2 oz. (1 cookie)	234
Iced Spice	1.1 oz. (1 cookie)	286
Lemon Drop	0.9 oz. (1 cookie)	94

	PORTION	SODIUM (MG.)
New Orleans Cake	2.1 oz. (1 piece)	365
Nutty Fudge	1 oz. (2 cookies)	100
Nutty Nougat	0.9 oz. (3 cookies)	39
Oatmeal	0.9 oz. (1 cookie)	86
Oatmeal Chocolate Chip	1.2 oz. (2 cookies)	226
Oatmeal Nut	1 oz. (2 cookies)	136
Old Fashioned Ginger	1 oz. (1 cookie)	203
Old Fashioned Molasses	1 oz. (1 cookie)	190
Peanut Butter	0.8 oz. (1 cookie)	90
Peanut Butter Chip	1.2 oz. (2 cookies)	206
Peanut Butter 'N Chips	1 oz. (1 cookie)	199
Pecan Crunch	1 oz. (2 cookies)	130
Pecan Ice Box	0.8 oz. (1 cookie)	68
Pineapple Filled	1 oz. (1 cookie)	169
Raspberry Filled	1 oz. (1 cookie)	86
Rocky Road	1 oz. (1 cookie)	47
Ruth's Golden Oatmeal	0.9 oz. (1 cookie)	70
Soft Sugar	0.9 oz. (1 cookie)	102
Sour Cream	0.9 oz. (1 cookie)	87
Strawberry Filled	0.9 oz. (1 cookie)	87
Sugar	1 oz. (1 cookie)	309
Sugar Lemon	1.1 oz. (1 cookie)	153

Carnation Breakfast Bars

Almond Crunch	1.5 oz. (1 bar)	175
Chocolate Chip	1.4 oz. (1 bar)	163
Chocolate Crunch	1.5 oz. (1 bar)	129
Peanut Butter Crunch	1.5 oz. (1 bar)	154

Carnation Slender Bars

Chocolate	2 bars	330
Chocolate Chips	2 bars	295
Chocolate Peanut Butter	2 bars	305
Lemon Yogurt	2 bars	295
Raspberry Yogurt	2 bars	295
Strawberry Yogurt	2 bars	295
Vanilla	2 bars	320

Duncan Hines

Almond Fudge Chocolate Chip	0.8 oz. (2 cookies)	90

	PORTION	SODIUM (MG.)
Butterscotch Chocolate	0.8 oz. (2 cookies)	70
Chocolate Chip	0.8 oz. (2 cookies)	70
Mint Chocolate Chip	0.8 oz. (2 cookies)	70
Peanut Butter	0.8 oz. (2 cookies)	90

Drake's

Oatmeal Cookies	3 cookies	200

Estee

Chocolate Chip	1 cookie	5
Coconut	1 cookie	5
Creme Filled Wafers, Assorted	1 wafer	5
Creme Filled Wafers, Chocolate	1 wafer	5
Creme Filled Wafers, Vanilla	1 wafer	5
Fudge	1 cookie	5
Lemon	1 cookie	5
Oatmeal Raisin	1 cookie	5
Sandwich Cookies, Duplex	1 cookie	10
Sandwich Cookies, Lemon	1 cookie	10
Snack Wafers, Chocolate Coated	1 wafer	10
Snack Wafers, Chocolate-Strawberry	1 wafer	5
Snack Wafers, Vanilla	1 wafer	5
Vanilla	1 cookie	5

Featherweight

Chocolate Chip Cookies	1 cookie	6
Chocolate Creme Wafers	1 wafer	14
Chocolate Crescent Cookies	1 cookie	6
Lemon Cookies	1 cookie	3
Peanut Butter Creme Wafers	1 wafer	14
Sandwich Creme Cookies	1 cookie	3
Vanilla Cookies	1 cookie	6
Vanilla Creme Wafers	1 wafer	14

Frito-Lay

Creme Filled Oatmeal Cake	2 oz.	70

	PORTION	SODIUM (MG.)
Nut Fudge Brownie	1.80 oz.	135
Peanut Butter Bar	1.75 oz.	125

Girl Scout

Chocolate Chunks	0.8 oz. (2 cookies)	73
Dosi Dos	0.9 oz. (2 cookies)	64
Medallions	1.1 oz. (4 cookies)	76
Pecan w/Praline Bottoms	1.1 oz. (4 cookies)	70
Samoas	1.1 oz. (2 cookies)	63
Tagalons	1.0 oz. (2 cookies)	85
Thin Mints	1 oz. (4 cookies)	88
Trefoils	0.9 oz. (4 cookies)	82

Grandma's

Old Time Molasses Cookie	1 oz.	155
Soft Raisin Cookie	1 oz.	87

Health Valley

Amaranth Jumbo Single Cookies	1 cookie	80
Carob Snaps	2 snaps	10
Cinnamon Animal Snaps	2 cookies	10
Cinnamon Honey Jumbo Cookies	1 cookie	35
Coconut Snaps	2 snaps	10
Date Pecan Cookies	2 cookies	30
Fruit Bakes, Apple	1 bar	23
Fruit Bakes, Date	1 bar	20
Fruit Bakes, Raisin	1 bar	18
Ginger Snaps	2 snaps	10
Honey Grahams	1 oz. (10 crackers)	155
Lemon Snaps	2 snaps	10
Oatmeal Cookies	2 cookies	40
Oatmeal Honey Jumbo Cookies	1 cookie	35
Peanut Butter Cookies	2 cookies	40
Peanut Butter Honey Jumbo Cookies	1 cookie	35
Peanut Butter Jumbo Single Cookie	1 cookie	80

PORTION   SODIUM (MG.)

Raisin Bran Cookies ....2 cookies....................................40
Raisin/Oatmeal Jumbo Single
  Cookie ....1 cookie....................................80
Vanilla Animal Snaps ....2 cookies....................................10

Hostess

Brownies, large ....2 oz. (1 brownie)..................120
Brownies, small ....1.3 oz. (1 brownie) 75

Keebler

C. C. Biggs Chocolate Chip
  Cookies ....0.8 oz. (2 cookies) 75
Coconut Chocolate Drop
  Cookies ....1.1 oz. (2 cookies) 115
Deluxe Grahams, Fudge
  Covered Graham Cracker......1.1 oz. (4 cookies) 110
Double Nutty Peanut Butter
  Sandwich.....................................0.9 oz. (2 cookies) 115
Elfwich Sandwich Cookies........0.8 oz. (2 cookies) 85
Fudge Stripes Cookies ....0.8 oz. (2 cookies) 75
Honey Grahams...........................0.5 oz. (4 crackers) 90
Old Fashioned Oatmeal ....1.3 oz. (2 cookies) 200
Pecan Sandies ....1 oz. (2 cookies)..................125
Pitter Patter Sandwich
  Cookies ....1.2 oz. (2 cookies) 175
Rich 'n Chips ....1.1 oz. (2 cookies) 100
Vanilla Wafers.............................1 oz. (1 cookie)....................115

Merico

Chocolate Chip.............................1.1 oz. (2 cookies) 70
Peanut Butter ....1.1 oz. (2 cookies) 70
Sugar ....1.1 oz. (2 cookies) 70

Mrs. Goodcookie

Chocolate Chip Cookies.............1.3 oz. (3 cookies) 40
Chocolate Fudge Cookies ....1.3 oz. (3 cookies) 53
Oatmeal Raisin Cookies............1.3 oz. (3 cookies) 93
Peanut Butter Cookies ....1.3 oz. (3 cookies) 67
Sugar Cookies ....1.3 oz. (3 cookies) 60

PORTION   SODIUM (MG.)

Nabisco

	PORTION	SODIUM (MG.)
Almond Windmill	1 oz. (3 cookies)	120
Apple Crisp	1 oz. (3 cookies)	125
Baker's Bonus Butterscotch Chip Cookies	1 oz. (2 cookies)	140
Baker's Bonus Cocoanut Bars	1 oz. (3½ bars)	140
Baker's Bonus Double Chips Fudge	1 oz. (2 cookies)	140
Baker's Bonus Oatmeal	1 oz. (1½ cookies)	120
Baker's Bonus Sugar Rings	1 oz. (2 cookies)	110
Barnum's Animal Crackers	1 oz. (11 crackers)	155
Biscos I Screams 'N You Screams Chocolate Creme Sandwich	1 oz. (2 cookies)	80
Biscos I Screams 'N You Screams Vanilla Creme Sandwich	1 oz. (2 cookies)	80
Biscos Sugar Wafers	1 oz. (8 wafers)	50
Biscos Triple Decker Peanut Butter Sugar Wafers	1 oz. (3 wafers)	130
Biscos Waffle Cremes	1 oz. (3 cookies)	45
Brown Edge Sandwich	1 oz. (2 cookies)	75
Brown Edge Wafers	1 oz. (5 wafers)	90
Bugs Bunny Grahams	1 oz. (9 crackers)	130
Butter Flavored Cookies	1 oz. (5½ cookies)	155
Cameo Creme Sandwich	1 oz. (2 cookies)	90
Chips Ahoy Chocolate Chip Cookies	1 oz. (2½ cookies)	100
Chocolate Chip Snaps	1 oz. (6½ snaps)	105
Chocolate Chocolate Chip Cookies	1 oz. (2½ cookies)	95
Chocolate Grahams	1 oz. (2½ cookies)	75
Chocolate Pinwheels Cakes	1 oz. (1 piece)	45
Chocolate Snaps	1 oz. (7½ snaps)	160
Cocoanut Chocolate Chip Cookies	1 oz. (2 cookies)	100
Cocoanut Macaroon Soft Cakes	1 oz. (1½ pieces)	80

PORTION   SODIUM (MG.)

Cookie Break Mixed Creme
Sandwich....................................1 oz. (3 cookies) 120
Cookie Break Vanilla Artificial
Flavor Creme Sandwich........1 oz. (3 cookies) 135
Country Cremes Natural
Coconut Custard
Sandwich....................................1 oz. (3 cookies) 95
Country Cremes Natural Flavor
Apple Cinnamon
Sandwich....................................1 oz. (3 cookies) 110
Country Cremes Natural Flavor
Vanilla Custard Sandwich ....1 oz. (3 cookies) 120
Creme Wafer Sticks ....1 oz. (3 wafers) 50
Devil's Food Cakes...................1 oz. (1½ pieces) 55
Famous Chocolate Wafers........1 oz. (4½ wafers) 235
Famous Cookie Assortment
Baronet Creme Sandwich ....1 oz. (2½ cookies) 90
Biscos Sugar Wafers ....1 oz. (3 wafers)....................55
Butter Flavored.......................1 oz. (5 cookies) 155
Cameo Creme Sandwich......1 oz. (2 cookies) 90
Kettle Cookie Sandwich ....1 oz. (4 cookies) 115
Lorna Doone Shortbread ......1 oz. (4 cookies) 160
Fancy Dip Grahams ....1 oz. (2½ cookies) 105
Fig Newtons Cakes...................1 oz. (2 pieces)....................125
Fig Wheats Whole Wheat Fig
Bars ....1 oz. (2 pieces) 80
Gaiety Fudge Chocolate
Sandwich....................................1 oz. (2 cookies) 130
Graham Crackers ....1 oz. (4 crackers) 210
Heyday Caramel Peanut
Logs ....1 oz. (1 piece) 65
Ideal Chocolate Peanut Bars...1 oz. (1½ pieces) 130
Lorna Doone Shortbread ....1 oz. (3½ cookies) 160
Mallomars Chocolate Cakes ....1 oz. (2 pieces) 50
Marshmallow Puffs ....1 oz. (1½ pieces) 60
Marshmallow Sandwich ....1 oz. (3½ cookies) 85
Marshmallow Twirls Cakes........1 oz. (1 piece) 45
Mayfair Assortment English
Style
Crown Creme Sandwich........1 oz. (2½ cookies) 185

PORTION   SODIUM (MG.)

Fancy Shortbread	1 oz. (6 cookies)	95
Filigree Creme Sandwich	1 oz. (2½ cookies)	100
Mayfair Creme Sandwich	1 oz. (2 cookies)	80
Tea Rose Creme	1 oz. (2½ cookies)	140
Tea Time Biscuit	1 oz. (5½ cookies)	130
Melt-a-Way Shortcake Cookies	1 oz. (2 cookies)	140
Mystic Mint Sandwich Cookies	1 oz. (1½ cookies)	95
Nana Banana Natural Flavor Creme Sandwich	1 oz. (2 cookies)	80
National Arrowroot Biscuit	1 oz. (6 cookies)	100
Nilla Wafers	1 oz. (7 wafers)	95
Nutter Butter Peanut Butter Sandwich	1 oz. (2 cookies)	110
Oatmeal, Cocoanut & Raisin Cookies	1 oz. (2½ cookies)	90
Old Fashioned Ginger Snaps	1 oz. (4 snaps)	215
Old Fashioned Granola Soft Snacks	1 oz. (1 piece)	125
Old Fashioned Oatmeal & Apple Soft Snacks	1 oz. (1 piece)	70
Old Fashioned Oatmeal & Raspberry Soft Snacks	1 oz. (1 piece)	75
Oreo Chocolate Sandwich Cookies	1 oz. (3 cookies)	210
Oreo Chocolate Sandwich Double Stuf	1 oz. (2 cookies)	165
Pantry Molasses Cookies	1 oz. (2 cookies)	140
Party Grahams Cookies	1 oz. (3 cookies)	115
Peanut Butter Fudge	1 oz. (2½ cookies)	100
Peanut Creme Patties	1 oz. (4½ cookies)	95
Pecan Shortbread Cookies	1 oz. (2 cookies)	85
Piccolo Crepe Cookies	1 oz. (6½ cookies)	50
Social Tea Biscuit	1 oz. (6 cookies)	115
Spiced Wafers	1 oz. (3½ wafers)	200
Striped Shortbread	1 oz. (3 cookies)	80
Swiss Creme Sandwich	1 oz. (3 wafers)	125
Twiddle Sticks Crunchy Creme Wafers	1 oz. (3 wafers)	45

PORTION   SODIUM (MG.)

Nature Valley

Chewy Granola Bar, Apple	1 bar	70
Chewy Granola Bar, Chocolate Chip	1 bar	80
Chewy Granola Bar, Peanut Butter	1 bar	80
Chewy Granola Bar, Raisin	1 bar	65
Granola and Fruit Bar, Apple	1 bar	150
Granola and Fruit Bar, Cherry	1 bar	165
Granola and Fruit Bar, Date	1 bar	140
Granola and Fruit Bar, Raspberry	1 bar	150
Granola Bar, Almond	0.8 oz (1 bar)	85
Granola Bar, Cinnamon	0.8 oz. (1 bar)	80
Granola Bar, Coconut	0.8 oz. (1 bar)	65
Granola Bar, Oats 'n Honey	0.8 oz. (1 bar)	70
Granola Bar, Peanut	0.8 oz. (1 bar)	85
Granola Bar, Peanut Butter	0.8 oz. (1 bar)	80
Granola Cluster, Apple Cinnamon	1 roll	100
Granola Cluster, Almond	1.3 oz. (1 roll)	100
Granola Cluster, Caramel	1.3 oz. (1 roll)	120
Granola Cluster, Chocolate	1 roll	110
Granola Cluster, Chocolate Chip	1 roll	120
Granola Cluster, Raisin	1.3 oz. (1 roll)	110

Penn-Dutch

Apple Filled Flips	2 cookies	290
Apricot Filled Flips	2 cookies	140
Banana Fudge Flips	2 cookies	140
Blueberry Flips	2 cookies	165
Date-Filled Oatmeal	2 cookies	270
Fruit & Honey Bars	2 cookies	390
Fudge Bars	2 cookies	240
Old Fashioned Dutch Molasses	2 cookies	380

PORTION   SODIUM (MG.)

	PORTION	SODIUM (MG.)
Old Fashioned Dutch Oatmeal	2 cookies	305
Original Health Cookies	2 cookies	205
Peanut Jumbles	2 cookies	165

Pepperidge Farm Cookies, Brownies and Bars

	PORTION	SODIUM (MG.)
Almond Supreme	0.9 oz. (2 cookies)	45
Apricot-Raspberry	1.1 oz. (3 cookies)	80
Blueberry	0.9 oz. (2 cookies)	53
Bordeaux	1 oz. (4 cookies)	93
Brownie Chocolate Nut	1.1 oz. (3 cookies)	80
Brownie Nut	0.9 oz. (1 cookie)	65
Brussels	1 oz. (3 cookies)	95
Brussels Mint	0.9 oz. (3 cookies)	80
Butter Assortment	1.0 oz. (4 cookies)	73
Cappucino	1 oz. (3 cookies)	60
Capri	1.1 oz. (2 cookies)	90
Champagne Assortment	0.9 oz. (4 cookies)	73
Chessmen	0.9 oz. (3 cookies)	80
Chocolate Chip	1.1 oz. (3 cookies)	85
Chocolate Chip, large	0.9 oz. (1 cookie)	80
Chocolate Chocolate Chip	1.1 oz. (3 cookies)	75
Chocolate Chunk Pecan	1 oz. (2 cookies)	60
Chocolate Laced Pirouettes Assortment	0.9 oz. (4 cookies)	60
Date Nut Granola	1.1 oz. (3 cookies)	95
Date Pecan	1.1 oz. (3 cookies)	60
Geneva	1.1 oz. (3 cookies)	65
Ginger Man	1 oz. (4 cookies)	75
Hazelnut	1.1 oz. (3 cookies)	110
Irish Oatmeal	1.0 oz. (3 cookies)	120
Lemon Nut Crunch	1.1 oz. (3 cookies)	75
Lido	1.2 oz. (2 cookies)	85
Marseilles Assortment	0.9 oz. (3 cookies)	75
Milano	1.2 oz. (3 cookies)	80
Mint Milano	0.9 oz. (3 cookies)	70
Molasses Crisps	1.1 oz. (3 cookies)	75
Nassau	1.1 oz. (2 cookies)	90
Oatmeal	0.9 oz. (1 cookie)	105
Orange	0.9 oz. (2 cookies)	43

PORTION   SODIUM (MG.)

Orange Milano	0.9 oz. (3 cookies)	70
Orleans	1.0 oz. (5 cookies)	50
Orleans Sandwich	1 oz. (3 cookies)	60
Peanut Butter Chip	1.1 oz. (3 cookies)	135
Peanut Butter Chip, large	0.9 oz. (1 cookie)	110
Pirouettes Assortment	0.9 oz. (4 cookies)	73
Raisin Bran	1.1 oz. (3 cookies)	80
Seville Assortment	0.7 oz. (2 cookies)	50
Shortbread	0.9 oz. (2 cookies)	85
Southport Assortment	1 oz. (2 cookies)	70
Strawberry	1.1 oz. (3 cookies)	70
Sugar	1.1 oz. (3 cookies)	115
Sugar, large	1.1 oz. (1 cookie)	100
Sunflower Raisin	1.1 oz. (3 cookies)	75
Tahiti	1 oz. (2 cookies)	50

Pepperidge Farm Snack Bars

Apple Nut	1.7 oz. (1 bar)	90
Apricot Raspberry	1.7 oz. (1 bar)	90
Blueberry	1.7 oz. (1 bar)	90
Brownie Nut	1.6 oz. (1 bar)	100
Chocolate Chip Macaroon	1.6 oz. (1 bar)	80
Coconut Macaroon	1.6 oz. (1 bar)	80
Date Nut	1.6 oz. (1 bar)	90
Raisin Spice	1.6 oz. (1 bar)	80

Pillsbury Figurines

Caramel Nut	1.8 oz. (2 bars)	285
Chocolate	1.8 oz. (2 bars)	230
Chocolate Caramel	1.8 oz. (2 bars)	180
Chocolate Mint	1.8 oz. (2 bars)	235
Chocolate Peanut Butter	1.8 oz. (2 bars)	240
Double Chocolate	1.8 oz. (2 bars)	230
Lemon Yogurt	1.8 oz. (2 bars)	180
Strawberry Yogurt	1.8 oz. (2 bars)	165
Vanilla	1.8 oz. (2 bars)	205

Pillsbury Milk Break Bars

Chocolate	1 bar	75

	PORTION	SODIUM (MG.)
Chocolate Mint	1 bar	80
Natural Flavor	1 bar	75
Peanut Butter	1 bar	115

Plus Products

Caro Chip Cookies	1 oz. (4 cookies)	100

Ralston Purina

Sugar Honey Graham Crackers	1 oz. (8 crackers)	140

Royal

Fig Cookies	2 oz. (1 cookie)	166
Granola Cookies	2 oz. (1 cookie)	137
Oatmeal Cookies	2 oz. (1 cookie)	161

Sara Lee

Chocolate Brownies	1.6 oz. (⅛ cake)	108

Stella D'Oro

Apple Pastry (dietetic)	0.8 oz. (1 pastry)	under 10
Coconut Cookies (dietetic)	1.2 oz. (3 cookies)	under 10
Egg Biscuits (dietetic)	0.8 oz. (2 pieces)	under 10
Fig Pastry (dietetic)	0.8 oz. (1 piece)	under 10
Kichel (dietetic)	1.1 oz. (20 pieces)	under 10
Love Cookies (dietetic)	0.8 oz. (1 cookie)	under 10
Peach Apricot Pastry (dietetic)	0.8 oz. (1 pastry)	under 10
Prune Pastry	0.8 oz. (1 pastry)	under 10
Royal Nuggets	0.7 oz. (75 pieces)	10
Sesame Cookies (dietetic)	0.9 oz. (3 cookies)	10

Tastykake

Butter Cookies	1 cookie	26
Chocolate Chip Cookies	1 cookie	48
Coated Mini-pretzel	1 pretzel	22
Coated Pretzel	1 pretzel	69

PORTION   SODIUM (MG.)

Fudge Bar	1 bar	144
Oatmeal Raisin Bar	1 bar	269
Vanilla Shortbread Cookie	1 cookie	23

Eggs and Egg Substitutes

Avoset

Second Nature Imitation Egg	1½ oz.	70

Fleischmann's

Egg Beaters	2 oz. (¼ cup)	90

Morningstar Farms

Scramblers	2 oz. (¼ cup)	150

Tillie Lewis

Eggstra, Cooked	½ package	105

USDA

Egg, Whole	1 egg	59
Egg White	1 white	50
Egg Yolk	1 yolk	9

Entrees and Dinners

(see also Fish and Seafood, page 311, Meat, page 339, and Poultry, page 367)

CANNED

Armour

Beef Stew	8 oz.	1190
Chili with Beans	7.5 oz.	1080
Chili without Beans	7.5 oz.	1140
Corned Beef Hash	7.5 oz.	1420
Sloppy Joe Beef	7.6 oz.	1700
Sloppy Joe Pork	7.6 oz.	1570

	PORTION	SODIUM (MG.)
Texas Chili with Beans	7.5 oz.	1310
Texas Chili without Beans	7.5 oz.	1550

Carnation

Chicken Sausage in Barbecue Sauce	5 oz. (1 can)	931
Chicken Sausage in Mustard Sauce	5 oz. (1 can)	927
Chicken Sausage in Pizza Sauce	5 oz. (1 can)	733
Chicken Sausage in Taco Sauce	5 oz. (1 can)	871

Chef Boy-ar-dee

ABC's & 1, 2, 3's in Sauce (15 oz. can)	7.5 oz.	989
ABC's & 1,2,3's in Sauce (40 oz. can)	7.5 oz.	848
ABC's & 1,2,3's w/Franks	7.5 oz.	1152
ABC's & 1,2,3's w/Meatballs	7.5 oz	1148
ABC's & 1,2,3's w/Mini Meatballs	7.5 oz.	1021
Beef Chili w/Beans	7.5 oz.	1005
Beef Chili w/o Beans	7.5 oz.	875
Beef Ravioli in Sauce (40 oz. can)	7.5 oz.	1191
Beef Ravioli in Tomato & Meat Sauce (15 oz. can)	7.5 oz.	1270
Beef-O-Getti	7.5 oz	1148
Beefaroni (15 oz. can)	7.5 oz.	1243
Beefaroni (40 oz. can)	7.5 oz	1392
Cannelloni	7.5 oz.	1060
Cheese Ravioli in Beef and Tomato Sauce	7.5 oz.	1205
Cheese Ravioli in Tomato Sauce	7.5 oz.	1116
Chicken Ravioli	7.5 oz.	1100
Chili Mac	7.5 oz.	1286
Corned Beef Hash	7.5 oz.	1210

PORTION   SODIUM (MG.)

Cosmic Kids in Chicken Sauce	7.5 oz.	1320
Cosmic Kids in Tomato Sauce	7.5 oz.	895
Cosmic Kids w/Meatballs	7.5 oz.	1115
Lasagna (15 oz. can)	7.5 oz.	1075
Lasagna (40 oz. can)	7.5 oz.	1372
Macaroni Shells in Tomato Sauce	7.5 oz.	1040
Manicotti	7.5 oz.	1116
Meat Balls in Brown Gravy	7.5 oz.	1402
Meat Ball Stew	7.5 oz.	1233
Meatball-a-roni	7.5 oz.	1116
Mini Bites	7.5 oz.	1101
Mini Ravioli, Beef (15 oz. can)	7.5 oz.	1138
Mini Ravioli, Beef (40 oz. can)	7.5 oz.	1137
Pac Man in Chicken Sauce	7.5 oz.	905
Pac Man in Tomato Sauce	7.5 oz.	830
Pac Man w/Meatballs	7.5 oz.	1140
Roller Coasters	7.5 oz.	1201
Sausage Ravioli in Tomato & Meat Sauce	7.5 oz.	785
Spaghetti & 8 Mini Meat Balls	7.5 oz.	1190
Spaghetti & Meatballs in Tomato Sauce (15 oz. can)	7.5 oz.	1010
Spaghetti & Meatballs in Tomato Sauce (40 oz. can)	7.5 oz.	1491
Spaghetti w/Beef in Tomato Sauce	7.5 oz.	1100
Spaghetti w/Tomato Sauce & Cheese (15 oz. can)	7.5 oz.	1040
Spaghetti w/Tomato Sauce & Cheese (40 oz. can)	7.5 oz.	975
Vegetables & Meat Balls in Gravy	7.5 oz.	715
Zooroni w/Meatballs in Sauce	7.5 oz.	904
Zooroni w/Tomato Sauce	7.5 oz.	829

PORTION   SODIUM (MG.)

Chef Boy-ar-dee, EZO Products

	PORTION	SODIUM (MG.)
ABC's & 1,2,3's in Sauce	7 oz.	885
ABC's & 1,2,3's w/Meatballs	7 oz.	995
Beef Ravioli	7 oz. (1 can)	1180
Beef Stew	7 oz. (1 can)	1120
Beefaroni	7 oz. (1 can)	965
Beef-O-Getti	7 oz. (1 can)	1005
Cannelloni	7.5 oz. (1 can)	1169
Cheese Ravioli in Tomato Sauce	7.5 oz. (1 can)	1010
Chef's Special Chili Beans in Meat Sauce	7.0 oz.	835
Chef's Special Mini Beef Ravioli	7.5 oz. (1 can)	1275
Chicken Stew	7 oz. (1 can)	1150
Chili Con Carne w/Beans	7 oz. (1 can)	960
Chili Mac	7 oz. (1 can)	1270
Corned Beef Hash	7 oz. (1 can)	1370
Hot Chili Con Carne w/Beans	7 oz. (1 can)	1140
Lasagna	7 oz. (1 can)	1115
Macaroni & Cheese	7 oz. (1 can)	1040
Macaroni Shells	7 oz. (1 can)	920
Noodles w/Chicken	7.0 oz.	780
Roller Coasters	7 oz. (1 can)	1070
Spaghetti & Meat Balls	7 oz. (1 can)	1075

Dennison's

	PORTION	SODIUM (MG.)
Chili Beans in Chili Gravy	7.5 oz.	705
Chili Con Carne w/Beans (15 oz. can)	7.5 oz.	875
Chili Con Carne w/Beans (30 oz. can)	7.5 oz.	840
Chili Con Carne w/Beans (40 oz. can)	7.5 oz.	1050
Chili Con Carne w/o Beans	7.5 oz.	1380
Chili Mac	7.5 oz.	1140
Hot Chili Con Carne w/Beans (15 oz. can)	7.5 oz.	910
Hot Chili Con Carne w/Beans (40 oz. can)	7.5 oz.	950

PORTION   SODIUM (MG.)

Tamalitos in Chili Gravy ....7.5 oz. 1395

Dennison's EZO

Chili Con Carne w/Beans ....7.0 oz. 1125
Hot Chili Con Carne w/Beans 7.0 oz. 945

Dia-Mel

Beef Ravioli.............................8 oz. 75
Beef Stew ....8 oz. 70
Chicken Stew ....8 oz. 65
Spaghetti and Meatballs............8 oz. 55

Featherweight

Beef Ravioli.............................8 oz. 75
Beef Stew ....7.5 oz. 50
Chicken Stew ....7.5 oz. 15
Chili w/Beans ....7.5 oz. 75
Spaghetti w/Meat Balls..............7.5 oz. 65
Stuffed Dumplings w/Chicken 7.5 oz. 85

Franco-American

Beef Ravioli in Meat Sauce......7.5 oz. 1095
Beef RavioliOs in Meat
  Sauce..................................7.5 oz. 1030
Beefy O's.................................7.5 oz. 1250
Cheese RavioliOs in Tomato
  Sauce..................................7.5 oz. 1160
Elbow Macaroni & Cheese ......7.4 oz. 960
Macaroni & Cheese ....7.4 oz. 960
PizzOs ....7.5 oz. 1060
Spaghetti in Meat Sauce ....7.5 oz. 1110
Spaghetti in Tomato Sauce
  w/Cheese...............................7.4 oz. 940
Spaghetti w/Meatballs in
  Tomato Sauce ....7.4 oz. 950
SpaghettiOs in Tomato &
  Cheese Sauce........................7.4 oz. 895
SpaghettiOs w/Meatballs in
  Tomato Sauce ....7.4 oz. 1035

PORTION   SODIUM (MG.)

	PORTION	SODIUM (MG.)
SpaghettiOs w/Sliced Franks in Tomato Sauce	7.4 oz.	1070
UFO's	7.5 oz.	780
UFO's with Meteors	7.5 oz.	790

Health Valley

	PORTION	SODIUM (MG.)
Chili Con Carne	4 oz.	330
Lentil Chili	½ cup	50
Mild Vegetarian Chili w/Beans	4 oz.	680
Mild Vegetarian Chili w/Beans, no salt	½ cup	20
Spicy Vegetarian Chili w/Beans	4 oz.	680
Spicy Vegetarian Chili w/Beans, no salt	4 oz.	25

Heinz

	PORTION	SODIUM (MG.)
Beans 'n' Franks	7¾ oz.	905
Beef Goulash	7½ oz.	920
Beef Stew	7½ oz.	1245
Chicken Stew w/Dumplings	7½ oz.	850
Chili Con Carne	7¾ oz.	1000
Chili Mac	7½ oz.	860
Hot Chili w/Beans	7¾ oz.	1140
Mac 'n' Beef in Tomato Sauce	7¼ oz.	850
Macaroni and Cheese	7½ oz.	1105
Noodles and Beef in Sauce	7½ oz.	825
Noodles and Chicken	7½ oz.	930
Noodles and Tuna	7½ oz.	950
Spaghetti in Tomato Sauce w/Cheese	7¾ oz.	1105
Spaghetti in Tomato Sauce w/Meat	7½ oz.	965
Spanish Rice	7¼ oz.	1045

La Choy

	PORTION	SODIUM (MG.)
Beef Chow Mein (14 oz.)	¾ cup	1000
Beef Chow Mein (Bi-Pack)	¾ cup prepared	750

PORTION   SODIUM (MG.)

Beef Pepper Oriental Chow Mein (14 oz.)	¾ cup	1345
Beef Pepper Oriental Chow Mein (Bi-Pack)	¾ cup prepared	1005
Chicken Chow Mein (14 ounce)	¾ cup	990
Chicken Chow Mein (Bi-Pack)	¾ cup prepared	635
Meatless Chow Mein (14 ounce)	¾ cup	720
Pork Chow Mein (Bi-Pack)	¾ cup prepared	1120
Shrimp Chow Mein (14 ounce)	¾ cup	855
Shrimp Chow Mein (Bi-Pack)	¾ cup prepared	820
Sukiyaki	¾ cup prepared	910
Sweet and Sour Oriental with Chicken	¾ cup	1335
Sweet and Sour Oriental with Pork	¾ cup	1455
Vegetable Chow Mein (Bi-Pack)	¾ cup prepared	790

Libby's

Chili w/Beans	7.5 oz.	810
Corned Beef Hash	7.5 oz.	1260
Sloppy Joe BBQ Sauce w/Beef	⅓ cup	190

Lido Club

Beef Ravioli	7.5 oz.	535
Spaghetti Rings & Little Meatballs	7.5 oz.	1105

Luck's

Chicken & Dumplings	7.5 oz.	626
Chili Con Carne w/Beans	7.5 oz.	985

Old El Paso

Tamales	3.5 oz.	326

PORTION   SODIUM (MG.)

Pasta Suprema

Curly Macaroni w/Mushroom Sauce	2.9 oz.	565
Fettuccine Alfredo	2.9 oz.	565
Rotini w/Herb 'N Butter Sauce	3.0 oz.	980
Shells w/Garlic Sauce	2.8 oz.	1005

Swanson

Beef Stew	7.6 oz.	890
Chicken a la King	5.3 oz.	695
Chicken & Dumplings	7.5 oz.	965
Chicken Stew	7.6 oz.	960

Van Camp's

Chili w/Beans	8 oz.	1155
Chili w/o Beans	8 oz.	1430
Chili Weenee	8 oz.	975
Skettee Weenee	8 oz.	1060

DRY

Appian Way

Pizza, Regular	3.1 oz.	510
Pizza, Thick Crust	5.3 oz.	980

Chef Boy-ar-dee Dinners and Pizza

Complete Cheese Pizza	¼ pkg. prepared	735
Complete Cheese Pizza in a Skillet	¼ pkg. prepared	610
Complete Hamburger Pizza	¼ pkg. prepared	775
Complete Pepperoni Pizza	¼ pkg. prepared	870
Complete Pepperoni Pizza in a Skillet	¼ pkg. prepared	620
Complete Sausage Pizza	¼ pkg. prepared	930
Lasagna Dinner	¼ pkg. prepared	1140
Pizza Mix	¼ pkg. prepared	575
Spaghetti & Meatball Dinner	¼ pkg. prepared	900

PORTION   SODIUM (MG.)

Spaghetti Dinner w/Condensed Meat Sauce	¼ pkg. prepared	595
Spaghetti Dinner w/Meat Sauce	¼ pkg. prepared	1155
Spaghetti Dinner w/Mushroom Sauce	¼ pkg. Prepared	1085
2 Complete Cheese Pizzas	⅛ pkg. prepared	650
2 Complete Pepperoni Pizzas	⅛ pkg. prepared	595

Fearn Soya

Bean Barley Stew	1.8 oz. dry	555
Blackbean Creole	1.9 oz. dry	761
Brazil Nut Burger	¼ cup dry	224
Breakfast Patty Mix	¼ cup dry	305
Sunflower Burger Mix	¼ cup dry	3
Tri-Bean Casserole	1.8 oz. dry	572

Hamburger Helper (plus ⅕ lb. hamburger)

Beef Noodle	⅕ pkg. prepared	970
Beef Romanoff	⅕ pkg. prepared	1095
Cheeseburger Macaroni	⅕ pkg. prepared	1025
Chili Tomato	⅕ pkg. prepared	1310
Hamburger Hash	⅕ pkg. prepared	920
Hamburger Pizza Dish	⅕ pkg. prepared	960
Hamburger Stew	⅕ pkg. prepared	945
Lasagne	⅕ pkg. prepared	925
Potato Stroganoff	⅕ pkg. prepared	965
Potatoes Au Gratin	⅕ pkg. prepared	890
Rice Oriental	⅕ pkg. prepared	1085
Spaghetti	⅕ pkg. prepared	1045
Tamale Pie	⅕ pkg. prepared	910

Kraft

American Style Spaghetti Dinner	1 cup prepared	575
Egg Noodle & Cheese Dinner	⅔ cup prepared	555
Egg Noodle w/Chicken Dinner	¾ cup prepared	960

PORTION   SODIUM (MG.)

Macaroni & Cheese Deluxe
  Dinner.....................................¾ cup prepared...................650
Macaroni & Cheese Dinner......¾ cup prepared...................655
Macaroni & Cheese Dinner—
  Family Size ....¾ cup prepared...................605
Spaghetti w/Meat Sauce
  Dinner.....................................¾ cup prepared...................765
Spiral Macaroni & Cheese
  Dinner.....................................¾ cup prepared...................770
Tangy Italian Style Spaghetti
  Dinner.....................................1 cup prepared 780

La Sauce

Chicken BBQ ....4.5 oz. 1175
Chicken Chinese Style ....4 oz. 1150
Chicken Italian Style ....3.9 oz. 625
Chicken Mild Mexican..............3.9 oz. 675
Chicken Sweet 'n Sour.............4.3 oz. 325

Mug-o-Lunch

Macaroni & Cheese Mix............1 pouch 695
Noodles & Beef-Flavored
  Sauce Mix ....1 pouch 825

Prince

Macaroni & Cheese Dinner......¾ cup prepared...................597

Ragu

Pizza Quick Mix for
  Homemade Pizza Crust ....¼ pizza 348

Swift

Homemade Stew Starter, Beef
  Stew w/Meat............................12 oz. prepared.................1385
Homemade Stew Starter, Beef
  Stew w/o Meat ....12 oz. prepared.................1280

Tuna Helper

Country Dumpling Noodles 'N
  Tuna ....⅕ pkg. prepared...............1020

PORTION   SODIUM (MG.)

Creamy Noodles 'N Tuna..........⅕ package prepared 880
Noodles, Cheese Sauce 'N
  Tuna ....⅕ pkg. prepared 745
Tuna Tetrazzini ....⅕ pkg. prepared 840

FROZEN

Armour Classic Lites

Beef Pepper Steak	10 oz.	1160
Chicken Burgundy	11.3 oz.	1060
Chicken Oriental	10 oz.	735
Filet of Cod Divan	13.3 oz.	1060
Seafood Natural Herbs	12 oz.	1410
Sliced Beef w/Broccoli	10.3 oz.	2120
Turf & Surf	10 oz.	640
Turkey Parmesan	11 oz.	980
Veal Pepper Steak	11 oz.	570

Armour Dinner Classics

Beef Burgundy	10.5 oz.	915
Beef Stroganoff	11.3 oz.	1200
Boneless Beef Short Ribs	10.5 oz.	1045
Chicken Fricassee	11.8 oz.	1190
Cod Almondine	12 oz.	1070
Lasagna	10 oz.	1045
Salisbury Steak	11 oz.	1580
Seafood Newburg	10.5 oz.	1385
Sirloin Tips	11 oz.	1110
Stuffed Green Peppers	12 oz.	1475
Swedish Meatballs	11.5 oz.	1645
Sweet and Sour Chicken	11 oz.	1440
Sweet and Sour Pork	11.5 oz.	965
Teriyaki Chicken	10.5 oz.	1145
Teriyaki Steak	10 oz.	1200
Veal Parmigiana	10.8 oz.	1515

Banquet

Beef	16 oz.	1731
Cheese Enchilada	21.3 oz.	4778

PORTION   SODIUM (MG.)

	PORTION	SODIUM (MG.)
Chicken & Dumplings	19 oz.	1923
Chopped Beef	18 oz.	1792
Fish & Chips	14 oz.	972
Fried Chicken	17 oz.	2780
Veal Parmagian	20 oz.	2123

Banquet Buffet Supper Main Dishes

Beef Enchilada	8 oz.	1477
Beef Stew	8 oz.	977
Cheese Manicotti & Meat	5.3 oz.	891
Gravy & Salisbury Steak	5.3 oz.	744
Gravy & Sliced Beef	8 oz.	890
Gravy & Sliced Turkey	5.3 oz.	576
Lasagna with Meat	8 oz.	1190
Macaroni & Beef	8 oz.	1336
Macaroni & Cheese	8 oz.	871
Meat Loaf	6 oz.	1032
Mostaccioli & Sauce	8 oz.	1421
Veal Parmagian	6.4 oz.	842

Banquet Cookin' Bag Entrees

BBQ Sauce w/Sliced Beef	4 oz.	929
Beef Enchilada	6 oz.	980
Chicken a la King	5 oz.	645
Creamed Chipped Beef	4 oz.	1082
Gravy & Sliced Turkey	5 oz.	471
Gravy & Sliced Beef	4 oz.	492
Macaroni & Cheese	5 oz.	583
Meat Loaf	5 oz.	894
Salisbury Steak	5 oz.	729
Veal Parmigian	5 oz.	853

Banquet Dinners

Beans & Franks	10.3 oz.	1377
Beef w/Gravy	10 oz.	1009
Beef Enchilada	12 oz.	1805
Cheese Enchilada	12 oz.	2166
Chopped Beef	11 oz.	1199
Fish	8.8 oz.	927

PORTION   SODIUM (MG.)

Fried Chicken	11 oz.	1831
Ham	10 oz.	1148
Italian Style	12 oz.	1783
Meat Loaf	11 oz.	1525
Mexican Style	12 oz.	1995
Mexican Style Combination	12 oz.	1978
Salisbury Steak	11 oz.	1333
Turkey	11 oz.	1416
Veal Parmigian	11 oz.	1310
Western	11 oz.	1548

Banquet Casseroles

Macaroni & Cheese	8 oz.	930
Spaghetti w/Meat Sauce	8 oz.	1242

Banquet Extra Helping Dinners

Chicken & Dressing	19 oz.	1817
Meat Loaf	19 oz.	2396
Salisbury Steak	19 oz.	2175
Turkey	19 oz.	2165

Banquet Meat Pies

Beef	8 oz.	964
Chicken	8 oz.	1027
Tuna	8 oz.	1305
Turkey	8 oz.	1111

Banquet Supreme Meat Pies

Beef	8 oz.	1245
Chicken	8 oz.	1325
Turkey	8 oz.	1370

Celeste Pizza

Cheese Pizza (7 oz.)	7 oz. (1 pie)	1310
Cheese Pizza (19 oz.)	4.8 oz. (¼ pie)	803
Chicago Style Combination	6 oz. (¼ pie)	1156
Chicago Style Pepperoni Deluxe	6 oz. (¼ pie)	1389

PORTION   SODIUM (MG.)

Chicago Style Sausage
  Deluxe ....6 oz. (¼ pie) 1153
Deluxe Pizza (9 oz.) ....9 oz. (1 pie) 1589
Deluxe Pizza (23.5 oz.) ....5.9 oz (¼ pie) 1049
Deluxe Pizza Sicilian Style ......6.5 oz. (¼ pie) 1190
Pepperoni Pizza (7.25 oz.) ......7.3 oz. (1 pie) 1776
Pepperoni Pizza (20 oz.) ....5 oz. (¼ pie) 1082
Sausage & Mushroom Pizza
  (9 oz.)......................................9 oz. (1 pie) 1576
Sausage & Mushroom Pizza
  (24 oz.) ....6 oz. (¼ pie) 1195
Sausage Pizza (8 oz.) ....8 oz. (1 pie) 1528
Sausage Pizza (22 oz.) ....5.5 oz. (¼ pie) 1227

Chef Boy-ar-dee Pizza & Lasagna

Cheese Pizza ....6.5 oz. (½ pie)......................925
Deluxe Pizza...............................7.8 oz. (½ pie) 1135
English Muffin Pizza,
  Cheese....................................6.6 oz. (2 pizzas) 1220
English Muffin Pizza,
  Pepperoni (7.3 oz.) ....7.3 oz. (1 pizza)..................1845
English Muffin Pizza,
  Pepperoni (14.5 oz.) ....7.3 oz. (2 pizzas) 1644
English Muffin Pizza,
  Sausage (7.8 oz.) ....7.3 oz. (1 pizza) 1815
English Muffin Pizza,
  Sausage (15.5 oz.) ....7.8 oz. (2 pizzas) 1549
Four Little Pizzas, Cheese.......5.3 oz. (2 pizzas) 795
Four Little Pizzas, Pepperoni 5.5 oz. (2 pizzas) 955
Four Little Pizzas, Sausage......5.3 oz. (2 pizzas) 840
Hamburger Pizza.......................7 oz. (½ pie) 1055
Lasagna.....................................6 oz 790
Pepperoni Pizza.........................6.5 oz. (½ pie) 1110
Premium Combo Italian
  Sausage & Peppers ....7.4 oz. (½ pie) 1275
Premium Combo Mushroom &
  Pepper ....7 oz. (½ pie) 1101
Premium Combo Peppers &
  Sausage..................................7.4 oz. (½ pie) 1438
Sausage Pizza ....7 oz. (½ pie) 1170

PORTION   SODIUM (MG.)

El Charrito Dinners

Beef Burrito Rolls	12 oz. (6 rolls)	1340
Beef Enchiladas	16 oz. (4 enchiladas)	1940
Beef Tacos	12 oz. (6 tacos)	1390
Beef/Bean Burritos	6.5 oz. (2 burritos)	910
Beef-Cheese Enchiladas	16 oz. (1 dinner)	2300
Cheese Enchilada Dinner	12 oz. (1 dinner)	1485
Chunky Beef Burritos	12 oz. (2 burritos)	1695
Enchilada Dinner	12 oz. (1 dinner)	1490
Enchiladas, Cheese Tacos	16 oz. (2 each)	2220
Mexican Style Dinner	14 oz. (1 dinner)	1940
Queso Dinner	13 oz. (1 dinner)	1840
Queso Rito	16 oz. (1 dinner)	2250
Rancho Rito Dinner	16 oz. (1 dinner)	1745
Saltillo Dinner	13 oz. (1 dinner)	1885

El Charrito Entrees

Green Chili Burrito	5 oz. (1 burrito)	800
Red Chili Burrito	5 oz. (1 burrito)	755

Fox Pizza

Pepperoni Deluxe	½ pizza	900

Green Giant Baked Entrees

Boneless Beef Ribs in BBQ Sauce w/Corn on Cob	1 package	1440
Chicken in BBQ Sauce w/Corn on Cob	1 package	885
Chicken in Herb Butter w/Stuffed Potato	1 package	965
Chicken Lasagna	12 oz.	1215
Enchilada Sonora Style	12 oz.	1245
Lasagna w/Meat Sauce	10.5 oz.	1515
Lasagna w/Meat Sauce, Single Serve	12 oz.	1660
Salisbury Steak w/Gravy	½ container	1095
Spinach Lasagna	12 oz.	1455
Stir-Fry Beef Teriyaki	10 oz.	780
Stir Fry Cashew Chicken	10 oz.	965

PORTION   SODIUM (MG.)

Stir Fry Chicken & Garden
  Vegetables....................10 oz. 705
Stir Fry Sweet and Sour
  Chicken ....10 oz. 585
Stuffed Cabbage Rolls w/Beef
  in Tomato Sauce ....½ container 800
Stuffed Green Peppers...............½ container 785
Swiss Steak in Gravy
  w/Stuffed Potato....................1 package..........................1235

Green Giant Single Pouch Boil 'N Bag Entrees

Beef Stew ....9 oz. 275
Lasagna.................................9.5 oz. 1145
Macaroni and Cheese ....9 oz. 1115
Salisbury Steak w/Creole
  Sauce....................................9 oz. 910
Stir Fry Shrimp Fried Rice........10 oz. 1130

Green Giant Twin Pouch Entrees

Beef Burgundy w/Rice and
  Carrots ....9 oz. 855
Beef Chow Mein w/Rice &
  Vegetables................................10 oz. 1050
Beef Stroganoff w/Noodles ......9 oz......................820
Chicken a la King w/Biscuits....9 oz. 1545
Chicken & Broccoli w/Rice in
  Cheese Sauce ....9.5 oz. 915
Chicken & Noodles
  w/Vegetables............................9 oz......................940
Chicken & Pea Pods in Sauce
  w/Rice & Vegetables...............10 oz. 995
Chicken Chow Mein w/Rice &
  Vegetables................................9 oz. 1075
Salisbury Steak w/Mashed
  Potatoes....................................11 oz. 1515
Steak & Green Peppers in
  Sauce w/Rice &
  Vegetables................................9 oz. 1170
Sweet & Sour Meatballs
  w/Rice & Vegetables...............9.4 oz. 820

PORTION   SODIUM (MG.)

Turkey Breast Slices w/White
  & Wild Rice Stuffing ....9 oz. 1225

Health Valley

Egg Roll.................................1 Roll 840
Lobster Roll...........................1 Roll 785
Nut Roll ....1 Roll 685
Shrimp Roll ....1 Roll 722
Teriyaki Roll ....1 Roll 675

La Choy Dinners

Beef Pepper Oriental ....12 oz. (1 dinner)................1985
Chicken Chow Mein ....12 oz. (1 dinner)................1740
Shrimp Chow Mein ....12 oz. (1 dinner)................1740

La Choy Egg Rolls

Chicken Egg Rolls....................1.3 oz. (3 rolls) 185
Lobster Egg Rolls ....1.3 oz. (3 rolls) 240
Meat & Shrimp Egg Rolls
  (6.5 oz.) ....1.3 oz. (3 rolls) 220
Meat & Shrimp Egg Rolls
  (7.5 oz.) ....1.5 oz. (6 rolls) 250
Shrimp Egg Rolls (6 oz.) ....3 oz. (1 roll) 575
Shrimp Egg Rolls (6.5 oz.) ......1.3 oz. (3 rolls) 210

La Choy Entrees

Beef Pepper Oriental ....6 oz. (⅔ cup).........................845
Chicken Chow Mein ....6 oz. (⅔ cup).........................690
Fried Rice with Pork ....4 oz. (¾ cup).........................885
Shrimp Chow Mein ....6 oz. (⅔ cup).........................800
Sweet and Sour Chicken ....6 oz. (⅔ cup).......................1005
Sweet and Sour Pork ....6 oz. (⅔ cup).......................1100

La Pizzeria Pizza

Cheese Pizza (9 in.)..................5.4 oz. 770
Combination Pizza (9 in.) ..6.8 oz. 1140
Pepperoni Pizza (12 in.)............5.3 oz. 920
Sausage Pizza (9 in.)................6.5 oz. 1020
Thick Crust Cheese (10 in.) ....6.2 oz. 970

PORTION   SODIUM (MG.)

Land O' Lakes

Bacon & Onion Pour-A-Quiche	4⅓ oz.	380
Ham Pour-A-Quiche	4⅓ oz.	360
Spinach & Onion Pour-A-Quiche	4⅓ oz.	365
3-Cheese Pour-A-Quiche	4⅓ oz.	385

Lean Living

Cheese Enchilada	10 oz.	460
Chicken a la King	9 oz.	280
Chicken Crepes Marco Polo	7 oz.	350
Spinach Lasagna	9 oz.	420
Spinach Mushroom Casserole	7½ oz.	410

Le Menu

Beef Sirloin Tips	11½ oz.	825
Breast of Chicken Parmigiana	11½ oz.	895
Chicken A La King	10¼ oz.	1170
Chopped Sirloin Beef	12¼ oz.	1115
Pepper Steak	11½ oz.	1045
Sliced Breast of Turkey with Mushrooms	11¼ oz.	1165
Sweet and Sour Chicken	11½ oz.	960
Yankee Pot Roast	11 oz.	830

Lifeline Healthy Gourmet Foods

Beef & Vegetables with Creole Sauce	7 oz.	224
Beef & Yogurt Sauce	7 oz.	440
Chicken Cacciatore	7 oz.	320
Chicken Fricassee	7 oz.	320
Italian Style Ziti in Tomato Sauce w/Cheese Flavored w/Beef	7 oz.	290
Mushroom Gravy w/Salisbury Steak	7 oz.	328

PORTION   SODIUM (MG.)

Sauce w/Swedish Style Meatballs	.7 oz.	156
Stuffed Cabbage Rolls	.7.5 oz.	320

Morton Boil-in-Bag Entrees

Beef Patty	.5 oz.	660
Chicken a la King	.5 oz.	510
Creamed Chipped Beef	.5 oz.	1370
Salisbury Steak	.5 oz.	690
Sliced Beef	.5 oz.	500
Sliced Chicken	.5 oz.	810

Morton Casseroles

Macaroni & Cheese	.8 oz.	900
Spaghetti & Meat	.8 oz.	760

Morton Country Table Dinners

Fried Chicken	.15 oz.	2180
Salisbury Steak	.15 oz.	1810
Sliced Beef	.14 oz.	1220
Sliced Turkey	.15 oz.	1722

Morton Dinners

Beans & Franks	.10.8 oz.	800
Beef	.10 oz.	840
Beef Enchilada	.11 oz.	1216
Boneless Chicken	.10 oz.	1350
Chicken Croquette	.10.3 oz.	930
Chicken 'n Dumpling	.11 oz.	1265
Chicken 'n Noodles	.10.3 oz.	685
Fish	.9 oz.	690
Fried Chicken	.11 oz.	1865
Fried Chicken (32 oz.)	.6.4 oz.	1585
Fried Chicken (22 oz.) Breast Portion	.5.5 oz.	1362
Ham	.10 oz.	375
Macaroni & Beef	.10 oz.	875
Macaroni & Cheese	.11 oz.	905
Meat Loaf	.11 oz.	1380

	PORTION	SODIUM (MG.)
Mexican Style	11 oz.	1242
Salisbury Steak	11 oz.	1465
Spaghetti & Meat Ball	11 oz.	905
Turkey	11 oz.	1260
Veal Parmigiana	11 oz.	1450
Western Style	11.8 oz.	1650

Morton Entrees

Fried Chicken	12 oz.	2545
Salisbury Steak	10.3 oz.	1372
Sliced Turkey	12.3	2032

Morton Family Meals

Beef Stew	8 oz.	495
Gravy/Salisbury Steaks	8 oz.	1220
Gravy/Sliced Beef	8 oz.	750
Gravy/Sliced Turkey	8 oz.	1120
Gravy/Turkey Croquettes	8 oz.	1360
Macaroni and Cheese	8 oz.	860
Mushroom Gravy/Beef Patties	8 oz.	1170
Onion Gravy/Beef Patties	8 oz.	1170
Tomato Sauce/Meat Loaf	8 oz.	1100
Veal Parmigiana	8 oz.	1180

Morton King-Size Dinners

Boneless Chicken	17 oz.	2112
Fried Chicken	17 oz.	2902
Salisbury Steak	19 oz.	2717
Turkey	19 oz.	2567
Veal Parmigiana	20 oz.	2430

Morton Pot Pies

Beef	8 oz.	1040
Chicken	8 oz.	1246
Tuna	8 oz.	1120
Turkey	8 oz.	1115

PORTION   SODIUM (MG.)

Morton Steak House Dinners

Beef Tenderloin	9.5 oz.	390
Chopped Sirloin	9.5 oz.	390
Rib Eye	9 oz.	390
Sirloin Strip	9.5 oz.	390
Sliced Turkey	5 oz.	640
Sloppy Joe	5 oz.	730
Veal Parmigiana	5 oz.	580

Mrs. Paul's

Eggplant Parmigiana	11 oz.	1810

Pepperidge Farm

Beef with Barbecue Sauce	8¾ oz. (1 deli)	640
Mexican Style	8¾ oz. (1 deli)	590
Pizza Style	8¾ oz. (1 deli)	780
Scrambled Egg, Canadian Bacon & Cheese in Pastry	7¾ oz. (1 deli)	670
Sliced Beef w/Brown Sauce, in Pastry	8¾ oz. (1 deli)	570
Turkey, Ham and Cheese in Pastry	7¾ oz. (1 deli)	765
Western Style Omelet	8¾ oz. (1 deli)	590

Royal Dragon Chinese Dimsum

Cocktail Spring Roll	3.5 oz.	670
Hargow	3.5 oz.	510
Potsticker	3.5 oz.	590
Shaomai	3.5 oz.	750
Spring Roll	3.5 oz.	670
Wonton	3.5 oz.	740

Stouffer's Entrees

Asparagus Souffle	4 oz.	440
Beef & Spinach Stuffed Pasta Shells	9 oz.	1315
Beef Chop Suey w/Rice	12 oz.	2040

	PORTION	SODIUM (MG.)
Beef Pie	10 oz.	1600
Beef Stew	10 oz.	1675
Beef Stroganoff w/Parsley Noodles	9.8 oz.	1300
Beef Teriyaki w/Rice & Vegetables	10 oz.	1450
Cheese Souffle	6 oz.	1360
Cheese Stuffed Pasta Shells w/Meat Sauce	9 oz.	1310
Chicken a la King w/Rice	9.5 oz.	900
Chicken Cacciatore w/Spaghetti	11.3 oz.	1135
Chicken Chow Mein w/o Noodles	8 oz.	1115
Chicken Crepes w/Mushroom Sauce	8.3 oz.	1040
Chicken Divan	8.5 oz.	830
Chicken Paprikash w/Egg Noodles	10.5 oz.	1325
Chicken Pie	10 oz.	1540
Chicken Stuffed Pasta Shells w/Cheese Sauce	9 oz.	1060
Chili Con Carne w/Beans	8.8 oz.	1265
Corn Souffle	4 oz.	510
Creamed Chicken	6.5 oz.	680
Creamed Chipped Beef	5.5 oz.	900
Escalloped Chicken & Noodles	5.8 oz.	720
Fettucini Alfredo	5 oz.	1194
Green Bean Mushroom Casserole	4¾ oz.	675
Green Pepper Steak w/Rice	10.5 oz.	1500
Ham & Asparagus Crepes	6.3 oz.	840
Ham & Swiss Cheese Crepes	7.5 oz.	905
Lasagna	10.5 oz.	1200
Linguini w/Clam Sauce	10.5 oz.	1010
Lobster Newburg	6.5 oz.	700
Macaroni & Beef w/Tomatoes	5.8 oz.	810
Macaroni & Cheese	6 oz.	780

PORTION   SODIUM (MG.)

Ratatouille	5 oz.	1320
Roast Beef Hash	5.8 oz.	760
Salisbury Steak w/Onion Gravy	6 oz.	1150
Scallops and Shrimp Mariner w/Rice	10.3 oz.	1120
Short Ribs Beef, Boneless, w/ Vegetable Gravy	5.8 oz.	560
Shrimp Newburg	6.5 oz.	555
Spaghetti w/Meat Sauce	14 oz.	1970
Spinach Crepes w/Cheddar Cheese Sauce	9.5 oz.	995
Spinach Souffle	4 oz.	600
Stuffed Green Pepper w/Beef, Tomato Sauce	7.8 oz.	960
Swedish Meatballs w/Parsley Noodles	11 oz.	1620
Tuna Noodle Casserole	5.8 oz.	670
Turkey Casserole w/Gravy & Dressing	9.8 oz.	1124
Turkey Pie	10 oz.	1735
Turkey Tetrazzini	6 oz.	620
Welsh Rarebit	5 oz.	660

Stouffer's Lean Cuisine

Beef and Pork Cannelloni w/Mornay Sauce	9.6 oz.	940
Cheese Cannelloni w/Tomato Sauce	9.1 oz.	885
Chicken Chow Mein w/Rice	11.3 oz.	1155
Chicken and Vegetables w/Vermicelli	12.8 oz.	1220
Filet of Fish Divan	12.4 oz.	785
Filet of Fish Florentine	9 oz.	815
Glazed Chicken w/Vegetable Rice	8.5 oz.	830
Meatball Stew	10 oz.	1260
Oriental Beef in Sauce w/Vegetables and Rice	8.6 oz.	1150

PORTION   SODIUM (MG.)

	PORTION	SODIUM (MG.)
Oriental Scallops and Vegetables w/Rice	11 oz.	1325
Salisbury Steak w/Italian Style Sauce and Vegetables	9.5 oz.	820
Spaghetti w/Beef and Mushroom Sauce	11.5 oz.	1455
Zucchini Lasagna	11 oz.	1000

Stouffer's Pizza

	PORTION	SODIUM (MG.)
Cheese	5.1 oz. (½ pie)	850
Deluxe	6.1 oz. (½ pie)	1150
Hamburger	6.1 oz. (½ pie)	1100
Mushroom	6 oz. (½ pie)	1755
Pepperoni	5.6 oz. (½ pie)	1190
Sausage	6 oz. (½ pie)	1320
Sausage & Mushroom	6.3 oz. (½ pie)	1220

Swanson Chunky Pies

	PORTION	SODIUM (MG.)
Beef	10 oz.	935
Chicken	10 oz.	850
Turkey	10 oz.	975

Swanson Hungry Man Dinners

	PORTION	SODIUM (MG.)
Boneless Chicken	17.5 oz.	1525
Chicken Parmigiana	20 oz.	2205
Chopped Beef Steak	17.3 oz.	2030
Fish 'n Chips	14.8 oz.	1280
Fried Chicken, Dark Portions	14 oz.	1640
Fried Chicken, White Portions	14 oz.	2060
Lasagna w/Meat	18.8 oz.	1395
Salisbury Steak	16.5	1565
Sliced Beef	16 oz.	1250
Turkey	18.5 oz.	2110
Veal Parmigiana	20 oz.	2075
Western Style	17.5 oz.	1900

Swanson Hungry Man Entrees

	PORTION	SODIUM (MG.)
Beef Enchilada	16 oz.	2010

PORTION   SODIUM (MG.)

Fried Chicken Breast		
Portions	11.8 oz.	1710
Fried Chicken Dark		
Portions	11 oz.	1340
Lasagna w/Meat	12.8 oz.	1250
Salisbury Steak	11.8 oz.	1330
Sliced Beef	12.3 oz.	1045
Turkey	13.3 oz.	1645

Swanson Hungry Man Meat Pies

Beef	16 oz.	1565
Chicken	16 oz.	1680
Steak Burger	16 oz.	1460
Turkey	16 oz.	1730

Swanson Main Course Entrees

Lasagna w/Meat in Tomato		
Sauce	13.3 oz.	1160
Macaroni & Cheese	12 oz.	1815
Salisbury Steak w/Gravy	10 oz.	1435
Steak & Green Pepper in		
Oriental Style Sauce	8.5 oz.	1105

Swanson Pies

Beef Pie	8 oz.	810
Chicken Pie	8 oz.	850
Macaroni & Cheese	7 oz.	880
Turkey Pie	8 oz.	890

Swanson Dinners

BBQ Flavored Fried Chicken	9.3 oz.	845
Bean and Beef Burrito	15.3 oz.	1635
Beans & Franks	12.5 oz.	1100
Beef	11.5 oz.	1085
Beef Enchiladas	15 oz.	1415
Chopped Sirloin Beef	11.5 oz.	915
Dark Meat Fried Chicken	10.3 oz.	1345
Fish 'n Chips	10.5 oz.	970

PORTION   SODIUM (MG.)

	PORTION	SODIUM (MG.)
Lasagna	13 oz.	780
Loin of Pork	11.3 oz.	635
Macaroni & Beef	12 oz.	925
Macaroni & Cheese	12.3 oz.	970
Meat Loaf	11 oz.	1010
Mexican Style Combination	16 oz.	1865
Noodles and Chicken	10.5 oz.	805
Polynesian Style	12 oz.	1355
Salisbury Steak	11 oz.	1055
Spaghetti and Meatballs	12.5 oz.	1065
Swiss Steak	10 oz.	830
Turkey	11.5 oz.	1295
Veal Parmigiana	12.8 oz.	1280
Western Style	12.3 oz.	1040
White Portions Fried Chicken	10.8 oz.	1425

Swanson Entrees and Breakfasts

	PORTION	SODIUM (MG.)
Beef Enchilada	11.3 oz.	1220
Chicken Nibbles	5 oz.	865
Fish 'n Chips	5.5 oz.	585
French Toast w/Sausages	6.5 oz.	820
Fried Chicken	7.3 oz.	1075
Gravy & Sliced Beef w/Potatoes	8 oz.	805
Meat Loaf w/Tomato Sauce	9 oz.	1020
Meatballs w/Brown Gravy	8.5 oz.	885
Omelets w/Cheese Sauce & Ham	7 oz.	1180
Salisbury Steak	5.5 oz.	635
Scrambled Eggs & Sausage w/Hash Potatoes	6.3 oz.	790
Spaghetti in Tomato Sauce w/Breaded Veal	8.3 oz.	915
Spanish Style Omelet	8.0 oz.	895
Turkey	8.8 oz.	1070

Swift International Entrees

	PORTION	SODIUM (MG.)
Chicken Cordon Bleu	6 oz.	925
Chicken Kiev	6 oz.	895

PORTION　SODIUM (MG.)

Chicken Lucerne	6 oz.	1005
Chicken Parmigiana	6 oz.	620
Chicken Romanoff	6 oz.	705
Chicken Royale	6 oz.	1040

Totino's Pizza

Canadian Bacon Party	½ pizza	875
Canadian Style Bacon Classic	⅓ pizza	1405
Cheese Party	½ pizza	635
Deluxe Cheese Classic	⅓ pizza	960
Deluxe Combination Classic	⅓ pizza	1435
Deluxe Sausage Classic	⅓ pizza	1200
Deluxe Pepperoni Classic	⅓ pizza	1600
Sausage & Pepperoni Party	½ pizza	1120
Hamburger Party	½ pizza	950
Mexican Style Party	½ pizza	880
Nacho Party	½ pizza	665
Pepperoni Party	½ pizza	1030
Sausage Party	½ pizza	1130

Weight Watchers

Beefsteak	9.8 oz.	958
Cannelloni Florentine	13 oz.	893
Cheese Pizza Pie	6 oz.	911
Chicken A La King	10 oz.	1275
Chicken Livers and Onions	9.3 oz.	495
Chicken New Orleans Style	11 oz.	1129
Chicken Oriental	12 oz.	1595
Chicken Parmigiana	7.8 oz.	1025
Chile Con Carne w/Beans	10 oz.	822
Deluxe Combination Pizza	7.3 oz.	691
Eggplant Parmigiana	7.8 oz.	1072
Fillet of Fish in Lemon Sauce	13.3 oz.	971
Fillet of Fish in Newburg Sauce	9.3 oz.	951
Flounder in Newburg Sauce	12.5 oz.	779
Flounder with Lemon Flavored Breadcrumbs	6.5 oz.	814
Haddock with Stuffing	7 oz.	547

	PORTION	SODIUM (MG.)
Lasagna with Veal	12.8 oz.	1602
Ocean Perch with Lemon Flavored Breadcrumbs	6.5 oz.	479
Sirloin of Beef in Mushroom Sauce	13 oz.	1608
Sliced Chicken in Celery Sauce	8.5 oz.	1000
Sliced Chicken with Gravy & Stuffing	14.8 oz.	1910
Sliced Turkey with Gravy & Stuffing	15.3 oz.	1962
Sole with Lemon Sauce	9.3 oz.	1179
Southern Fried Chicken Patty	6.8 oz.	835
Sweet 'N Sour Chicken	9.5 oz.	563
Turkey Tetrazzini	13 oz.	1443
Veal Patty Parmigiana	9 oz.	1081
Veal Sausage Pizza	6.8 oz.	872
Veal Stuffed Pepper	11.8 oz.	1002
Vegetable Supreme Pizza	7.3 oz.	727
Ziti Macaroni	12.5 oz.	1257

Fast Food Restaurants

Arby's

	PORTION	SODIUM (MG.)
Apple Turnover	3 oz.	240
Arby's Sauce	1 oz. (1 serving)	325
Bac'n Cheddar Deluxe	7.9 oz. (1 serving)	1375
Bacon & Egg Croissant	4.5 oz.	550
Beef & Cheese	6 oz. (1 serving)	1745
Blueberry Turnover	3 oz.	255
Broccoli & Cheese Potato	12 oz.	480
Butter Croissant	2 oz.	225
Cherry Turnover	3 oz.	254
Chicken Breast Sandwich	7.3 oz. (1 serving)	1323
Chicken Club Sandwich	15 oz.	1300
Chicken Salad Croissant	15 oz.	725
Deluxe Potato	11 oz.	480
Chocolate Shake	14 oz. (1 serving)	290

PORTION   SODIUM (MG.)

French Dip	5.5 oz. (1 serving)	1111
French Fries	2.5 oz. (1 serving)	39
Ham 'N Cheese	8.3 oz. (1 serving)	1745
Ham & Swiss Croissant	4 oz.	995
Horsey Sauce	1 oz. (1 serving)	350
Jamocha Shake	14 oz. (1 serving)	265
Junior Beef	3 oz. (1 serving)	530
Mushroom & Cheese Potato	10.6 oz.	640
Mushroom & Swiss Croissant	4 oz.	630
Potato Cakes (2)	3.5 oz. (1 serving)	476
Roast Beef	5 oz. (1 serving)	880
Roast Beef Deluxe	8.25 oz. (1 serving)	1288
Roast Beef, Super	9.75 oz. (1 serving)	1420
Roasted Chicken Breast	5 oz.	930
Sausage & Egg Croissant	5.75 oz.	745
Sub (no dressing)	9.5 oz.	1766
Taco Potato	15 oz.	1060
Vanilla Shake	14 oz. (1 serving)	275

Burger Chef

Big Shef	1 serving	840
Biscuit Sandwich w/Sausage	1 serving	1313
Cheeseburger	1 serving	641
Double Cheeseburger	1 serving	835
French Fries, large	1 serving	46
French Fries, regular	1 serving	33
Fun Meal	1 serving	513
Hamburger	1 serving	480
Hash Rounds	1 serving	349
Mushroom Burger	1 serving	744
Scrambled Eggs & Bacon Platter	1 serving	1108
Scrambled Eggs & Sausage Platter	1 serving	1411
Shake, chocolate	1 serving	378
Shake, vanilla	1 serving	325
Sunrise w/Bacon	1 serving	978
Sunrise w/Sausage	1 serving	1412
Super Shef	1 serving	1088
Top Shef	1 serving	1007

PORTION   SODIUM (MG.)

Burger King

	PORTION	SODIUM (MG.)
Apple Pie	3 oz. (1 serving)	385
Bacon Double Cheeseburger	7.1 oz. (1 serving)	985
Cheeseburger	4.4 oz. (1 serving)	705
Double Beef Whopper	11.9 oz. (1 serving)	1015
Double Beef Whopper w/Cheese	12.9 oz. (1 serving)	1295
Double Cheeseburger	6.3 oz. (1 serving)	865
Double Hamburger	5.8 oz. (1 serving)	585
French Fries, regular	2.4 oz. (1 serving)	230
Hamburger	3.9 oz. (1 serving)	560
Onion Rings, regular	2.7 oz. (1 serving)	450
Shake, chocolate	10 oz. (1 serving)	280
Shake, vanilla	10 oz. (1 serving)	320
Specialty Chicken Sandwich	7.3 oz. (1 serving)	775
Specialty Ham & Cheese Sandwich	7.8 oz. (1 serving)	1550
Veal Parmagiana	8.5 oz. (1 serving)	805
Whaler Sandwich	7.2 oz. (1 serving)	745
Whaler Sandwich w/Cheese	7.7 oz. (1 serving)	885
Whopper	9.2 oz. (1 serving)	975
Whopper Junior	5.1 oz. (1 serving)	545
Whopper Junior w/Cheese	5.6 oz. (1 serving)	685
Whopper w/Cheese	10.2 oz. (1 serving)	1260

Carl's Jr.

	PORTION	SODIUM (MG.)
American Cheese	0.7 oz. (1 serving)	120
Apple Turnover	4 oz. (1 serving)	470
Bacon	0.4 oz. (2 strips)	220
Bacon & Cheese Omelette	4.6 oz. (1 serving)	660
Blue Cheese Dressing	1 oz. (2 tbl.)	115
California Omelette	5.4 oz. (1 serving)	550
California Roast Beef Sandwich	6.4 oz. (1 serving)	505
Carbonated Beverages, regular	16 oz. (1 serving)	45
Carrot Cake	3 oz. (1 serving)	375
Charbroiler Steak	2.9 oz. (1 serving)	215
Charbroiler Steak Sandwich	8.2 oz. (1 serving)	700

	PORTION	SODIUM (MG.)
Cheese Omelette	4.4 oz. (1 serving)	440
Chili Cheese Dog	5.0 oz. (1 serving)	740
Chili Dog	4.5 oz. (1 serving)	640
Crispirito	7.6 oz. (1 serving)	1050
English Muffin w/Butter & Jelly	2.2 oz. (1 serving)	245
Famous Star Hamburger	8 oz. (1 serving)	705
Filet of Fish Sandwich	8.6 oz. (1 serving)	790
French Fries, regular	3 oz. (1 serving)	460
Happy Star Hamburger	4.5 oz. (1 serving)	670
Hashed Brown Potatoes	3.5 oz. (1 serving)	260
Hearty Chicken Sandwich	9.2 oz. (1 serving)	1070
Hot Cakes w/Syrup & Butter	6.3 oz. (1 serving)	530
Hot Chocolate	6 oz. (1 cup)	120
Iced Tea, regular	16 oz. (1 serving)	0
Low-Cal Italian Dressing	1 oz. (2 tbl.)	180
Old Time Star Hamburger	6.8 oz. (1 serving)	625
Onion Rings	3.1 oz. (1 serving)	75
Orange Juice	6 oz. (1 cup)	0
Original Hot Dog	4.9 oz. (1 serving)	880
Salad, regular	10 oz. (1 serving)	695
Sausage	1 oz. (1 patty)	235
Scrambled Eggs	2.9 oz. (1 serving)	110
Shakes, regular	1 serving	350
Sunrise Sandwich w/Bacon	1 oz. (1 serving)	780
Sunrise Sandwich w/Sausage	1.6 oz. (1 serving)	790
Super Star Hamburger	11 oz. (1 serving)	785
Sweet Roll w/Butter	3.9 oz. (1 serving)	450
Swiss Cheese	0.7 oz.	125
Thousand Island Dressing	1 oz. (2 tbl.)	160
Western Bacon Cheeseburger	7.8 oz. (1 serving)	1330

Church's

Dark Chicken Portion	3.5 oz.	475
White Chicken Portion	3.5 oz.	498

Dairy Queen

Banana Split	13.5 oz.	150
Buster Bar	5.2 oz.	175

PORTION   SODIUM (MG.)

	PORTION	SODIUM (MG.)
Chocolate Dipped Cone, large	8.2 oz.	145
Chocolate Dipped Cone, regular	5.5 oz.	100
Chocolate Dipped Cone, small	3.2 oz.	55
Chocolate Malt, large	20.7 oz.	360
Chocolate Malt, regular	14.7 oz.	260
Chocolate Malt, small	10.2 oz.	180
Chocolate Shake, large	20.7 oz.	360
Chocolate Shake, regular	14.7 oz.	260
Chocolate Shake, small	10.2 oz.	180
Chocolate Sundae, large	8.7 oz.	165
Chocolate Sundae, regular	6.2 oz.	120
Chocolate Sundae, small	3.7 oz.	75
Cone, large	7.5 oz.	115
Cone, regular	5 oz.	80
Cone, small	3 oz.	45
Dairy Queen Sandwich	2 oz.	40
Dilly Bar	3 oz.	50
Double Delight	9 oz.	150
Float	14 oz.	85
Freeze	14 oz.	180
Hot Fudge Brownie Delight	9.4 oz.	225
Mr. Misty, large	15.5 oz.	under 10
Mr. Misty Float	14.5 oz.	95
Mr. Misty Freeze	14.5 oz.	140
Mr. Misty Kiss	3 oz.	under 10
Parfait	10 oz.	140
Peanut Buster Parfait	11 oz.	250
Strawberry Shortcake	11 oz.	215

Dairy Queen/Brazier

	PORTION	SODIUM (MG.)
Chicken Sandwich	7.7 oz.	870
Double Hamburger	7.4 oz.	660
Double Hamburger w/Cheese	8.4 oz.	980
Fish Sandwich	6 oz.	875
Fish Sandwich w/Cheese	6.2 oz.	1035
French Fries, regular	2.5 oz.	115
French Fries, large	4 oz.	185

	PORTION	SODIUM (MG.)
Hot Dog	3.5 oz.	830
Hot Dog w/Cheese	4 oz.	990
Hot Dog w/Chili	4.5 oz.	985
Onion Rings	3 oz.	140
Single Hamburger	5.2 oz.	630
Single Hamburger w/Cheese	5.7 oz.	790
Super Hot Dog	6.2 oz.	1365
Super Hot Dog w/Cheese	6.9 oz.	1605
Super Hot Dog w/Chili	7.7 oz.	1595
Triple Hamburger	9.6 oz.	690
Triple Hamburger w/Cheese	10.6	1010

Hardee's

Bacon & Egg Biscuit	4 oz. (1 serving)	823
Bacon Cheeseburger	8.6 oz. (1 serving)	1074
Big Cookie	1.9 oz. (1 serving)	258
Big Deluxe	8.7 oz.	1083
Fisherman's Filet	6.9 oz.	1013
Big Roast Beef	5.9 oz.	1770
Biscuit	2.9 oz.	650
Biscuit w/Egg	5.6 oz.	819
Biscuit w/Jelly	3.5 oz.	653
Cheeseburger	4.1 oz.	789
Chef Salad	11.5 oz. (1 serving)	517
Chicken Fillet	6.8 oz.	360
Egg, Fried, medium	1.8 oz.	169
French Fries, large	4 oz.	192
French Fries, small	2.5 oz.	121
Ham Biscuit	3.8 oz.	1415
Ham Biscuit w/Egg	6.5 oz.	1584
Hamburger	3.9 oz.	682
Hot Dog	4.2 oz.	744
Hot Ham & Cheese	5.2 oz.	1067
Jelly	0.6 oz.	3
Mushroom 'N' Swiss	7.2 oz. (1 serving)	1051
Roast Beef Sandwich	5 oz.	1030
Sausage Biscuit	3.9 oz.	864
Sausage Biscuit w/Egg	5.7 oz.	1033
Shrimp 'N' Pasta Salad	11.8 oz. (1 serving)	941
Steak Biscuit	4.7 oz.	804

PORTION   SODIUM (MG.)

	PORTION	SODIUM (MG.)
Steak Biscuit w/Egg	5.7 oz.	973
Turkey Club	6.8 oz. (1 serving)	1185

Jack in the Box

	PORTION	SODIUM (MG.)
Apple Turnover	4.2 oz.	350
Bacon Cheeseburger Supreme	1 serving	1307
Bleu Cheese Dressing	1 serving	735
Breakfast Jack Sandwich	4.3 oz.	1035
Buttermilk House Dressing	1 serving	555
Cheeseburger	3.8 oz.	875
Cheese Nachos	1 serving	1154
Chicken Strips Dinner	1 serving	1213
Chicken Supreme	1 serving	1582
Club Pita	1 serving	953
French Fries	2.8 oz.	130
French Toast Breakfast	6.3 oz.	1130
Hamburger	3.4 oz.	565
Hot Ham N' Cheese	1 serving	1705
Jumbo Jack Hamburger	8.7 oz.	1135
Jumbo Jack Hamburger w/Cheese	9.6 oz.	1665
Moby Jack Sandwich	5 oz.	835
Onion Rings	3 oz.	320
Pancake Breakfast	8.2 oz.	1670
Sausage Crescent	1 serving	1012
Scrambled Eggs Breakfast	9.4 oz.	1110
Shake, Chocolate	11.2 oz.	295
Shake, Strawberry	11.5 oz.	270
Shake, Vanilla	11.1 oz.	265
Shrimp Salad	1 serving	460
Sirloin Steak Dinner	1 serving	969
Supreme Crescent	1 serving	1053
Supreme Nachos	1 serving	1782
Swiss & Bacon Burger	1 serving	1354
Taco, regular	2.9 oz.	460
Taco, super	5.1 oz.	970
Taco Salad	1 serving	1436
Thousand Island Dressing	1 serving	560

PORTION   SODIUM (MG.)

Kentucky Fried Chicken

	PORTION	SODIUM (MG.)
Chicken Breast Filet Sandwich	5.5 oz.	1093
Cole Slaw	3.2 oz.	225
Combination Dinner, Extra Crispy	13 oz.	1529
Combination Dinner, Original Recipe	12 oz.	1536
Corn	4.8 oz.	11
Dark (meat) Dinner, Extra Crispy	13 oz.	1480
Dark (meat) Dinner, Original Recipe	12 oz.	1441
Drumstick, Extra Crispy	2.0 oz.	263
Drumstick, Original Recipe	1.7 oz.	207
Gravy	0.5 oz.	57
Keel, Extra Crispy	3.7 oz.	584
Keel, Original Recipe	3.4 oz.	631
Kentucky Fries	3.5 oz.	174
Mashed Potatoes	3.0 oz.	268
Roll	0.7 oz.	118
Side Breast, Extra Crispy	3.0 oz.	564
Side Breast, Original Recipe	2.4 oz.	558
Thigh, Extra Crispy	3.8 oz.	549
Thigh, Original Recipe	3.1 oz.	566
White (meat) Dinner, Extra Crispy	12 oz.	1544
White (meat) Dinner, Original Recipe	11 oz.	1528
Wing, Extra Crispy	1.9 oz.	312
Wing, Original Recipe	1.5 oz.	302

McDonald's

	PORTION	SODIUM (MG.)
Apple Pie	3 oz. (1 serving)	398
Big Mac	7.2 oz. (1 serving)	1010
Cheeseburger	4 oz. (1 serving)	767
Cherry Pie	3.1 oz. (1 serving)	427
Chicken McNugget Sauce, Barbecue	1.1 oz. (1 serving)	309

PORTION SODIUM (MG.)

	PORTION	SODIUM (MG.)
Chicken McNugget Sauce, Honey	0.5 oz. (1 serving)	2
Chicken McNugget Sauce, Hot Mustard	1 oz. (1 serving)	259
Chicken McNugget Sauce, Sweet & Sour	1.1 oz. (1 serving)	186
Chicken McNuggets	3.9 oz. (1 serving)	525
Chocolaty Chip Cookies	2.4 oz. (1 serving)	313
Cone	4.1 oz (1 serving)	109
Egg McMuffin	4.9 oz. (1 serving)	885
English Muffin w/Butter	2.2 oz. (1 serving)	318
Filet-o-Fish	4.9 oz. (1 serving)	781
French Fries, regular	2.4 oz. (1 serving)	109
Ham Biscuit	4.2 oz. (1 serving)	1949
Hamburger	3.6 oz. (1 serving)	520
Hash Brown Potatoes	1.9 oz. (1 serving)	325
Hotcakes w/Butter, Syrup	7.5 oz. (1 serving)	1070
McChicken Sandwich	6 oz. (1 serving)	990
McDonaldland Cookies	2.4 oz. (1 serving)	358
Quarter Pounder	5.8 oz. (1 serving)	735
Quarter Pounder w/Cheese	6.8 oz. (1 serving)	1236
Sausage	1.9 oz. (1 serving)	615
Sausage Biscuit	4.8 oz. (1 serving)	1380
Scrambled Eggs	3.5 oz. (1 serving)	205
Shake, Chocolate	10.2 oz (1 serving)	300
Shake, Strawberry	10.2 oz. (1 serving)	207
Shake, Vanilla	10.2 oz. (1 serving)	201
Sundae, Caramel	5.8 oz. (1 serving)	195
Sundae, Hot Fudge	5.8 oz. (1 serving)	175
Sundae, Strawberry	5.8 oz. (1 serving)	96

Pizza Hut

	PORTION	SODIUM (MG.)
Pepperoni, Thick 'N Chewy	2 slices (¼ medium pie)	900
Pepperoni, Thin 'N Crispy	2 slices (¼ medium pie)	1000
Pork/Mushroom, Thick 'N Chewy	2 slices (¼ medium pie)	1000
Pork/Mushroom, Thin 'N Crispy	2 slices (¼ medium pie)	1200
Standard Cheese, Thick 'N Chewy	2 slices (¼ medium pie)	800

PORTION   SODIUM (MG.)

	PORTION	SODIUM (MG.)
Standard Cheese, Thin 'N Crispy	2 slices (¼ medium pie)	900
Super Cheese, Thick 'N Chewy	2 slices (¼ medium pie)	1000
Super Cheese, Thin 'N Crispy	2 slices (¼ medium pie)	1100
Super Pepperoni, Thick 'N Chewy	2 slices (¼ medium pie)	1200
Super Pepperoni, Thin 'N Crispy	2 slices (¼ medium pie)	1200
Super Pork/Mushroom, Thick 'N Chewy	2 slices (¼ medium pie)	1200
Super Pork/Mushroom, Thin 'N Crispy	2 slices (¼ medium pie)	1400
Super Supreme, Thin 'N Crispy	2 slices (¼ medium pie)	1500
Supreme, Thin 'N Crispy	2 slices (¼ medium pie)	1200

Roy Rogers

	PORTION	SODIUM (MG.)
Apple Danish	2.5 oz.	255
Bacon Cheeseburger	6.3 oz.	1536
Bacon 'n Tomato Dressing	2 Tbl.	150
Biscuit	2.2 oz.	575
Bleu Cheese Dressing	2 Tbl.	153
Breakfast Crescent Sandwich	4.5 oz.	867
Breakfast Crescent Sandwich w/bacon	4.7 oz.	1035
Breakfast Crescent Sandwich w/ham	5.8 oz.	1192
Brownie	2.3 oz.	150
Caramel Sundae	5.1 oz.	193
Cheese Danish	2.5 oz.	260
Cheeseburger	6.3 oz.	1404
Cherry Danish	2.5 oz.	242
Chicken Breast	4.4 oz.	601
Chicken Breast and Wing	6 oz.	867
Chicken Leg	1.7 oz.	162
Chicken Thigh	3.5 oz.	505
Chicken Thigh and Leg	5.1 oz.	667
Chicken Wing	1.5 oz.	266

	PORTION	SODIUM (MG.)
Cole Salw	3.5 oz.	261
Crescent Roll	2.5 oz.	547
Egg and Biscuit Platter	5.8 oz.	734
Egg and Biscuit Platter w/bacon	6.1 oz.	957
Egg and Biscuit Platter w/ham	7 oz.	1156
Egg and Biscuit Platter w/sausage	7.1 oz.	1059
French Fries	3 oz.	165
French Fries, large	4 oz.	221
Fried Chicken, breasts	1 serving	601
Fried Chicken, thighs	1 serving	505
Fried Chicken wings	1 serving	266
Hamburger	5 oz.	495
Hot Fudge Sundae	5.3 oz.	186
Hot Topped Potato (Plain)	8 oz.	65
Hot Topped Potato w/bacon 'n cheese	8.7 oz.	778
Hot Topped Potato w/ broccoli 'n cheese	11 oz.	523
Hot Topped Potato w/oleo	8.3 oz.	161
Hot Topped Potato w/sour cream 'n chives	10.5 oz.	138
Hot Topped Potato w/taco beef 'n cheese	12.6 oz.	726
Hot Chocolate	8 oz.	210
Italian Dressing, Lo-cal	2 Tbl.	100
Macaroni	3.5 oz.	603
Pancakes w/butter and syrup	1 serving	511
Pancake Platter w/bacon	6.1 oz.	1065
Pancake Platter w/ham	7 oz.	1264
Pancake Platter w/sausage	7.1 oz.	1167
Potato Salad	3.5 oz.	696
Ranch Dressing	2 Tbl.	100
Roast Beef	5.4 oz.	785
Roast Beef, large	6.4 oz.	1044
Roast Beef w/Cheese	6.4 oz.	1694
Roast Beef w/Cheese, large	7.4 oz.	1953
RR Bar Burger	7.3 oz.	1826

PORTION   SODIUM (MG.)

	PORTION	SODIUM (MG.)
Shake, Chocolate	11.2 oz.	290
Shake, Strawberry	11 oz.	261
Shake, Vanilla	10.8 oz.	282
Strawberry Shortcake	7.2 oz.	674
Strawberry Sundae	5 oz.	99
1,000 Island Dressing	2 Tbl.	150

Rustler

	PORTION	SODIUM (MG.)
Beef Patty Platter, large	8 oz.	533
Beef Patty Platter, small	4 oz.	499
Cheese	2 slices	160
Filet Mignon Platter	1 serving	518
Jello, cherry	1 serving	39
Pudding, chocolate	1 serving	146
Rib-eye Steak Platter	1 serving	494
Steak & Crab Platter	1 serving	900
Strip Steak Platter	1 serving	515
T-Bone Steak Platter	1 serving	529
Trailboss Sandwich	1 serving	183
Westerner Sandwich	1 serving	234

Wendy's

	PORTION	SODIUM (MG.)
Bacon Cheeseburger on white bun	5 oz.	860
Chicken Sandwich on multi-grain bun	5 oz.	500
Chili	8 oz.	1070
Double Cheeseburger	11.4 oz.	1414
Double Hamburger	7 oz.	575
French Fries	3.5 oz.	95
Frosty Dairy Dessert	12 fl. oz.	220
Kids' Meal Hamburger	2 oz.	265
Multi-grain Bun	1.7 oz.	220
Pick-Up Window Side Salad	18 oz.	540
Potato, Bacon & Cheese	12.3 oz.	1180
Potato, Broccoli & Cheese	12.9 oz.	430
Potato, Cheese	12.3 oz.	450
Potato, Chicken A La King	12.6 oz.	820
Potato, Chili & Cheese	14 oz.	610

	PORTION	SODIUM (MG.)
Potato, Plain	8.8 oz.	60
Potato, Sour Cream & Chives	11 oz.	230
Potato, Stroganoff & Sour Cream	14.3 oz.	910
Single Cheeseburger	8.5 oz.	1085
Single Hamburger	4 oz.	410
Single Hamburger on multi-grain bun	4 oz.	290
Taco Salad	12.6 oz.	1100
Triple Cheeseburger	14.1 oz.	1848
Triple Hamburger	12.7 oz.	1217
White Bun	1.8 oz.	266

Fats, Oils, and Salad Dressings

FATS

Butter

Land O Lakes

Butter, Lightly Salted Sweet Cream	0.5 oz. (1 tbl.)	115
Butter, Lite Salt, Sweet Cream, Whipped	0.3 oz. (1 tbl.)	75
Butter, Unsalted Sweet Cream	0.5 oz. (1 tbl.)	2
Butter, Unsalted Sweet Cream, Whipped	0.3 oz. (1 tbl.)	1

USDA

Butter	0.2 oz. (1 pat)	41
Butter	0.5 oz. (1 tbl.)	123

Lard

USDA

Lard	1 tbl.	0

Margarine

Blue Bonnet

Diet	0.5 oz. (1 tbl.)	110
Soft	0.5 oz. (1 tbl.)	110
Spread	0.5 oz. (1 tbl.)	110
Stick	0.5 oz. (1 tbl.)	110
Whipped Soft	0.3 oz. (1 tbl.)	70
Whipped Stick	0.3 oz. (1 tbl.)	70

Chiffon

Soft Regular Margarine	0.5 oz. (1 tbl.)	105
Soft Unsalted Margarine	0.5 oz. (1 tbl.)	under 10
Soft Whipped Margarine	0.3 oz. (1 tbl.)	80
Stick Margarine	0.5 oz. (1 tbl.)	110

Country Morning Blend

Lightly Salted Soft (Tub)	1 tbl.	85
Lightly Salted Stick	1 tbl.	115
Unsalted Soft (Tub)	1 tbl.	1
Unsalted Stick	1 tbl.	1

Dalewood

Lard Print Margarine	0.5 oz. (1 tbl.)	129

Empress

Corn Oil Stick Margarine	0.5 oz. (1 tbl.)	124
60% Corn Oil Tub Spread	0.5 oz. (1 tbl.)	136
Soft Corn Oil Tub	0.5 oz. (1 tbl.)	121
Tub Margarine	0.5 oz. (1 tbl.)	127
Vegetable Print Margarine	0.5 oz. (1 tbl.)	116

Fleischmann's

Diet	0.5 oz. (1 tbl.)	110
Soft	0.5 oz. (1 tbl.)	110
Soft Whipped	0.3 oz. (1 tbl.)	70
Spread	0.5 oz. (1 tbl.)	110
Stick	0.5 oz. (1 tbl.)	110
Unsalted Parve	0.5 oz. (1 tbl.)	0
Unsalted Stick	0.5 oz. (1 tbl.)	1

PORTION   SODIUM (MG.)

Holiday

Margarine	0.5 oz. (1 tbl.)	170

Land O Lakes

Margarine, Regular Corn Oil Stick	0.5 oz. (1 tbl.)	115
Margarine, Regular Soy Stick	0.5 oz. (1 tbl.)	115
Soy Soft Tub	0.5 oz. (1 tbl.)	115

Mazola

Margarine	0.5 oz. (1 tbl.)	115
Margarine, Diet Imitation	0.5 oz. (1 tbl.)	130
Margarine, Unsalted	0.5 oz. (1 tbl.)	1

Miracle (Kraft)

Corn Oil Margarine	0.5 oz. (1 tbl.)	75
Margarine	0.5 oz. (1 tbl.)	75

Nucoa

Margarine	0.5 oz. (1 tbl.)	165
Margarine, Soft	0.5 oz. (1 tbl.)	155

Parkay (Kraft)

Corn Oil Margarine	0.5 oz. (1 tbl.)	115
Corn Oil Margarine, Soft	0.5 oz. (1 tbl.)	115
Light Corn Oil Spread	0.5 oz. (1 tbl.)	120
Light Spread	0.5 oz. (1 tbl.)	110
Margarine	0.5 oz. (1 tbl.)	95
Margarine, Soft	0.5 oz. (1 tbl.)	110
Margarine, Soft, Diet	0.5 oz. (1 tbl.)	70
Squeeze Margarine	0.5 oz. (1 tbl.)	110
Whipped Margarine (Cup)	0.5 oz. (1 tbl.)	75

Scotch Buy

Lard Print Margarine	0.5 oz. (1 tbl.)	129
Print Spread 60% Oil	0.5 oz. (1 tbl.)	126

PORTION   SODIUM (MG.)

60% Oil Tub Spread ....0.5 oz. (1 tbl.) 146
Soybean Tub Margarine ....0.5 oz. (1 tbl.) 133
Vegetable Print Margarine ....0.5 oz. (1 tbl.) 131

Weight Watchers

Reduced-Calorie Margarine
  (Tub) ....2 tsp. 60
Reduced-Calorie Margarine
  Sticks ....1.5 tsp. 55

Shortening
Crisco

Butter Flavor Crisco ....1 tbl. 0
Crisco.......................................1 tbl. 0

NuMade

Shortening....................................0.5 oz. (1 tbl.) 379

Scotch Buy

Shortening....................................0.5 oz. (1 tbl.) 379

OILS
USDA

Corn Oil ....1 tbl. 0
Cottonseed Oil ....1 tbl. 0
Olive Oil ....1 tbl. 0
Peanut Oil ....1 tbl. 0
Safflower Oil ....1 tbl. 0
Sesame Oil ....1 tbl. 0
Soybean Oil ....1 tbl. 0
Sunflower Oil ....1 tbl. 0

SALAD DRESSINGS
Bottled

Aristocrat

French Style, Low Fat, Low
  Calorie, Low Salt ....0.5 oz. (1 tbl.) 5

PORTION   SODIUM (MG.)

Italian Garlic, Low Fat, Low
   Calorie, Low Salt ....0.5 oz. (1 tbl.) 2

Bama

Salad Dressing.............................1 tbl......................................120

Catalina

French ....0.5 oz. (1 tbl.) 180
French, Reduced Calorie ....0.5 oz. (1 tbl.) 125

Dia-Mel

Blue Cheese.................................1 tbl..30
Creamy Cucumber ....1 tbl..30
Creamy Garlic ....1 tbl..10
Creamy Italian ....1 tbl..10
Diet Whip.....................................1 tbl......................................100
French ....1 tbl..20
Italian ....1 tbl..10
Red Wine Vinegar ....1 tbl..10
Tahiti..1 tbl..10
Thousand Island ....1 tbl..30
Yogurt Buttermilk ....1 tbl..15

Eastern

Bleu Cheese.................................1 tbl..55
Cucumber ....1 tbl......................................135
Honey French..............................1 tbl......................................108
Italian ....1 tbl......................................140
Poppy Seed ....1 tbl..18
Slaw...1 tbl..55
Spicy Oil & Vinegar.....................1 tbl..56
Thousand Island ....1 tbl......................................100
Wine & Cheese ....1 tbl......................................140

Estee

Caesar...1 tbl......................................150
Herb Garden................................1 tbl......................................150
Italian ....1 tbl......................................150
Onion and Cucumber ....1 tbl..80

PORTION   SODIUM (MG.)

Featherweight

Low Sodium Imitation French	1 tbl.	4
2-Calorie Low Sodium Dressing	1 tbl.	6

Hellmann's

Spin Blend Salad Dressing	1 tbl.	110

Henri's

Bacon 'N Tomato French	1 tbl.	145
Blue Cheese	1 tbl.	195
Blue Cheese, reduced calorie	1 tbl.	205
Buttermilk Farms	1 tbl.	135
Chopped Chive	1 tbl.	130
Creamy Dill	1 tbl.	135
Creamy Garlic	1 tbl.	145
Creamy Garlic, reduced calorie	1 tbl.	210
Cucumber 'N Onion	1 tbl.	210
Cucumber 'N Onion, reduced calorie	1 tbl.	210
French	1 tbl.	115
French Style, reduced calorie	1 tbl.	135
Hearty Beefsteak French	1 tbl.	95
Russian	1 tbl.	90
Slaw	1 tbl.	180
Smoky Bits	1 tbl.	80
Sweet 'N Saucy French	1 tbl.	105
Tas-Tee	1 tbl.	105
Thousand Island	1 tbl.	140
Thousand Island, reduced calorie	1 tbl.	175
Yogonaise, reduced calorie	1 tbl.	125
Yogowhip, reduced calorie	1 tbl.	105

Kraft

Bacon & Buttermilk	0.5 oz. (1 tbl.)	125
Bacon & Tomato	0.5 oz. (1 tbl.)	130

PORTION   SODIUM (MG.)

Bacon & Tomato, reduced calorie	0.5 oz. (1 tbl.)	150
Buttermilk & Chives Creamy	0.5 oz. (1 tbl.)	120
Buttermilk Creamy	0.5 oz. (1 tbl.)	120
Buttermilk Creamy, reduced calorie	0.5 oz. (1 tbl.)	125
Chunky Blue Cheese	0.5 oz. (1 tbl.)	230
Chunky Blue Cheese, reduced calorie	0.5 oz. (1 tbl.)	240
Coleslaw	0.5 oz. (1 tbl.)	195
Creamy Bacon, reduced calorie	0.5 oz. (1 tbl.)	150
Creamy Cucumber	0.5 oz. (1 tbl.)	200
Creamy Cucumber, reduced calorie	0.5 oz. (1 tbl.)	230
Creamy Italian, reduced calorie	0.5 oz. (1 tbl.)	125
Creamy Italian w/Real Sour Cream	0.5 oz. (1 tbl.)	120
Creamy Onions & Chives	0.5 oz. (1 tbl.)	145
French	0.5 oz. (1 tbl.)	125
French, reduced calorie	0.5 oz. (1 tbl.)	145
Golden Blend Italian	0.5 oz. (1 tbl.)	150
Golden Caesar	0.5 oz. (1 tbl.)	175
Italian, reduced calorie	0.5 oz. (1 tbl.)	210
Oil & Vinegar	0.5 oz. (1 tbl.)	220
Oil Free Italian	0.5 oz. (1 tbl.)	215
Russian	0.5 oz. (1 tbl.)	125
Russian, reduced calorie	0.5 oz. (1 tbl.)	195
Thousand Island	0.5 oz. (1 tbl.)	150
Thousand Island, reduced calorie	0.5 oz. (1 tbl.)	140
Zesty Italian	0.5 oz. (1 tbl.)	275

Marie's

Avocado Goddess	0.5 oz. (1 tbl.)	88
Blue Cheese	0.5 oz. (1 tbl.)	64
Blue Cheese, reduced calorie	0.5 oz. (1 tbl.)	192
Creamy Cucumber	0.5 oz. (1 tbl.)	143

PORTION　SODIUM (MG.)

Italian Garlic	0.5 oz. (1 tbl.)	108
Italian with Cheese	0.5 oz. (1 tbl.)	95
Ranch	0.5 oz. (1 tbl.)	79
Roquefort	0.5 oz. (1 tbl.)	95
Russian	0.5 oz. (1 tbl.)	86
Thousand Island	0.5 oz. (1 tbl.)	103

Miracle Whip

Miracle Whip Salad Dressing	1 tbl.	85

NuMade

Blue Cheese	0.5 oz. (1 tbl.)	133
Caesar	0.5 oz. (1 tbl.)	216
Creamy Cucumber	0.5 oz. (1 tbl.)	203
French, reduced calorie	0.5 oz. (1 tbl.)	138
Green Goddess	0.5 oz. (1 tbl.)	138
Italian	0.5 oz. (1 tbl.)	333
Italian, reduced calorie	0.5 oz. (1 tbl.)	198
Red Wine Vinegar & Oil	0.5 oz. (1 tbl.)	232
Russian	0.5 oz. (1 tbl.)	129
Salad Dressing	0.5 oz. (1 tbl.)	100
Savory French	0.5 oz. (1 tbl.)	220
Thousand Island	0.5 oz. (1 tbl.)	156
Thousand Island, reduced calorie	0.5 oz. (1 tbl.)	143
Zesty French	0.5 oz. (1 tbl.)	220

Philadelphia

Cucumber	1 tbl.	105
Garlic & Chive	1 tbl.	115
Italian Herb	1 tbl.	120
Toasted Onion	1 tbl.	150

Roka

Blue Cheese	1 tbl.	175
Blue Cheese, reduced calorie	1 tbl.	285

Salad Life

Cheese & Herb	1 tbl.	130

	PORTION	SODIUM (MG.)
Cheese & Herbs, no salt	1 tbl	15
French	1 tbl	59
Italian	1 tbl	114
Real French, no salt	1 tbl	3
Real Italian, no salt	1 tbl	3
Thousand Island, unsalted	1 tbl	62

Scotch Buy

Salad Dressing	0.5 oz. (1 tbl.)	100

Seven Seas

Buttermilk Recipe	1 tbl	130
Buttermilk Recipe Country Spice	1 tbl	135
Caesar	1 tbl	260
Capri	1 tbl	130
Chunky Blue Cheese	1 tbl	195
Creamy Bacon	1 tbl	205
Creamy French	1 tbl	265
Creamy Italian	1 tbl	255
Creamy Onion 'N Chive	1 tbl	190
Creamy Parmesan	1 tbl	140
Creamy Russian	1 tbl	115
Family Style French	1 tbl	135
Green Goddess	1 tbl	140
Herbs & Spices	1 tbl	160
Mild Italian	1 tbl	170
Red Wine Vinegar & Oil	1 tbl	265
Thousand Island	1 tbl	155
Viva Italian!	1 tbl	320
Viva Parmesan	1 tbl	205

Weight Watchers

Creamy Italian	1 tbl	123
Russian/Thousand Island	1 tbl	114

Western

Coleslaw	1 tbl	158
Creamy Cucumber	1 tbl	108

	PORTION	SODIUM (MG.)
Creamy French	1 tbl.	102
Creamy Garlic	1 tbl.	212
Creamy Italian	1 tbl.	133
French	1 tbl.	102
Fruit Salad	1 tbl.	111
Oil & Vinegar	1 tbl.	46
Russian	1 tbl.	165
Salad Dressing	1 tbl.	104
Thousand Island, Pourable	1 tbl.	137
Western	1 tbl.	105

Wishbone

Caesar	1 tbl.	195
California Onion	1 tbl.	155
Chunky Blue Cheese	1 tbl.	150
Creamy Cucumber	1 tbl.	125
Creamy Garlic	1 tbl.	170
Creamy Italian	1 tbl.	170
Deluxe French	1 tbl.	75
Garlic French	1 tbl.	155
Green Goddess	1 tbl.	165
Italian	1 tbl.	285
Russian	1 tbl.	145
Sweet 'n Spicy French	1 tbl.	150
Thousand Island	1 tbl.	135

Mixes

Good Seasons

Low Calorie Italian	1 tbl.	160

Hain

No Oil Bleu Cheese	1 tbl.	135
No Oil Buttermilk	1 tbl.	90
No Oil Caesar	1 tbl.	170
No Oil French	1 tbl.	160
No Oil Garlic Cheese	1 tbl.	130
No Oil Herb	1 tbl.	100
No Oil Italian	1 tbl.	145
No Oil 1,000 Island	1 tbl.	120

PORTION SODIUM (MG.)

Weight Watchers

Blue Cheese	1 tbl. prepared	108
Creamy Italian	1 tbl. prepared	224
French Style	1 tbl. prepared	164
Italian	1 tbl. prepared	175
Russian	1 tbl. prepared	128
Thousand Island	1 tbl. prepared	265

Fish and Seafood

(See also Entrees and Dinners, page 264)

CANNED

Bumble Bee

Salmon, Keta	3 oz. (⅖ cup)	412
Salmon, Pink	3 oz. (⅖ cup)	418
Salmon, Red Sockeye	3 oz. (⅖ cup)	350
Tuna, Chunk Light in Oil, undrained	3 oz. (½ cup)	302
Tuna, Chunk Light in Water, undrained	3 oz. (½ cup)	287
Tuna, Solid White in Oil, undrained	3 oz. (⅖ cup)	355
Tuna, Solid White in Water, undrained	3 oz. (⅖ cup)	285

Chicken of the Sea

Tuna, Albacore Chunk White in Water, Dietetic, Low Sodium	3 oz. drained	33
Tuna, Albacore White in Oil	3 oz. drained	536
Tuna, Albacore White in Water	3 oz. drained	400
Tuna, Light in Oil	3 oz. drained	408
Tuna, Light in Water	3 oz. drained	344

Del Monte

Salmon, Pink	3 oz. undrained	495

PORTION   SODIUM (MG.)

Salmon, Red..............................3 oz. undrained 495
Sardines in Tomato Sauce........3 oz. undrained 405

Doxsee

Clam Juice ....4 fl. oz. 285
Clams (Minced & Chopped) ....6.5 oz. 1130

Featherweight

Salmon....................................3 oz...................................57
Sardines in Oil ....3 oz...................................104
Sardines in Tomato Sauce........3 oz...................................104
Sardines in Water ....3 oz...................................104
Tuna ....3 oz...................................44

Gorton's

Minced Clams...........................3.25 oz. (½ can) 400

Health Valley

Tuna ....3 oz....................................376
Tuna, Dietetic ....3 oz....................................43

Snow's

Clam Juice ....3 fl. oz. 470
Minced Clams...........................6.5 oz. 920

The Spreadables

Tuna Salad ....2 oz....................................208

Star Kist

Bonito, Chunk in Oil ....3 oz. drained 338
Bonito, Solid in Oil ....3 oz. drained 338
Tuna, Chunk Light in Oil ....3 oz. drained 338
Tuna, Chunk Light in Water......3 oz. drained 338
Tuna, Chunk Light in Water
   (Diet Pack) ....3 oz. drained 43
Tuna, Chunk White in Oil..........3 oz. drained 338
Tuna, Chunk White in Water
   (Diet Pack) ....3 oz. drained 43

PORTION   SODIUM (MG.)

Tuna, Grated Light Flakes
   in Oil......................................3 oz. drained 338
Tuna, Select Chunk Lite in
   Spring Water w/60% Less
   Salt ....3 oz. drained 180
Tuna, Solid Light in Oil..............3 oz. drained 338
Tuna, Solid Light in Tonno Olive
   Oil ....3 oz. drained 338
Tuna, Solid Light in Water ....3 oz. drained 338
Tuna, Solid White in Oil ....3 oz. drained 338
Tuna, Solid White in Water
   (import Albacore) ....3 oz. drained 338
Tuna, Solid White in Water
   (local Albacore) ....3 oz. drained 338

Underwood

Sardines (Mustard Sauce) ....3 oz. (1 can) 680
Sardines (Soy Bean Oil) ....3 oz. (1 can) 186
Sardines (Tomato Sauce)..........3 oz. (1 can) 680

FRESH

USDA

Bass, Black Sea, raw ....4 oz..76
Bluefish, baked w/butter............4 oz..116
Bluefish, breaded, fried ....4 oz..164
Catfish, raw..............................4 oz..67
Clams, hard, raw ....4 oz..232
Clams, soft, raw.......................4 oz..40
Cod, broiled w/butter................4 oz..124
Crab, steamed ....4 oz..419
Eel, raw ....4 oz..89
Flounder, baked w/butter ....4 oz..268
Haddock, breaded, fried ....4 oz..200
Halibut, broiled w/butter ....4 oz..152
Herring, smoked.......................1 oz. 1745
Lingcod, raw ....4 oz..67
Lobster, boiled ....4 oz..283
Lox, regular..............................2 oz. 1100
Mackerel, raw ....4 oz..53

	PORTION	SODIUM (MG.)
Mullet, breaded, fried	4 oz.	111
Mussels, raw	4 oz.	324
Ocean Perch, fried	4 oz.	171
Oysters, fried	4 oz.	232
Oysters, raw	4 oz.	151
Pollock, creamed	4 oz.	125
Pompano, cooked	4 oz.	64
Rockfish, oven-steamed	4 oz.	76
Salmon, broiled w/butter	4 oz.	132
Scallops, raw	4 oz.	289
Scallops, steamed	4 oz.	300
Shad, baked w/butter	4 oz.	88
Shrimp, fried	4 oz.	212
Shrimp, raw	4 oz.	183
Snapper, raw	4 oz.	75
Trout, Lake, raw	4 oz.	89

FROZEN

Gorton's

	PORTION	SODIUM (MG.)
Baked Stuffed Scrod	6 oz. (1 fillet)	650
Batter Fried Fish Fillets	3 oz. (1 fillet)	440
Batter Fried Fish Sticks	3 oz. (3 sticks)	505
Crunchy Fish Fillets	3.4 oz. (2 fillets)	680
Crunchy Fish Sticks	2 oz. (3 sticks)	325
Filet of Haddock in Lemon Butter Sauce	6 oz. (1 fillet)	735
Filet of Sole in Cheese Sauce	7 oz. (1 fillet)	900
Filet of Sole in Lemon Butter Sauce	6 oz. (1 fillet)	760
Lightly Battered Tempura Fillets	3 oz. (1 fillet)	405
Lightly Battered Tempura Sticks	3 oz. (3 sticks)	465
Lightly Breaded Fillets	3 oz. (1 fillet)	320
Lightly Breaded Sticks	3 oz. (3 sticks)	285
Potato Crisp Fish Fillets	3.4 oz. (2 fillets)	770
Potato Crisp Fish Sticks	2.4 oz. (3 sticks)	305
Shrimp Scampi	6 oz.	415
Stuffed Flounder	6.5 oz. (1 fillet)	1075

PORTION   SODIUM (MG.)

Mrs. Paul's

Batter Dipped Fish Fillets	6 oz. (2 fillets)	830
Buttered Fish Fillets	5 oz. (2 pieces)	780
Catfish Fillets	3.6 oz. (1 fillet)	243
Catfish Fingers	4 oz.	260
Combination Seafood Platter	9 oz.	1340
Crispy Crunchy Fish Fillets	4.5 oz. (2 fillets)	650
Crispy Crunchy Fish Sticks	3 oz. (4 sticks)	455
Crispy Crunchy Flounder Fillets	4 oz. (2 fillets)	800
Crispy Crunchy Haddock Fillets	4 oz. (2 fillets)	800
Crispy Crunchy Perch Fillets	4 oz. (2 fillets)	480
Crunchy Light Batter Fish Fillets	4.5 oz. (2 fillets)	855
Crunchy Light Batter Fish Sticks	3.5 oz. (4 pieces)	795
Crunchy Light Batter Haddock Fillets	4.5 oz. (2 fillets)	935
Deviled Crab Miniatures	3.5 oz.	195
Deviled Crabs	3 oz. (1 cake)	385
Fish Cake Thins	5 oz. (2 pieces)	1020
Fish Cakes	4 oz. (2 cakes)	669
Fish Dijon	8.5 oz.	925
Fish Florentine	9 oz.	1025
Fish Mornay	10 oz.	665
Fish Parmesan	5 oz.	540
French Fried Scallops	3.5 oz.	545
Fried Clams in a Light Batter	2.5 oz.	385
Fried Shrimp	3 oz.	525
Light and Natural Fish Fillet	6 oz. (1 fillet)	770
Light and Natural Flounder Fillets	6 oz. (1 fillet)	975
Light and Natural Haddock Fillets	6 oz. (1 fillet)	960
Light and Natural Sole Fillets	6 oz. (1 fillet)	700
Scallops Mediterranean	11 oz.	775

PORTION   SODIUM (MG.)

Seafood Newburg	8.5 oz.	610
Shrimp Oriental	11 oz.	940
Shrimp Primavera	11 oz.	1185
Supreme Light Batter Fish Fillets	3.6 oz. (1 piece)	505

Van De Kamp

Fish Fillets, Batter Dipped	6 oz.	704
Fish Fillets, Country Seasoned	4.8 oz.	668
Fish Fillets, Light 'n Crispy	4 oz.	722
Fish Kabobs, Batter Dipped	4 oz	577
Fish Kabobs, Country Seasoned	4 oz.	490
Fish 'n Chips, Batter Dipped	8 oz.	551
Fish Portions, Batter Dipped	3 oz.	510
Fish Sticks, Batter Dipped	5 oz.	982
Fish Sticks, Light 'n Crispy	3.8 oz.	769
Haddock, Batter Dipped	4 oz.	529
Halibut, Batter Dipped	4 oz.	618
Sole, Batter Dipped	4 oz.	578

Freeze-Dried Foods

Mountain House

Apples	1 oz., recon.	0
Beans & Beef Franks	8 oz., recon.	102
Beef, cooked, diced	3 oz., recon.	20
Beef & Rice w/Onions	8 oz., recon.	270
Beef Chop Suey	8 oz., recon.	128
Beef Patties, raw	3 oz., recon.	19
Beef Steak, raw	3 oz., recon.	19
Beef Stew	8 oz., recon.	71
Beef Stroganoff	8 oz., recon.	122
Blueberries	1 oz., recon	0
Carrots	1 oz., recon.	1
Cheese Omelette	1 oz., recon.	62
Chicken, cooked, diced	3 oz., recon.	19

	PORTION	SODIUM (MG.)
Chicken Chop Suey	8 oz., recon.	119
Chicken Salad	1 oz., recon.	65
Chicken Stew	8 oz., recon.	190
Chili Mac w/Beef	8 oz., recon	158
Chili w/Beans	8 oz., recon.	274
Corn	1 oz., recon.	0
Eggs w/Butter	1 oz., recon.	52
Eggs w/Real Bacon Bits	1 oz., recon.	47
Fish Steaks, raw	3 oz., recon.	11
Green Beans	1 oz., recon.	0
Green Peas	1 oz., recon.	8
Mexican Omelette	1 oz., recon.	43
Noodles & Chicken	8 oz., recon.	256
Peaches	1 oz., recon.	0
Pears	1 oz., recon.	0
Pork Chops, raw	3 oz., recon.	18
Potatoes & Beef	8 oz., recon.	154
Rice & Chicken	8 oz., recon.	249
Sausage Patties, cooked	3 oz., recon.	148
Shrimp Creole	8 oz., recon.	206
Strawberries	1 oz., recon.	0
Tuna Salad	1 oz., recon.	43
Vegetable Beef Stew	8 oz., recon.	74

Fruit

CANNED

Del Monte Lite

Apricots, Halves, unpeeled	½ cup	under 10
Chunky Mixed Fruit	½ cup	under 10
Fruit Cocktail	½ cup	under 10
Peaches, Yellow Cling Halves or Slices	½ cup	under 10
Pears, Bartlett Halves or Slices	½ cup	under 10

Del Monte Regular

Applesauce	½ cup	under 5

PORTION   SODIUM (MG.)

Apricot, Halves, unpeeled ....½ cup............................under 10
Apricots, Whole, peeled ....½ cup............................under 10
Cherries, Dark Sweet, Pitted ....½ cup............................under 10
Cherries, Dark Sweet w/Pits ....½ cup............................under 10
Cherries, Light Sweet w/Pits ....½ cup............................under 10
Chunky Mixed Fruits ....½ cup............................under 10
Figs, Whole...............................½ cup............................under 10
Fruit Cocktail............................½ cup............................under 10
Fruit for Salad ....½ cup............................under 10
Grapefruit Sections ....½ cup............................under 10
Mandarin Oranges.....................5.5 oz. ....under 10
Mixed Fruit ....5 oz. ....under 10
Peaches, Freestone, Halves
    or Slices...........................½ cup............................under 10
Peaches, Spiced w/Pits ....3.5 oz. ....under 10
Peaches, Yellow Cling Halves
    or Slices...........................½ cup............................under 10
Pears, Halves or Slices ....½ cup............................under 10
Pineapple, Crushed or Chunk
    or Slices, w/Juice................½ cup............................under 10
Pineapple, Crushed or Chunk
    or Slices, w/Syrup ....½ cup............................under 10
Tropical Fruit Salad ....½ cup............................under 10

Diet Delight

Applesauce ....½ cup.....................................2
Apricots ....½ cup.....................................5
Cherries ....½ cup.....................................4
Fruit Cocktail............................½ cup.....................................5
Fruits for Salad ....½ cup.....................................6
Grapefruit Sections ....½ cup.....................................6
Mandarin Oranges.....................½ cup.....................................4
Peaches, Cling ....½ cup.....................................6
Peaches, Freestone ....½ cup.....................................7
Pears ....½ cup.....................................7
Pineapple..................................½ cup.....................................8
Purple Plums ....½ cup.....................................3

Health Valley

Apricot Halves ....½ cup 10

PORTION   SODIUM (MG.)

Fruit Mix	½ cup	10
Peaches	½ cup	10
Pears	½ cup	10

Kraft

Fruit Salad	½ cup	5
Grapefruit Sections, unsweetened	½ cup	0

Libby

Apricots, Heavy Syrup	½ cup	8
Apricots, Lite	½ cup	under 5
Cherries	½ cup	5
Fruit Cocktail, Heavy Syrup	½ cup	10
Fruit Cocktail, Lite	½ cup	under 5
Fruits for Salad, Heavy Syrup	½ cup	8
Fruits for Salad, Lite	½ cup	under 5
Grapefruit	½ cup	5
Grapefruit, Lite	½ cup	under 5
Mixed Fruits, Chunky	½ cup	10
Mixed Fruits, Chunky Lite	½ cup	under 5
Peaches, Chunky	½ cup	10
Peaches, Cling, Heavy Syrup	½ cup	10
Peaches, Lite	½ cup	under 5
Pears, Chunky	½ cup	8
Pears, Heavy Syrup	½ cup	8
Pears, Lite	½ cup	under 5
Pineapple, Lite	½ cup	under 5
Plums	½ cup	5

Luck's

Fried Apples	4 oz.	40

Stokely-Van Camp

Applesauce	½ cup	35
Apricot Halves	½ cup	23
Fruit Cocktail	½ cup	15
Fruit for Salad	½ cup	15
Peach Halves	½ cup	23

PORTION   SODIUM (MG.)

Peaches, Sliced	½ cup	20
Pear Halves	½ cup	15
Pears, Sliced	½ cup	10
Purple Plums	½ cup	25

DRIED

Del Monte

Apples, uncooked	2 oz.	50
Apricots, uncooked	2 oz.	under 10
Currants, Zante	2.4 oz. (½ cup)	under 10
Mixed Dried Fruits	2 oz.	10
Peaches, uncooked	2 oz.	under 10
Prunes, Pitted, uncooked	2 oz.	under 10
Prunes w/Pits, uncooked	2 oz.	under 10
Raisins, Golden Seedless	3 oz.	under 10
Raisins, Thompson Seedless	3 oz.	15

USDA

Apples, Sulfured	½ cup	39
Apricots	½ cup	6
Currants	½ cup	5
Dates	10 fruits	1
Figs	1 fruit	2
Peaches	½ cup	5
Pears	½ cup	5
Prunes	5 large	2
Prunes, cooked	½ cup	4
Raisins, Seedless	½ cup	9

FRESH

USDA

Apples, raw or baked	1 fruit	2
Apricots	3 fruits	1
Banana	1 fruit	2
Blackberries (Boysenberries)	½ cup	1
Blueberries	½ cup	1
Cantaloupe Melon	½ fruit	24
Casaba Melon	⅕ fruit	34
Cherries	½ cup	1

	PORTION	**SODIUM (MG.)**
Cranberries	½ cup	1
Currants	½ cup	2
Figs	1 fruit	2
Grapefruit	½ fruit	1
Grapes, Thompson Seedless	10 fruits	1
Honeydew Melon	⅕ fruit	28
Kumquat	1 fruit	1
Lemon	1 fruit	1
Mango	1 fruit	1
Nectarine	1 fruit	1
Orange	1 fruit	1
Papaya	1 fruit	8
Peach	1 fruit	1
Pear	1 fruit	1
Pineapple	½ cup	1
Plum	1 fruit	1
Raspberries	½ cup	1
Strawberries	⅔ cup	1
Tangelo	1 fruit	1
Tangerine	1 fruit	1
Watermelon	1/16 fruit	8

FROZEN

Birds Eye

Mixed Fruit (in pouch)	5 oz.	5
Peaches (in pouch)	5 oz.	9
Red Raspberries (in pouch)	5 oz.	0
Strawberries (in pouch)	5 oz.	5
Strawberry Halves	5 oz.	5
Whole Strawberries	5 oz.	0

Mrs. Paul's

Apple Fritters	4 oz. (2 fritters)	770

Stokely-Pictsweet

Red Raspberries	5 oz.	10
Strawberry Halves	5 oz.	10

Stouffer's

Escalloped Apples	4 oz.	50

Gelatin and Pudding

GELATIN

Cooked

Dia-Mel

Gel Mixes (all varieties) ....½ cup prepared..................80

D-Zerta

Low Calorie Gelatin (average
  for all flavors) ....½ cup prepared.....................5

Estee

Gel Mix ....½ cup prepared.....................5

Featherweight

Gelatin (All Flavors) ....½ cup prepared.....................2

Jell-O

Black Raspberry	½ cup prepared	40
Cherry	½ cup prepared	75
Concord Grape	½ cup prepared	40
Lemon	½ cup prepared	80
Lime	½ cup prepared	60
Orange-Pineapple	½ cup prepared	70
Wild Strawberry	½ cup prepared	80
Other Flavors	½ cup prepared	55

Royal

All Flavors ....½ cup prepared..................90
Sweet As You Please
  Gelatin ....½ cup prepared..................90

Safeway

Blackberry ....½ cup prepared..................59
Black Cherry ....½ cup prepared..................68

PORTION   SODIUM (MG.)

Black Raspberry	½ cup prepared	69
Cherry	½ cup prepared	68
Grape	½ cup prepared	69
Lemon	½ cup prepared	69
Lime	½ cup prepared	76
Mixed Fruit	½ cup prepared	69
Orange	½ cup prepared	70
Pineapple-Orange	½ cup prepared	69
Raspberry	½ cup prepared	32
Strawberry	½ cup prepared	69
Strawberry-Banana	½ cup prepared	69

USDA

Gelatin, unflavored	1 tsp.	0

Ready to Serve

Dia-Mel

RTE Gel-a-thin	4 oz.	12

Estee

Gel, Ready-to-Eat	½ cup	5

PUDDING

Cooked

Dia-Mel

Butterscotch	½ cup prepared	80
Chocolate	½ cup prepared	80
Lemon	½ cup prepared	20
Vanilla	½ cup prepared	80

D-Zerta

Butterscotch Reduced Calorie Pudding	½ cup prepared	115
Chocolate Reduced Calorie Pudding	½ cup prepared	115
Vanilla Reduced Calorie Pudding	½ cup prepared	105

PORTION   SODIUM (MG.)

Estee

Chocolate	½ cup prepared	65
Lemon	½ cup prepared	5
Vanilla	½ cup prepared	75

Jell-O

Americana Golden Egg Custard	½ cup prepared	220
Americana Rice Pudding	½ cup prepared	155
Americana Tapioca Pudding, Chocolate	½ cup prepared	170
Americana Tapioca Pudding, Vanilla	½ cup prepared	170
Banana Cream Pudding/Pie Filling	⅙ pie filling prepared	165
Butterscotch Pudding/Pie Filling	½ cup prepared	245
Chocolate Fudge Pudding/Pie Filling	½ cup prepared	170
Chocolate Pudding/Pie Filling	½ cup prepared	170
Coconut Cream Pudding/Pie Filling	⅙ pie filling prepared	140
French Vanilla Pudding/Pie Filling	½ cup prepared	200
Lemon Pudding/Pie Filling	⅙ filling prepared	90
Milk Chocolate Pudding/Pie Filling	½ cup prepared	170
Vanilla Pudding/Pie Filling	½ cup prepared	200

Royal

Banana Pudding	½ cup prepared	225
Butterscotch Pudding	½ cup prepared	225
Chocolate Pudding	½ cup prepared	145
Chocolate Tapioca Pudding	½ cup prepared	145
Custard Pudding/Pie Filling	½ cup prepared	130
Dark 'N Sweet Pudding	½ cup prepared	145
Flan Pudding/Pie Filling	½ cup prepared	130
Key Lime Pudding/Pie Filling	½ cup prepared	115
Lemon Pudding/Pie Filling	½ cup prepared	115

PORTION   SODIUM (MG.)

Vanilla Pudding	½ cup prepared	225
Vanilla Tapioca Pudding	½ cup prepared	225

Safeway

Banana	½ cup prepared	298
Butterscotch	½ cup prepared	312
Chocolate	½ cup prepared	356
Lemon	½ cup prepared	220
Pistachio	½ cup prepared	296
Vanilla	½ cup prepared	307

Instant
Jell-O Instant Pudding

Banana Cream	½ cup prepared	440
Butter Pecan	½ cup prepared	440
Butterscotch	½ cup prepared	475
Chocolate	½ cup prepared	515
Chocolate Fudge	½ cup prepared	480
Coconut Cream	½ cup prepared	355
French Vanilla	½ cup prepared	435
Lemon	½ cup prepared	385
Milk Chocolate	½ cup prepared	505
Pineapple Cream	½ cup prepared	390
Pistachio	½ cup prepared	425
Vanilla	½ cup prepared	420

Royal Instant Pudding

Banana	½ cup prepared	245
Butterscotch	½ cup prepared	245
Chocolate	½ cup prepared	365
Coconut	½ cup prepared	320
Coffee	½ cup prepared	245
Dark 'N Sweet	½ cup prepared	365
Lemon	½ cup prepared	245
Pistachio Nut	½ cup prepared	320
Vanilla	½ cup prepared	245

Ready to Serve

Del Monte

Banana Pudding Cup	5 oz	285

PORTION   SODIUM (MG.)

Butterscotch Pudding Cup	5 oz.	285
Chocolate Fudge Pudding Cup	5 oz.	260
Chocolate Pudding Cup	5 oz.	280
Tapioca Pudding Cup	5 oz.	250
Vanilla Pudding Cup	5 oz.	285

General Mills Pudding

Chocolate	5 oz. (½ cup)	260
Rice	4.2 oz. (½ cup)	150
Tapioca	4.3 oz. (½ cup)	170

Hunt's

Banana	5 oz. (1 can)	220
Butterscotch	5 oz. (1 can)	240
Chocolate	5 oz. (1 can)	160
Chocolate Fudge	5 oz. (1 can)	170
Chocolate Marshmallow	5 oz. (1 can)	160
German Chocolate	5 oz. (1 can)	160
Lemon	5 oz. (1 can)	80
Rice	5 oz. (1 can)	230
Tapioca	5 oz. (1 can)	170
Vanilla	5 oz. (1 can)	190

Jell-O Pudding Pops

Banana	1 bar	65
Chocolate	1 bar	100
Chocolate-Vanilla Swirl	1 bar	80
Vanilla	1 bar	65

Rich's Pudding

Banana, artificial flavor	3 oz.	117
Butterscotch, artificial flavor	4.5 oz.	190
Chocolate	4.5 oz.	204
Chocolate Sundi	3.5 oz.	129
Lemon Flavor	3 oz.	113
Vanilla, artificial flavor	4.5 oz.	242

Sego Spoon Up Pudding

Banana, artificial flavor	4 oz. (½ container)	174

PORTION   SODIUM (MG.)

	PORTION	SODIUM (MG.)
Butterscotch, artificial flavor	4 oz. (½ container)	174
Chocolate Flavored	4 oz. (½ container)	189
Chocolate Fudge, artificial flavor	4 oz. (½ container)	189
Chocolate Marshmallow, artificial flavor	4 oz. (½ container)	189
Vanilla, artificial flavor	4 oz. (½ container)	174

Swiss Miss

	PORTION	SODIUM (MG.)
Butterscotch Pudding	4 oz. (½ container)	175
Chocolate Custard	4 oz. (½ container)	149
Chocolate Malt Pudding	4 oz. (½ container)	175
Chocolate Pudding	4 oz. (½ container)	176
Chocolate Pudding Sundae	4 oz. (½ container)	166
Double Rich Pudding	4 oz. (½ container)	173
Egg Custard	4 oz. (½ container)	180
Rice Pudding	4 oz. (½ container)	296
Tapioca Pudding	4 oz. (½ container)	170
Vanilla Pudding	4 oz. (½ container)	175
Vanilla Pudding Sundae	4 oz. (½ container)	166

Grains

USDA

	PORTION	SODIUM (MG.)
Barley, Pearled	1 cup cooked	6
Buckwheat Groats	1 cup cooked	under 10
Bulgur	1 cup cooked	under 10
Cornmeal	1 cup cooked	under 10
Millet	1 cup cooked	under 10
Rice, Brown	1 cup cooked	10
Rice, White	1 cup cooked	4
Rice, White, Quick	1 cup cooked	13
Rice, White, Regular	1 cup cooked	6
Soy Grits	1 cup cooked	under 10
Wheat Germ, Toasted	1 oz. (¼ cup)	1
Whole Wheat Berries	1 cup cooked	under 10

Gravies and Sauces

(see also Condiments, page 240 and Seasonings, page 373)

GRAVIES

Durkee

Au Jus Gravy Mix	¼ cup prepared	229
Au Jus Roastin' Bag	¼ pkg. dry	657
Brown Gravy Mix	¼ cup prepared	260
Brown Gravy w/Mushrooms Mix	¼ cup prepared	351
Brown Gravy w/Onions Mix	¼ cup prepared	339
Chicken Gravy Mix	¼ cup prepared	428
Chicken Gravy Roastin' Bag	¼ pkg. dry	900
Chicken Italian Style Roastin' Bag	¼ pkg. dry	904
Creamy Chicken Gravy Mix	¼ cup prepared	382
Creamy Chicken Style Roastin' Bag	¼ pkg. dry	632
Homestyle Gravy Mix	¼ cup prepared	208
Meatloaf Roastin' Bag	¼ pkg. dry	868
Mushroom Gravy mix	¼ cup prepared	293
Onion Gravy Mix	¼ cup prepared	239
Onion Pot Roast Roastin' Bag	¼ pkg. dry	716
Pork Gravy Mix	¼ cup prepared	544
Pork Gravy Roastin' Bag	¼ pkg. dry	645
Pot Roast and Stew Roastin' Bag	¼ pkg. dry	741
Sparerib Sauce Roastin' Bag	¼ pkg. dry	546
Swiss Steak Gravy Mix	¼ cup prepared	371
Swiss Steak Roastin' Bag	¼ pkg. dry	752
Turkey Gravy Mix	¼ cup prepared	253

Franco-American

Au Jus Gravy	2 oz.	290

	PORTION	SODIUM (MG.)
Beef Gravy	2 oz.	315
Brown Gravy w/Onions	2 oz.	340
Chicken Giblet Gravy	2 oz.	320
Chicken Gravy	2 oz.	320
Mushroom Gravy	2 oz.	320
Pork Gravy	2 oz.	350
Turkey Gravy	2 oz.	300

French's

Au Jus Gravy Mix	2 oz. (¼ cup prepared)	265
Brown Gravy Mix	2 oz. (¼ cup prepared)	280
Gravy for Chicken Mix	2 oz. (¼ cup prepared)	300
Gravy for Pork Mix	2 oz. (¼ cup prepared)	280
Gravy for Turkey Mix	2 oz. (¼ cup prepared)	380
Home Style Gravy Mix	2 oz. (¼ cup prepared)	335
Mushroom Gravy Mix	2 oz. (¼ cup prepared)	305
Onion Gravy Mix	2 oz. (¼ cup prepared)	350

Heinz

Brown Homestyle Gravy	2 oz.	180
Chicken Homestyle Gravy	2 oz.	275
Mushroom Homestyle Gravy	2 oz.	220
Onion Homestyle Gravy	2 oz.	220
Pork Homestyle Gravy	2 oz.	280
Turkey Homestyle Gravy	2 oz.	280

La Choy

Brown Gravy	¼ cup	612

Pillsbury

Brown Gravy Mix	0.2 oz. dry (¼ cup prepared)	305
Chicken Gravy Mix	0.2 oz. dry (¼ cup prepared)	230
Home Style Gravy Mix	0.2 oz. dry (¼ cup prepared)	300

Weaver

Chicken Croquette Gravy	2 oz.	340

PORTION   SODIUM (MG.)

Weight Watchers

Brown Gravy Mix	0.1 oz., dry	335
Brown Gravy Mix w/Mushrooms	0.1 oz., dry	374
Brown Gravy Mix w/Onions	0.1 oz., dry	370
Chicken Flavored Gravy Mix	0.1 oz., dry	522

SAUCES

Appian Way

Pizza Topping	1.5 oz.	185

Aunt Millie's

Spaghetti Sauce, Italian Sausage	4 oz.	322
Spaghetti Sauce, Marinara	4 oz.	327
Spaghetti Sauce, Meatless	4 oz.	322
Spaghetti Sauce, Peppers/Mushrooms	4 oz.	343
Spaghetti Sauce, Peppers/ Onions	4 oz.	322
Spaghetti Sauce, Peppers/ Sausage	4 oz.	306
Spaghetti Sauce w/Meat	4 oz.	322

Chef Boy-ar-dee (Cans)

Pizza Sauce w/Cheese	1 oz.	146
Pizza Sauce w/Pepperoni Slices	1 oz.	196
Pizza Sauce w/Sausage	1 oz.	230
Spaghetti Sauce (meatless)	4 oz.	704
Spaghetti Sauce w/Meat (8 oz. can)	4 oz.	650
Spaghetti Sauce w/Meat (29 oz. can)	4 oz.	706
Spaghetti Sauce w/Meat (15 oz. can)	4 oz.	755
Spaghetti Sauce w/Mushrooms (8 oz. can)	4 oz.	650

PORTION   SODIUM (MG.)

Spaghetti Sauce w/Mushrooms (15 oz. can)	4 oz.	790
Spaghetti Sauce w/Mushrooms (30 oz. can)	4 oz.	615

Chef Boy-ar-dee (Dry Mix)

Spaghetti Sauce Mix Complete w/Tomato Base and Cheese	4 oz., prepared	740

Chef Boy-ar-dee (Glass Jars)

Pizza Magic Pepperoni Flavor Sauce	1 oz.	153
Pizza Magic Pepperoni Style Sauce	1 oz.	159
Pizza Magic Sausage Flavor Sauce	1 oz.	163
Pizza Sauce w/Cheese	3.8 oz.	565
Portovista Sauce Regular (14 oz. can)	3.5 oz.	705
Portovista Sauce Regular (28 oz. can)	3.5 oz.	720
Portovista Sauce w/Meat (14 oz. can)	3.5 oz.	725
Portovista Sauce w/Meat (28 oz. can)	3.5 oz.	715
Portovista Sauce w/Mushrooms (14 oz. can)	3.5 oz.	770
Portovista Sauce w/Mushrooms (28 oz. can)	3.5 oz.	760
Spaghetti Sauce, Marinara	4 oz.	755
Spaghetti Sauce, Meatless (16 oz. jar)	4 oz.	790
Spaghetti Sauce, Meatless (29 oz. jar)	4 oz.	705
Spaghetti Sauce w/Ground Beef (16 oz. jar)	4 oz.	605
Spaghetti Sauce w/Ground Beef (29 oz. jar)	4 oz.	820

PORTION   SODIUM (MG.)

Spaghetti Sauce w/Mushrooms
  (16 oz. jar) ....4 oz....................655
Spaghetti Sauce w/Mushrooms
  (29 oz. jar) ....4 oz....................690

Contadina

Heavy Tomato Puree ....4 oz........................22
Italian Sauce.........................4 oz......................556
Pizza Sauce ....4 oz......................675
Sweet 'n Sour Sauce ....4 oz......................450
Swiss Steak Sauce ....4 oz......................638
Tomato Paste ....4 oz........................46
Tomato Sauce ....4 oz......................600

Del Monte

Enchilada Sauce, Hot.............4 oz. (½ cup)..............1090
Enchilada Sauce, Mild ....4 oz. (½ cup)..............1150
Tomato Paste ....4 oz........................73
Tomato Paste, No Salt
  Added..................................4 oz........................73
Tomato Sauce ....4 oz......................665
Tomato Sauce, No Salt
  Added..................................4 oz........................25
Tomato Sauce w/Onions ....4 oz......................575
Tomato Sauce w/Tomato Bits 4 oz......................525

Durkee

A la King Sauce, dry mix..........½ cup prepared..................692
Cheese Sauce, dry mix ....¼ cup prepared..................223
Enchilada Sauce Mix ....1 oz. dry 284
Hearty Beef Simmer Sauce......½ oz. (¼ package)..............845
Hollandaise Sauce, dry mix......3 tbl. prepared 137
Homestyle Simmer Sauce ....½ oz. (½ package)............2161
Italian Simmer Sauce ....1 oz. (¼ package) 1041
Meat Marinade, dry mix ....2 tbl. prepared..................1026
Mushroom Simmer Sauce ....½ oz. (¼ package)............384
Savory Tomato Sauce................½ oz. (¼ package)............1003
Sour Cream, dry mix ....2½ tbl. prepared 170
Spaghetti Sauce, dry mix..........½ cup prepared..................787

PORTION   SODIUM (MG.)

Spaghetti Sauce with
  Mushrooms ....½ cup prepared..................582
Stroganoff, dry mix ....⅓ cup prepared..................290
Sweet & Sour Sauce Mix.........½ cup prepared..................527
White Sauce, dry mix ....⅓ cup prepared..................266

Featherweight

Spaghetti Sauce.......................½ cup.....................................8

French's

Cheese Sauce Mix ....¼ cup prepared....................425
Hollandaise Sauce Mix.............3 tbl. prepared 290
Sour Cream Sauce Mix ....2½ tbl. prepared 130
Spaghetti Sauce Mix, Italian
  Style ....5 oz. prepared 900
Spaghetti Sauce Mix, Thick
  Homemade Style ....7 oz. prepared..................1455
Spaghetti Sauce w/Mushrooms
  Mix ....5 oz. prepared...................1045
Stroganoff Sauce Mix ....⅓ cup prepared..................490
Sweet 'N Sour Sauce Mix ....½ cup prepared..................135
Teriyaki Sauce Mix ....2 tbl. prepared...................1180

Health Valley

Bellissimo Pasta Sauce ....4 oz.....................................425
Bellissimo Pasta Sauce, no
  salt added ....4 oz......................................85
Tomato Sauce, Natural.............4 oz. (½ cup)......................585
Tomato Sauce, Natural, no
  salt ....4 oz. (½ cup) 33

Hunt's

Italian Tomato Sauce ....4 oz.....................................520
Manwich Sauce, Mexican ....2.5 oz. 470
Manwich Sauce, Regular..........2.5 oz. 410
Prima Salsa Meat Sauce..........4 oz.....................................635
Prima Salsa Mushroom
  Sauce....................................4 oz.....................................560
Prima Salsa Regular Sauce ....4 oz.....................................600

PORTION   SODIUM (MG.)

	PORTION	SODIUM (MG.)
Spaghetti Sauce Regular, No Salt Added	4 oz.	30
Tomato Paste	4 oz.	300
Tomato Paste, Italian Style	4 oz.	1050
Tomato Paste, No Salt Added	4 oz.	50
Tomato Puree	4 oz.	180
Tomato Sauce	4 oz.	670
Tomato Sauce, No Salt Added	4 oz.	25
Tomato Sauce Special	4 oz.	320
Tomato Sauce w/Bits	4 oz.	690
Tomato Sauce w/Cheese	4 oz.	800
Tomato Sauce w/Herbs	4 oz.	500
Tomato Sauce w/Mushrooms	4 oz.	710
Tomato Sauce w/Onions	4 oz.	670

La Choy

	PORTION	SODIUM (MG.)
Sweet & Sour Sauce	4 oz. (½ cup)	640

Prego

	PORTION	SODIUM (MG.)
Spaghetti Sauce	4 oz.	670
Spaghetti Sauce Flavored w/Meat	4 oz.	680
Spaghetti Sauce w/Mushrooms	4 oz.	640
Spaghetti Sauce, No Salt Added	4 oz.	25

Ragu

	PORTION	SODIUM (MG.)
Chunky Garden Style	4 oz.	435
Extra Thick and Zesty Plain Spaghetti Sauce	4 oz.	1110
Extra Thick and Zesty Spaghetti Sauce Flavored w/Meat	4 oz.	1110
Extra Thick and Zesty Spaghetti Sauce w/Mushrooms	4 oz.	1110

PORTION   SODIUM (MG.)

Homestyle Spaghetti Sauce, Plain	4 oz.	470
Homestyle Spaghetti Sauce Flavored w/Meat	4 oz.	470
Homestyle Spaghetti Sauce w/Mushrooms	4 oz.	470
Marinara Sauce	4 oz.	770
Pizza Quick Sauce, All Flavors	1.7 oz. (3 Tbsp.)	1110
Pizza Quick Sauce, Chunky Styles	1.9 oz. (3 Tbsp.)	1110
Plain Spaghetti Sauce	4 oz.	770
Spaghetti Sauce Flavored w/Meat	4 oz.	770
Spaghetti Sauce w/Mushrooms	4 oz.	770

Snow's

Welsh Rarebit Cheese Sauce	½ cup	460

Stokely-Van Camp

Tomato Sauce	4 oz.	825

Weight Watchers

Lemon-Butter Flavored Sauce	0.2 oz. dry	118

Western

Ham Glaze	4 oz.	89
Sweet 'N Sour Sauce	4 oz.	1700

Ice Cream and Frozen Desserts

ICE CREAM
Breyers

Butter Almond	4 oz. (½ cup)	42
Butter Pecan	4 oz. (½ cup)	42
Chocolate	4 oz. (½ cup)	30

PORTION   SODIUM (MG.)

Chocolate Almond	4 oz. (½ cup)	28
Coffee	4 oz. (½ cup)	44
Mint Chocolate Chip	4 oz. (½ cup)	42
Neapolitan	4 oz. (½ cup)	36
Peach	4 oz. (½ cup)	31
Southern Pecan Butterscotch	4 oz. (½ cup)	47
Strawberry	4 oz. (½ cup)	34
Vanilla	4 oz. (½ cup)	45
Vanilla Fudge Twirl	4 oz. (½ cup)	41

Foremost

Vanilla	4 oz. (½ cup)	38

Land O Lakes

Vanilla	4 oz. (½ cup)	60

Sealtest

Banana	4 oz. (½ cup)	62
Chocolate	4 oz. (½ cup)	83
Coconut	4 oz. (½ cup)	69
Coffee	4 oz. (½ cup)	77
French Vanilla	4 oz. (½ cup)	76
Lemon	4 oz. (½ cup)	77
Pineapple	4 oz. (½ cup)	64
Vanilla	4 oz. (½ cup)	78

ICE MILK AND SHERBET

Foremost

Ice Milk, Vanilla	4 oz. (½ cup)	51
Sherbet, all flavors	4 oz. (½ cup)	26

Land O Lakes

Ice Milk, Vanilla	4 oz. (½ cup)	50
Sherbet, fruit flavors	4 oz. (½ cup)	25

Sealtest Light 'n Lively Ice Milk

Caramel Nut	4 oz. (½ cup)	122

	PORTION	SODIUM (MG.)
Cherry-Pineapple	4 oz. (½ cup)	72
Chocolate	4 oz. (½ cup)	82
Coffee	4 oz. (½ cup)	87
Lemon Chiffon	4 oz. (½ cup)	90
Neapolitan	4 oz. (½ cup)	81
Orange-Pineapple	4 oz. (½ cup)	79
Peach	4 oz. (½ cup)	73
Strawberry	4 oz. (½ cup)	74
Toffee Crunch	4 oz. (½ cup)	80

YOGURT, FROZEN

Colombo

Frozen Soft Serve	8 oz.	211

Dannon

Danny Flip	5 oz.	20
Danny-in-a-Cup	8 oz.	47
Danny-on-a-Stick	2.5 oz.	15
Danny Parfait	4 oz.	15
Danny Sampler	3 oz.	30
Danny-Yo	3.5 oz.	25

Sealtest

Black Cherry	4 oz. (½ cup)	47
Lemon	4 oz. (½ cup)	52
Peach	4 oz. (½ cup)	45
Red Raspberry	4 oz. (½ cup)	45
Strawberry	4 oz. (½ cup)	46
Vanilla	4 oz. (½ cup)	52

OTHER

Sugar Lo

Frozen Dairy Desserts	4 oz. (½ cup)	47

Weight Watchers

Chocolate Mint Treat	3 fl. oz. (1 bar)	70
Chocolate Treat	3 fl. oz. (1 bar)	70

PORTION   SODIUM (MG.)

Dietary Frozen Dessert/ Imitation Ice Milk (all flavors)	4.9 fl. oz.	70
Orange Vanilla Treat	3 fl. oz. (1 bar)	70
Snack Cups Frozen Dessert (all flavors)	5 fl. oz. (1 cup)	70

Jams and Jellies

Dia-Mel

All varieties	1 tbl.	3

Diet Delight

Apple Spread	1 tsp.	3
Apricot/Pineapple Spread	1 tsp.	2
Blackberry Spread	1 tsp.	6
Grape Spread	1 tsp.	6
Strawberry Spread	1 tsp.	2

Kraft

Grape Jelly, reduced calorie	1 tsp.	10
Jam & Jelly (all varieties)	1 tsp.	5
Preserves (all varieties)	1 tsp.	0
Strawberry Preserves, reduced calorie	1 tsp.	5

Safeway

All Jams & Jellies	1 tbl.	0
All Preserves, except Plum & Red Raspberry	1 tbl.	0
Plum Preserves	1 tbl.	5
Red Raspberry Preserves	1 tbl.	5

Smucker's

Artificially Sweetened Imitation Grape	1 tbl.	under 10
Artificially Sweetened Imitation Strawberry	1 tbl.	under 10
Grape Jelly	1 tbl.	under 10

PORTION   SODIUM (MG.)

Tillie Lewis

	PORTION	SODIUM (MG.)
Apple Fruit Spread	1 tbl.	under 10
Apricot-Pineapple Fruit Spread	1 tbl.	under 10
Grape Fruit Spread	1 tbl.	under 10
Peach Fruit Spread	1 tbl.	under 10
Strawberry Fruit Spread	1 tbl.	under 20

Meat and Meat Substitutes

(see also Entrees and Dinners, page 264)

MEAT
Canned
Armour

	PORTION	SODIUM (MG.)
Banner Sausage	2.6 oz.	560
Beef Tripe	6 oz.	280
Chopped Beef	3 oz.	1250
Chopped Ham	3 oz.	1280
Deviled Ham	1.5 oz.	380
Deviled Treet	1.5 oz.	390
Lunch Tongue	3 oz.	1400
Pork Brains in Milk Gravy	2.8 oz.	410
Potted Meat (3 oz. size)	1.5 oz.	460
Potted Meat (5.5 oz. size)	1.83 oz.	560
Sliced Dried Beef	1 oz.	1220
Smoked Vienna Sausage	2 oz.	400
Treet	3 oz.	1190
Vienna Sausage in Barbecue Sauce	2.5 oz.	580
Vienna Sausage in Beef Stock	2 oz.	380

Libby's

	PORTION	SODIUM (MG.)
Corned Beef	2.3 oz.	720
Potted Meat	1.8 oz.	320
Vienna Sausage in Beef Broth	2 oz. (3½ links)	330

PORTION   SODIUM (MG.)

Oscar Mayer

Sandwich Spread	2 oz.	550

The Spreadables

Ham Salad	2 oz.	354

Swift Premium

Canned Ham	3.5 oz.	945
Hostess Ham	3.5 oz.	1265

Underwood

Corned Beef Spread	2 oz.	538
Deviled Ham	2 oz.	569
Liverwurst Spread	2 oz.	475
Roast Beef Spread	2 oz.	429

Fresh

USDA

Beef, Cooked, Lean	4 oz.	73
Brain, Raw	4 oz.	140
Heart, Beef, Braised	4 oz.	116
Heart, Calf, Braised	4 oz.	128
Kidney Beef, Braised	4 oz.	284
Lamb, Cooked, Lean	4 oz.	77
Liver, Calf, Fried	4 oz.	132
Liver, Pork, Simmered	4 oz.	56
Pork, Fresh, Cooked, Lean	4 oz.	79
Rabbit, Flesh, Cooked	4 oz.	70
Rabbit, Leg, Raw	4 oz.	40
Sweetbreads, Calf, Cooked	4 oz.	128
Tongue, Beef, Braised	4 oz.	68
Tripe, Commercial	4 oz.	52

Processed

Armour

Armour 1877 Canadian Bacon	2.0 oz.	850

PORTION   SODIUM (MG.)

	PORTION	SODIUM (MG.)
Armour Golden Star Canned Ham	3.0 oz.	970
Armour Golden Star Nugget	3.0 oz.	970
Armour Star Canned Chopped Ham	3.0 oz.	1100
Armour Star Canned Ham	3.0 oz.	970
Armour Star Sliced Bacon	0.9 oz.	180
Armour Star Spiced Luncheon Meat with Chicken	3.0 oz.	1040
Armour Star Thick Sliced Bacon	1.3 oz.	265
Barbecue Loaf	2 oz.	760
Beef Bologna	2 oz.	570
Beef Franks	1.6 oz. (1 link)	455
Bologna	2 oz.	570
Chopped Ham	2 oz.	770
Cooked Salami	2 oz.	600
German Sausage-Beef	2 oz.	770
Giant All Meat Franks	1.6 oz. (1 link)	455
Giant All Meat Franks	1.9 oz. (1 link)	565
Giant Beef Bologna	2 oz.	570
Giant Beef Franks	2 oz. (1 link)	565
Giant Great 8 Beef Franks	2.0 oz. (1 link)	565
Giant Great 8 Meat Franks	2.0 oz. (1 link)	565
Giant Olive Loaf	2 oz.	470
Giant P&P Loaf	2 oz.	700
Giant Plain Bologna	2 oz.	570
Hard Salami	2 oz.	1050
Italian Hard Salami	2 oz.	1040
Italian Pepperoni	2 oz.	1000
Meat Bologna	2 oz.	570
N/C Liverwurst	2 oz.	650
Old Fashioned Loaf	2 oz.	640
Sliced Genoa Salami	2 oz.	950
Sliced Pepperoni	2 oz.	1000
Sliced Salami	2 oz.	1050
Spiced Luncheon Meat	2 oz.	600
Summer Sausage-Cheese	2 oz.	760

PORTION   SODIUM (MG.)

Eckrich

	PORTION	SODIUM (MG.)
Banquet Loaf (Beef Smorgas Pac)	1.5 oz. (2 slices)	500
Bar-B Loaf	2 oz. (2 slices)	740
Beef Bologna	2 oz. (2 slices)	560
Beef Bologna, (Beef Smorgas Pac)	1.5 oz. (2 slices)	460
Beef Franks (12 oz.)	1.2 oz. (1 frank)	380
Beef Franks (1 lb.)	1.6 oz. (1 frank)	480
Beef Franks, Jumbo	2 oz. (1 frank)	620
Beef Pickle Loaf (Beef Smorgas Pac)	1.5 oz.	520
Beer Salami, sliced	2 oz. (2 slices)	700
Bologna, Beef (12 oz.)	2 oz. (2 slices)	540
Bologna, Lunch (chub)	2 oz.	580
Bologna, Sandwich	2 oz. (2 slices)	620
Bologna, Sliced	2 oz. (2 slices)	580
Bologna w/Cheese	1.3 oz (2 slices)	580
Bologna (Smorgas Pac)	1.5 oz. (2 slices)	460
Bologna (Smorgas Pac)	2 oz. (2 slices)	620
Braunschweiger Chub	2 oz. (2 slices)	800
Canadian Style Bacon	1 oz. (1 slice)	460
Cheese Franks	2 oz. (1 frank)	650
Cheese Smoked Sausage	2 oz.	500
Chopped Ham (Smorgas Pac)	1.5 oz (2 slices)	500
Cotto Beef Salami	2.7 oz. (2 slices)	960
Cotto Salami	(2 slices)	680
Franks (12 oz.)	1.2 oz. (1 frank)	360
Franks (1 lb.)	1.6 oz. (1 frank)	470
Franks, Jumbo	2 oz. (1 frank)	630
Garlic Bologna, Sliced	2 oz. (2 slices)	580
German Brand Bologna	2 oz. (2 slices)	700
German Brand Bologna Chub	2 oz. (2 slices)	720
Gourmet Loaf	2 oz. (2 slices)	780
Gourmet Loaf (Beef Smorgas Pac)	1.5 oz. (2 slices)	600
Ham, Chopped	2 oz. (2 slices)	660
Ham, Cooked, Sliced	2.4 oz. (2 slices)	940
Ham, Smoked, Sliced Sweet	1.5 oz. (2 slices)	540

PORTION   SODIUM (MG.)

	PORTION	SODIUM (MG.)
Ham & Cheese Loaf	2 oz. (2 slices)	700
Ham Loaf	2 oz. (2 slices)	660
Ham Smok-Y-Links	1.7 oz. (2 links)	560
Hard Salami, Sliced	2 oz.	1200
Honey Style Loaf	2 oz. (2 slices)	705
Honey Style Loaf (Smorgas Pac)	1.5 oz. (2 slices)	560
Honey Style Loaf (Smorgas Pac)	2 oz. (2 slices)	740
Hot Links Smoked Sausage	2.7 oz. (1 link)	640
Imported Danish Ham	2 oz.	780
Italian Sausage Links	1 link	562
Macaroni Cheese Loaf	2 oz. (2 slices)	740
Maple Flavor Smok-Y-Links Sausage	1.7 oz. (2 links)	400
Minced Roll Sausage	2 oz. (2 slices)	680
New England Brand Sausage	2 oz. (2 slices)	880
Old Fashion Loaf	2 oz. (2 slices)	660
Old Fashion Loaf (Smorgas Pac)	1.5 oz. (2 slices)	500
Old Fashion Loaf (Smorgas Pac)	2 oz. (2 slices)	680
Olive Loaf	2 oz. (2 slices)	740
Peppered Loaf	2 oz. (2 slices)	780
Pickle Loaf	2 oz. (2 slices)	640
Pickle Loaf (Smorgas Pac)	2 oz. (2 slices)	640
Polska Kielbasa, 16 or 20 oz.	2 oz.	520
Polska Kielbasa Skinless Links, 12 oz.	2 oz. (2 links)	500
Polska Kielbasa Skinless Links, 1 lb.	2 oz. (1 link)	490
Ring Bologna	2 oz.	560
Salami (Chub), Cooked	2 oz.	720
Skinless Smoked Sausage	2 oz.	500
Slender Sliced Beef	2 oz.	1120
Slender Sliced Corned Beef	2 oz.	680
Slender Sliced Smoked Ham	2 oz.	720
Slender Sliced Pastrami	2 oz.	720
Slender Sliced Smoked Pork	2 oz.	700

PORTION   SODIUM (MG.)

Smoked Sausage with Cheese	2.7 oz. (1 link)	670
Smok-Y-Links, Skinless	1.7 oz. (2 links)	410
Smok-Y-Links, Beef	1.7 oz. (2 links)	403
Smoked Sausage	2 oz.	530
Smoked Sausage, Beef	2 oz.	520
Smoked Sausage Skinless Links	2 oz. (1 link)	490
Smoky Tang Summer Sausage, Sliced	2 oz.	700
Thick Sliced Beef Bologna	2 slices	800
Thick Sliced Bologna, 12 oz.	2 slices	980
Thick Sliced Bologna, 1 lb.	1.8 oz. (1 slice)	510
Thin Sliced Bologna	2 oz. (2 slices)	640
Thin Sliced Beef Bologna	2 oz. (2 slices)	640

Featherweight

Cooked Ham	3 oz.	60
Corned Beef Loaf	2.5 oz.	60
Pork Frankfurters	1 frank	20

Frito-Lay

Beef Jerky	1 oz.	652
Hot Sausage	1 oz.	575
Smoked Beef Sticks	1 oz.	430
Summer Sausage	1 oz.	543

Homestead

Bologna	2 oz. (2 slices)	419
Canadian Bacon	1.2 oz.	551
Chip Chop Loaf	2 oz. (2 slices)	608
Cooked Salami	2 oz. (2 slices)	434
Dutch Loaf	2 oz. (2 slices)	489
Ham, Cooked	1.2 oz. (2 slices)	296
Ham & Cheese Loaf	2 oz. (2 slices)	447
Honey Loaf	2 oz. (2 slices)	457
Italian Loaf	2 oz. (2 slices)	480
Kielbassi Loaf	2 oz. (2 slices)	451

	PORTION	SODIUM (MG.)
Kielbassi Sausage	2 oz.	441
Pepper Loaf	2 oz. (2 slices)	530
Pickle & Pimiento Loaf	2 oz. (2 slices)	424
Pizza Loaf	2 oz. (2 slices)	468
Pride Ham	2 oz.	494
Smoked Sausage	2 oz.	428

Jefferson

Jumbo Franks	2 oz. (1 frank)	538
Ring Bologna	2 oz.	751
Sliced Big Bologna	2 oz.	722
Sliced Summer Sausage	2 oz.	725
Wieners	1.6 oz. (1 wiener)	465

Klement's

Barbeque Loaf	2 oz.	619
Beef, Cooked	2 oz.	574
Beef, Cooked, Reduced Salt	2 oz.	350
Blood and Tongue	2 oz.	318
Bologna, Ring (Beef)	2 oz.	583
Bologna, Ring (Beef & Pork)	2 oz.	605
Bologna, Slicing (Beef & Pork)	2 oz.	605
Bratwurst, Cooked, Natural Casing	2 oz.	510
Bratwurst, Cooked, Skinless	2 oz.	555
Braunschweiger Liver Sausage	2 oz.	579
Corned Beef, Pressed & Formed	2 oz.	628
Corned Beef, Top Round	2 oz.	446
Cotto Salami	2 oz.	691
Dutch Loaf	2 oz.	573
Farm Sausage	2 oz.	561
Good Chews	2 oz.	838
Ham, Cooked	2 oz.	625
Honey Loaf	2 oz.	716
Italian Sausage (Skinless)	2 oz.	601
Jellied Beef Loaf	2 oz.	559
Kishka, Dark	2 oz.	345

	PORTION	SODIUM (MG.)
Liver Sausage	2 oz	579
Meat Loaf	2 oz	432
Mortadela	2 oz	604
N.C. Coarse Wieners	2 oz	584
New England Ham	2 oz	645
Old Fashioned Loaf	2 oz	636
Olive Loaf	2 oz	690
Party Loaf (Ham & Cheese)	2 oz	608
Pepper Loaf	2 oz	661
Pickle and Pimento Loaf	2 oz	596
Polska Kielbasa	2 oz	561
Sandwich Loaf, Jumbo	2 oz	605
Sausage, Fresh (all)	2 oz	512
Short Polish	2 oz	561
Smoked Bratwurst (Grillwurst)	2 oz	561
Smoked Butts	2 oz	557
Smoked Ham	2 oz	601
Smoked Sausage	2 oz	561
Sulze or Head Cheese	2 oz	333
Summer Sausage	2 oz	650
Thuringer	2 oz	650
Top Round, Cooked, Choice	2 oz	190
Turkey Breast	2 oz	522
Wieners (Franks, Sheep Casing, Skinless)	2 oz	605
Yachtwurst	2 oz	651

Klement's Old Fashioned Sausage

	PORTION	SODIUM (MG.)
BBQ Loaf	2 oz. (2 slices)	740
Chicken Roll	2 oz. (2 slices)	290
Cooked Beef	2 oz. (2 slices)	740
Cooked Top Round Choice	2 oz. (2 slices)	190
Corn Beef	2 oz. (2 slices)	740
Ham, Cooked	2 oz. (2 slices)	740
Ham, Smoked	2 oz. (2 slices)	740
Honey Loaf	2 oz. (2 slices)	740
Jellied Beef Loaf	2 oz. (2 slices)	740
Pepper Ham Loaf	2 oz. (2 slices)	740
Yachtwurst	2 oz. (2 slices)	740

PORTION   SODIUM (MG.)

Olde Smithfield

	PORTION	SODIUM (MG.)
Bacon	2 oz.	380
Bologna (Beef)	2 oz.	412
Bologna (Meat)	2 oz.	400
Cooked Salami	2 oz.	518
Franks	1.2 oz.	338
Franks, large	2 oz.	474
Ham, Buffet	2 oz.	500
Ham, Cooked	2 oz.	476
Ham, Pressed	2 oz.	392
Ham, Smoked	2 oz.	538
Hot Pork Sausage	2 oz.	309
Mild Pork Sausage	2 oz.	309
Spiced Luncheon Meat	2 oz.	420

Oscar Mayer

	PORTION	SODIUM (MG.)
Bacon	0.4 oz. cooked (2 slices)	228
Bar-B-Q Loaf	2 oz. (2 slices)	738
Beef Cotto Salami	1.6 oz. (2 slices)	558
Beef Franks	1.6 oz. (1 frank)	466
Beef Honey Roll Sausage	1.6 oz. (2 slices)	608
Beef Salami for Beer	1.6 oz. (2 slices)	468
Beef Smokies	1.5 oz. (1 sausage)	455
Beef Summer Sausage	1.6 oz. (2 slices)	634
Bologna	1.6 oz. (2 slices)	482
Bologna, Beef	1.6 oz. (2 slices)	478
Bologna & Cheese	1.6 oz. (2 slices)	494
Braunschweiger (Liver Sausage)	2 oz.	660
Canadian Style Bacon	2 oz. (2 slices)	786
Cheese Smokies	1.5 oz. (1 sausage)	450
Corn Dogs (after heating)	4 oz. (1 sausage)	1252
Cotto Salami	1.6 oz. (2 slices)	490
Ham, Chopped	2 oz. (2 slices)	746
Ham, Cooked (smoked; sectioned & formed)	1.5 oz. (2 slices)	590
Ham & Cheese Loaf	2 oz. (2 slices)	730
Ham Roll Sausage	1.6 oz. (2 slices)	540
Hard Salami	0.6 oz. (2 slices)	334

PORTION   SODIUM (MG.)

Head Cheese	2 oz. (2 slices)	704
Honey Loaf	2 oz. (2 slices)	746
Jubilee Canned Ham	2 oz.	682
Jubilee Ham Slice (smoked; sectioned & formed)	2 oz.	746
Jubilee Ham, Smoked boneless	2 oz.	772
Jubilee Ham Steaks	2 oz. (1 slice)	741
Lean 'N Tasty Beef Breakfast Strips	0.6 oz. cooked (2 strips)	404
Lean 'N Tasty Pork Breakfast Strips	0.7 oz. cooked (2 strips)	440
Little Friers Pork Sausage	1.4 oz. cooked	446
Liver Cheese (pork fat wrap)	2.7 oz. (2 slices)	912
Luncheon Meat	2 oz. (2 slices)	716
Luxury Loaf	2 oz. (2 slices)	656
New England Brand Sausage	1.6 oz. (2 slices)	590
Old Fashioned Loaf	2 oz. (2 slices)	686
Olive Loaf	2 oz. (2 slices)	800
Peppered Loaf	2 oz. (2 slices)	798
Pickle & Pimento Loaf	2 oz. (2 slices)	764
Picnic Loaf	2 oz. (2 slices)	640
Salami for Beer	1.6 oz. (2 slices)	564
Smokie Links Sausage	1.5 oz. (1 link)	396
Southern Brand Pork Sausage Patties	1 oz. cooked (1 patty)	282
Summer Sausage	1.6 oz. (2 slices)	680
Wieners	1.6 oz. (1 wiener)	514
Wieners w/Cheese	1.6 oz. (1 wiener)	551

Swift

Beef Sizzlean	0.5 oz. (2 strips), cooked	260
Country Recipe Brown 'N Serve Links	1.5 oz. (2 links), cooked	375
Original Pork Brown 'N Serve Links	1.5 oz. (2 links), cooked	375

PORTION   SODIUM (MG.)

Original Pork Brown 'N Serve
  Patties ....1.7 oz. (2 patties), cooked 365
Pork Sizzlean ....0.6 oz. (2 strips),
                                   cooked................................380

MEAT SUBSTITUTES, Canned
Worthington

Chili...6.7 oz. (⅔ cup) 940
Choplets......................................3.2 oz. (2 slices) 770
Cutlets ....3.2 oz. (1½ slices) 430
FriChik ....3.2 oz. (2 pieces) 720
Numete ....2.4 oz. (½-in. slice) 480
Prime Stakes ....3.2 oz. (1 piece) 570
Protose.......................................2.7 oz. (½-in. slice) 560
Saucettes...................................2.4 oz. (2 links) 440
Skallops, Vegetable...................3 oz. (½ cup)......................475
Soyameat, Beef-like, Sliced......2 oz. (2 slices)......................450
Super-Links................................1.7 oz. (1 link) 480
Vegetable Steaks.......................3.2 oz. (2½ pieces) 400
Vegetarian Burger ....4 oz. (½ cup) 850
Vegetarian Country Stew ....9.5 oz. 410
Veja-Links ....2.2 oz. (2 links) 500
Worthington 209 (Turkey-like
  Slices) ....2.2 oz. (2½ slices) 560

Frozen

Morningstar Farms

Breakfast Links...........................2.4 oz. (3 links) 630
Breakfast Patties ....2.7 oz. (2 patties) 885
Breakfast Strips ....0.9 oz. (3 strips) 335
Grillers ....2.3 oz. (1 griller) 330

Processed

Worthington

Beef Style Roll ....2.5 oz. 940
Bolono Roll ....1.5 oz. (2 slices) 730
Corned Beef Style Roll..............2.5 oz. 1055

	PORTION	SODIUM (MG.)
Corned Beef Style Slices	2 oz. (4 slices)	710
Dinner Roast	2 oz.	495
Fillets	3 oz. (2 pieces)	920
FriPats	2.3 oz. (1 piece)	470
Granburger	1.2 oz. dry (6 tbl.)	985
Prosage Links	2.4 oz. (3 links)	535
Prosage Patties	2.7 oz. (2 patties)	765
Prosage Roll	2.5 oz. (2 slices)	590
Salami, Meatless	1.5 oz. (2 slices)	645
Smoked Beef Style Roll	2.5 oz.	1050
Smoked Beef Style Slices	2 oz. (6 slices)	990
Stakelets	2.5 oz. (1 piece)	570
Stakes Au Sauce	7 oz.	960
Stripples	1.2 oz. (4 strips)	550
Tuno Roll	2 oz.	390
Veelets Parmesano	7 oz.	1125
Veja-Links	2.2 oz. (2 links)	735
Wham Roll	2.4 oz.	1155
Wham Slices	2.4 oz. (3 slices)	1155

Milk, Cream, and Sour Cream

CREAM

Fresh

Land O Lakes

Half and Half	1 tbl.	5
Gourmet Heavy Whipping Cream	1 tbl.	5
Whipping Cream	1 tbl.	5

USDA

Half & Half	1 tbl.	6
Heavy Whipping Cream	1 tbl.	6
Light Table Cream	1 tbl.	6
Light Whipping Cream	1 tbl.	5
Medium Cream, 25% Fat	1 tbl.	6

PORTION   SODIUM (MG.)

Substitutes, Dry

Carnation

Coffee Mate ....2 tsp. 8

Cremora

Non-Dairy Creamer ....1 tsp. 5

N-Rich

Coffee Creamer, Nutritional ......2 tsp. 23

Pet

Non-Dairy Creamer ....2 tsp. 10

Substitutes, Liquid

Avoset

High Poly Milk Free Creamer 1 tbl. 6
Non-Dairy Creamer ....1 tbl. 8

Lucerne

Cereal Blend (Non-Dairy)..........1 tbl. 6
Coffee Tone........................1 tbl. 5

Rich's

Coffee Rich ....1 tbl. 5
Poly Rich ....1 tbl. 5
Richwhip (liquid)...................½ tbl. 10

CREAM TOPPINGS

Fresh

Kraft

Real Cream Topping ....¼ cup 10

Substitutes

Avoset

20% Cream Topping ....1 tbl. 4

PORTION   SODIUM (MG.)

Birds Eye

Cool Whip Extra Creamy
  Dairy Recipe Whip
  Topping ....1 tbl. 2

Cool Whip

Non-Dairy Whipped Topping ....1 tbl. 1
Dover Farms Whipped
  Topping ....1 tbl. 2

Dream Whip

Whipped Topping Mix ....1 tbl. 5

D-Zerta

Reduced Calorie Whipped
  Topping Mix..............................1 tbl. 5

Kraft

Whipped Topping for
  Dessert....................................¼ cup 10

La Creme

Whipped Topping........................1 tbl. 5

Rich's

Richwhip (pressurized) ....1 tbl..................................10
Richwhip (pre-whipped) ....1 tbl. 5

MILK

Canned

Carnation

Evaporated Lowfat Milk ....4 oz................................138
Evaporated Milk ....4 oz................................133
Evaporated Skim Milk...............4 oz................................140

Eagle

Sweetened Condensed Milk ....⅓ cup 120

PORTION   SODIUM (MG.)

Pet

Evaporated Milk ....½ cup 119
99 Evaporated Skimmed Milk....½ cup 118

USDA

Condensed, sweetened ....1 fl. oz. 49
Evaporated, skim.....................1 fl. oz. 37

Dry

Carnation
Instant Nonfat Dry Milk.............8 oz. recon. 125

Fearn Soya
Nonfat Dry Milk ....3 tbl., dry 155

Land O Lakes
Flash Instant Nonfat Dry
  Milk.....................................8 oz. recon. 125

Pet
Instant Nonfat Dry Milk.............5 tbl., dry...........................92

Sanalac
Instant Nonfat Dry Milk.............8 oz. recon. 125

Fresh

Foremost
Bulgarian Buttermilk ....8 oz. (1 cup) 230
Buttermilk.................................8 oz. (1 cup) 227
Profile Nonfat Milk....................8 oz. (1 cup) 125
So-Lo 2% Milk ....8 oz. (1 cup) 139
So-Lo Chocolate Lowfat
  Milk.....................................8 oz. (1 cup) 198
Vitamin D Milk ....8 oz. (1 cup) 121

Land O Lakes
Buttermilk.................................8 oz. (1 cup) 255

	PORTION	SODIUM (MG.)
Chocolate Milk	8 oz. (1 cup)	150
Chocolate Milk, 1% fat	8 oz. (1 cup)	150
Chocolate Skim Milk	8 oz. (1 cup)	155
Lowfat Milk, 1% fat	8 oz. (1 cup)	125
Lowfat Milk, 2% fat	8 oz. (1 cup)	120
Milk	8 oz. (1 cup)	120
Skim Milk	8 oz. (1 cup)	125

Sealtest

Chocolate Milk	8 oz. (1 cup)	113
Cultured Lowfat Buttermilk	8 oz. (1 cup)	221
Vitamin A & D Lowfat Milk	8 oz. (1 cup)	116

Weight Watchers

Skim Milk	8 oz. (1 cup)	118

USDA

Buttermilk, Cultured, Salted	1 cup	257
Chocolate, Lowfat, 2% fat	1 cup	150
Dry, Nonfat, Instantized	1 cup	125
Low Sodium, Whole	1 cup	6
Lowfat, 1% fat	1 cup	123
Lowfat, 1% fat, protein fortified	1 cup	143
Lowfat, 2% fat	1 cup	122
Lowfat, 2% fat, protein fortified	1 cup	145
Skim	1 cup	126
Skim, protein fortified	1 cup	144
Whole, 3.3% fat	1 cup	120

Substitutes

Health Valley

Soy Moo, Plain (Dry)	1 cup	110
Soy Moo, Carob (Dry)	1 cup	110
Soy Moo, Plain (Frozen)	1 cup	55
Soy Moo, Carob (Frozen)	1 cup	30

PORTION   SODIUM (MG.)

Magnolia

Sweetened Condensed Filled
  Dairy Blend ....⅓ cup 120

Pet

Dairymate Evaporated Filled
  Milk.......................................½ cup 120

SOUR CREAM

Breakstone
Sour Cream ....3 tbl..23

Foremost

Sour Cream ....3 tbl..21

Land O Lakes

Sour Cream ....3 tbl..15

USDA

Sour Cream, Cultured................1 tbl. 6
Sour Cream, Half & Half ....1 tbl. 6

Noodles, Potatoes, and Rice

**(see also Entrees and Dinners, page 264,
& Vegetables, Beans, and Tofu, page 400)**

NOODLES

Dry

Betty Crocker International Noodle Mixes

Fettuccine Alfredo ....¼ pkg. prepared 490
Parisienne ....¼ pkg. prepared 570
Romanoff..¼ pkg. prepared 705
Stroganoff ....¼ pkg. prepared 605

PORTION   SODIUM (MG.)

USDA

Macaroni	1 cup cooked	2
Spaghetti	1 cup cooked	2

Frozen

Stouffer's

Noodles Romanoff	4 oz	675

POTATOES
Canned

Del Monte

Sliced and Whole	½ cup	355

Dry

American Beauty

Mashed Potatoes (Flakes)	½ cup prepared	340

Betty Crocker

Chicken 'n Herb	⅙ package	585
Creamed Potatoes	½ cup prepared	415
Hash Browns w/Onions	½ cup prepared	460
Hickory Smoke Cheese	⅙ package	650
Julienne Potatoes w/Mild Cheese Sauce	½ cup prepared	570
Potato Buds Mashed Potatoes	½ cup prepared	355
Potatoes Au Gratin	½ cup prepared	605
Scalloped Potatoes	½ cup prepared	570
Sour Cream 'n Chive Potatoes	½ cup prepared	535

French's

Big Tate Mashed Packaged Potatoes	½ cup prepared	410
Cheese Scalloped Potato Casserole	½ cup prepared	540
Idaho Mashed, Packaged Potatoes	½ cup prepared	365

PORTION   SODIUM (MG.)

Potato Pancakes Mix ....three 3-in pancakes 490
Scalloped Potato Casserole,
  Crispy Top.................................½ cup prepared...................520
Sour Cream & Chives ....½ cup prepared...................660
Tangy Au Gratin Potatoes ....½ cup prepared...................525

Pillsbury

Hungry Jack Mashed (Flakes)....½ cup prepared...................380

Fresh

USDA

Potato, Baked or Boiled............1 medium 5
Sweet Potato, Baked or Boiled
  in Skin ....1 potato 20

Frozen

Birds Eye

Cottage Fries ....2.8 oz. 15
Crinkle Cuts ....3 oz...35
Farm Style Wedges...................3 oz...25
French Fries ....3 oz...25
Hash Browns ....4.0 oz......................................55
Shoestrings ....3.3 oz......................................50
Shredded Hash Browns ....3 oz...20
Steak Fries ....3 oz...25
Tasti Fries French Fried
  Potatoes..................................2.5 oz. 270
Tasti Puffs Potato Puffs ....2.5 oz. 400
Tiny Taters Potato Bits ....3.2 oz. 280
Whole Peeled Potatoes ....3.2 oz. 6

Green Giant

Potatoes and Sweet Peas in
  Bacon Cream Sauce...............½ cup 400
Sliced Potatoes in Butter
  Sauce.......................................½ cup 470
Stuffed Baked Potato
  w/Cheese Flavored
  Topping ....½ potato 520

PORTION   SODIUM (MG.)

Stuffed Baked Potato
  w/Sour Cream & Chives ......½ potato 580

Heinz

Self-Sizzling Crinkles ....3 oz.............................30
Self-Sizzling Fries ....3 oz.............................30
Self-Sizzling Shoestrings ....3 oz.............................40

Mrs. Paul's

Candied Yams ....4 oz............................105
Candied Yams 'N Apples ....4 oz.............................49
Potato and Cheese
  Pierogies ....5 oz. (3 pierogies) 720

Ore-Ida

Cottage Fries ....3 oz.............................40
Country Style Dinner Fries .....3 oz.............................50
Crispers...................................3 oz............................560
Crispy Crowns ....3 oz............................370
Crispy Crowns w/Onions ....3 oz............................480
Golden Crinkles ....3 oz.............................40
Golden Fries.............................3 oz.............................40
Golden Patties ....2.5 oz..........................340
Home Style Potato Wedges ....3 oz.............................20
Home Style Potato Planks........3 oz.............................30
Home Style Potato Slices ....3 oz.............................40
Home Style Potato Thins..........3 oz.............................40
O'Brien Potatoes ....3 oz.............................50
Pixie Crinkles ....3 oz.............................40
Shoe Strings.............................3 oz.............................40
Shredded Hash Browns ....6 oz.............................50
Small Whole Peeled
  Potatoes................................3 oz.............................40
Southern Style Hash
  Browns..................................3 oz.............................60
Tater Tots.................................3 oz............................550
Tater Tots w/Bacon Flavor ....3 oz............................720
Tater Tots w/Onions..................3 oz............................600

PORTION   SODIUM (MG.)

Stouffer's

Potatoes au Gratin	3.8 oz.	480
Scalloped Potatoes	4 oz.	450
Yams & Apples	5 oz.	225

RICE

Canned

Featherweight

Spanish Rice	7.5 oz.	32

La Choy

Chicken Fried Rice	4 oz.	1153
Chinese Style Fried Rice	4 oz.	1225
Fried Rice	¾ cup	965

Van Camp

Spanish Rice	8 oz.	1265

Dry

Minute Rice

Beef (Rib Roast)	½ cup prepared	720
Chicken (Drumstick)	½ cup prepared	685
Chinese Fried Rice	½ cup prepared	635
Long Grain & Wild Rice	½ cup prepared	570
Minute Rice, w/o salt & butter	⅔ cup prepared	2

Uncle Ben's

Brown, Long Grain, w/o butter or salt	4 oz. prepared (⅔ cup)	5
Brown & Wild Seasoned, w/o butter	4 oz. prepared (½ cup)	474
Converted, White, Enriched, Long Grain, w/o butter or salt	4 oz. prepared (⅔ cup)	2
Long Grain & Wild Seasoned, Fast Cooking	3.5 oz. prepared (½ cup)	387

PORTION   SODIUM (MG.)

Long Grain & Wild Seasoned,
w/o butter ....3.7 oz. prepared (½ cup)....420
Quick, White, Enriched, Long
Grain, w/o butter or salt ....4.5 oz. prepared (⅔ cup) 13

USDA

Brown ....1 cup cooked 10
White, Parboiled ....1 cup cooked 4
White, Quick ....1 cup cooked 13
White, Regular ....1 cup cooked 6

Frozen

Birds Eye

Rice & Peas w/Mushrooms ......2.3 oz. 320

Birds Eye Rice Dishes

Chinese Fried Style Rice ....3.6 oz. 425
French Style Rice ....3.6 oz. 635
Italian Style Rice ....3.6 oz. 385
Northern Italian Style Rice ....3.6 oz. 495
Oriental Style Rice ....3.6 oz. 460
Spanish Style Rice ....3.6 oz. 495

Green Giant Rice Dishes

Italian Blend White Rice
& Spinach in Cheese
Sauce ....½ cup 460
Long Grain White & Wild
Rice ....½ cup 565
Rice Medley ....½ cup 280
Rice 'n Broccoli in Flavored
Cheese Sauce ....½ cup 405
Rice Pilaf ....½ cup 520
Rice with Herb Butter Sauce ..½ cup 405

Nuts and Nut Butter

NUTS

Fisher

	Portion	Sodium (mg.)
Almonds, dry roasted, salted	1 oz.	45
Almonds, oil roasted, salted	1 oz.	56
Almonds, raw	1 oz.	1
Black Walnuts, raw	1 oz.	1
Brazil Nuts, oil roasted, salted	1 oz.	57
Cashews, dry roasted, salted	1 oz.	57
Cashews, oil roasted, salted	1 oz.	57
English Walnuts, raw	1 oz.	1
Filberts, oil dipped, salted	1 oz.	57
Peanuts, blanched, dry roasted, salted	1 oz.	101
Peanuts, blanched dry roasted unsalted	1 oz.	1
Peanuts, blanched, oil roasted, salted	1 oz.	119
Peanuts, in shell, roasted, salted	1 oz.	64
Peanuts, redskin, oil roasted, salted	1 oz.	119
Pecans, dry roasted, salted	1 oz.	100
Pecans, oil dipped, salted	1 oz.	60
Pecans, raw	1 oz.	0
Pistachios, in shell, roasted, salted	1 oz.	50
Pistachios, shelled, roasted, salted	1 oz.	100
Sunflower seeds, in shell, roasted, salted	1 oz.	58
Sunflower seeds, shelled, dry roasted, salted	1 oz.	108
Sunflower seeds, shelled, oil roasted, salted	1 oz.	108

Franklin

	Portion	Sodium (mg.)
Almonds	1 oz.	231

PORTION   SODIUM (MG.)

	PORTION	SODIUM (MG.)
Cashews	1 oz.	156
Club Mix	1 oz.	265
Crunch N Munch	1.25 oz.	185
Crunch N Munch Peanuts w/jackets	1 oz.	182
Crunch N Munch Peanuts w/o jackets	1 oz.	250
Filberts	1 oz.	185
Mixed Nuts	1 oz.	308
Peanuts w/jackets	1 oz.	182
Peanuts w/o jackets	1 oz.	158

Frito-Lay

	PORTION	SODIUM (MG.)
Cashews	1 oz.	115
Dry Roasted Peanuts	1 oz.	220
Salted in the Shell Peanuts	1 oz.	265
Salted Peanuts	1 oz.	170
Sunflower Kernels	1 oz.	170
Sunflower Seeds	1 oz.	50

Laura Scudder

	PORTION	SODIUM (MG.)
Almonds, Dry Roast Smoke Flavor	1 oz.	191
Cashews, shelled, salted, oil roasted	1 oz.	57
Dry Roast Cashews	1 oz.	261
Dry Roast Snackin Peanuts	1 oz.	247
Dry Roast Sunflower Nuts	1 oz.	215
Goober, roasted in shell	1 oz.	1
Pistachios, roasted in shell	1 oz.	0
Spanish Peanuts, shelled, salted, oil roasted	1 oz.	156
Sunflower Nuts, roasted in shell	1 oz.	5
Sunflower Nuts, shelled, salted, oil roasted	1 oz.	125
Virginia Peanuts, shelled, salted, oil roasted	1 oz.	183

PORTION   SODIUM (MG.)

Planters

Almonds, dry roasted	1 oz.	220
Cashews, dry roasted	1 oz.	220
Cashews, unsalted	1 oz.	under 10
Cashews, vacuum packed	1 oz.	220
Mixed Nuts, dry roasted	1 oz.	220
Mixed Nuts, unsalted	1 oz.	under 10
Mixed Nuts w/Peanuts	1 oz.	220
Mixed Nuts w/o Peanuts	1 oz.	220
Natural Pistachios	1 oz.	220
Old Fashioned Peanut Candy	1 oz.	120
Old Fashioned Peanuts	1 oz.	220
Peanuts, Cocktail	1 oz.	220
Peanuts, dry roasted	1 oz.	220
Peanuts, unsalted	1 oz.	under 10
Pecans	1 oz.	220
Spanish Peanuts	1 oz.	220
Sunflower Nuts, unsalted	1 oz.	under 10

USDA

Pine Nuts	1 oz.	3

PEANUT BUTTER

Bama

Creamy Peanut Butter	2 tbl.	160
Crunchy Peanut Butter	2 tbl.	135

Country Pure

Chunky Peanut Butter	1 oz. (2 tbl.)	135
Creamy Peanut Butter	1 oz. (2 tbl.)	137

Dia-Mel

Peanut Butter	2 tbl.	2

Elam's

Natural Peanut Butter w/Defatted Wheat Germ	2 tbl.	5

PORTION   SODIUM (MG.)

Health Valley

Chunky Peanut Butter	1 tbl.	30
Chunky Peanut Butter, unsalted	1 tbl.	1
Creamy Peanut Butter	1 tbl.	15
Creamy Peanut Butter, unsalted	1 tbl.	1

Jif

Creamy Peanut Butter	2 tbl	155
Crunchy Peanut Butter	2 tbl	130

Koeze

Old Fashioned Peanut Butter, no salt added	1 oz.	15

Laura Scudder

Old Fashioned Peanut Butter	1 oz. (2 tbl.)	214
Peanut Butter, homogenized	1 oz. (2 tbl.)	150

NuMade

Chunky Peanut Butter	1 oz. (2 tbl.)	142
Creamy Peanut Butter	1 oz. (2 tbl.)	131

Peter Pan

Crunchy Peanut Butter	1 oz. (2 tbl.)	190
Peanut Butter, low sodium	1 oz. (2 tbl.)	10
Creamy Peanut Butter	1 oz. (2 tbl.)	190

Planters

Creamy Peanut Butter	1 oz. (2 tbl.)	190
Crunchy Peanut Butter	1 oz. (2 tbl.)	190

Scotch Buy

Chunky Peanut Butter	1 oz. (2 tbl.)	113
Creamy Peanut Butter	1 oz. (2 tbl.)	162

PORTION   SODIUM (MG.)

Skippy

Creamy Peanut Butter ....1 oz. (2 tbl.) 150
Super Chunk Peanut Butter ....1 oz. (2 tbl.) 130

Smucker's

Goober Grape ....2 oz...............................154
Natural Peanut Butter ....1 oz. (2 tbl.) 135

TAHINI

USDA

Tahini ....1 tbl....................................12

Pancakes and Waffles

FROZEN

Aunt Jemima

Blueberry Pancake Batter ....4 oz. (three 4-in. pancakes)..698
Buttermilk Pancake Batter ....4 oz. (three 4-in. pancakes)..733
Cinnamon Swirl French Toast....3 oz. (2 slices).....................359
French Toast............................3 oz. (2 slices).....................430
Jumbo Apple & Cinnamon
   Waffles.................................1.3 oz. (1 waffle) 243
Jumbo Blueberry Waffles ....1.3 oz. (1 waffle) 243
Jumbo Buttermilk Waffles..........1.3 oz. (1 waffle) 267
Jumbo Original Waffles..............1.3 oz. (1 waffle) 261
Pancake Batter...........................4 oz. (three 4-in pancakes)....857

Eggo

Blueberry Artificially Flavored
   Waffles ....1.4 oz. (1 waffle) 260
Homestyle Waffles.....................1.4 oz. (1 waffle) 265
Strawberry Artificially Flavored
   Waffles ....1.4 oz. (1 waffle) 265

Swanson

Pancakes & Blueberries in
   Sauce.................................7 oz................................800
Pancakes & Sausages ....6 oz....................................950

PORTION   SODIUM (MG.)

MIXES

Arrowhead Mills

Buckwheat Pancake & Waffle Mix	four 5-in pancakes	502
Triticale Pancake and Waffle Mix	four 5-in. pancakes	495

Aunt Jemima

Buckwheat Pancake & Waffle Mix	1.1 oz. dry (three 4-in. pancakes)	429
Buttermilk Pancake & Waffle Mix	1.8 oz. dry (three 4-in. pancakes)	847
Complete Pancake & Waffle Mix	1.9 oz. dry (three 4-in. pancakes)	643
Original Pancake & Waffle Mix	1.1 oz. dry (three 4-in. pancakes)	482

Betty Crocker

Buttermilk Pancake & Waffle Mix	1.6 oz. dry (three 4-in. pancakes)	810
Complete Buttermilk Pancake Mix	2 oz. dry (three 4-in. pancakes)	580

Dia-Mel

Pancake Mix	three 3-in. pancakes	70

Featherweight

Pancake Mix	3 pancakes	70

Health Valley

Plain Pancake Mix	1 oz. dry	165
Buttermilk Pancake Mix	1 oz. dry	165

Log Cabin

Complete Pancake & Waffle Mix	three 4-in. pancakes	635

PORTION   SODIUM (MG.)

Regular Pancake & Waffle
  Mix ....three 4-in. pancakes...........515

Pillsbury Hungry Jack

Blueberry Pancake Mix..............three 4-in. pancakes...........815
Buttermilk Pancake Mix ....three 4-in. pancakes...........570
Complete Bulk Pancake Mix ....three 4-in. pancakes...........730
Complete Buttermilk Bulk
  Pancake Mix.............................three 4-in. pancakes...........730
Complete Packets Pancake
  Mix ....three 4-in. pancakes...........675
Extra Lights Pancake Mix ....three 4-in. pancakes...........485
Golden Blend Complete
  Pancake Mix.............................three 4-in. pancakes...........915
Golden Blend Pancake
  Mix ....three 4-in. pancakes...........500
Panshakes...................................three 4-in. pancakes...........880

Soy-o

Buckwheat Pancake Mix ....2.5 oz. dry (five
                                                        4-in. pancakes) 713
Low Sodium Pancake Mix........2.2 oz. dry (five
                                                        4-inch. pancakes) 4

Poultry

(see also Entrees and Dinners, page 264)

CHICKEN
Canned
Armour

Chicken Vienna Sausage in
  Beef Stock................................2 oz..545

Health Valley

Chicken ....2.5 oz. 225
Chicken, no salt.........................2.5 oz. 38

PORTION   SODIUM (MG.)

The Spreadables

Chicken Salad	2 oz	209

Swanson

Chicken Spread	2 oz	280
Chunk Chicken	2.5 oz	270
Chunk Style Mixin' Chicken	2.5 oz	225
Chunk White Chicken	2.5 oz.	235

Underwood

Chunky Chicken Spread	2 oz	479

Fresh

USDA

Breast, roasted, w/skin	3.5 oz. (½ breast)	69
Drumstick, roasted, w/skin	3.6 oz. (2 drumsticks)	94

Frozen

Banquet

Breaded Chicken Nuggets	3 oz	169
Breaded Chicken Patties	3 oz	151
Breaded Chicken Sticks	3 oz	166
Fried Chicken	6.4 oz	1201
Fried Chicken Breast Portions	4.4 oz.	772
Fried Chicken Thighs & Drumsticks	5 oz	892
Fried Chicken Wing Portions	6.75 oz	1226

Morton

Fried Chicken	6.4 oz.	1585
Fried Chicken (Breast Portion)	5.5 oz.	1238

Swanson Fried Chicken Parts

Fried Breast Portions	4.5 oz.	865
Fried Chicken, assorted pieces	3.3 oz.	655
Fried Chicken Nibbles (Wing Sections)	3.3 oz.	645

PORTION   SODIUM (MG.)

	PORTION	SODIUM (MG.)
Fried Thighs and Drumsticks	3.3 oz.	550
Take-out Style Fried Chicken, assorted pieces	3.3 oz.	730

Weaver

Breasts, Batter Dipped	3.5 oz.	240
Breasts, Dutch Frye	3.7 oz.	410
Chicken au Gratin	6 oz.	720
Chicken Croquettes	2 oz.	380
Chicken Croquettes, w/gravy	2.0 oz.	530
Chicken Turnovers	4 oz.	640
Crispy Light Variety Pack	2.9 oz.	400
Crispy Sticks	3.0 oz.	360
Good 'n Wholesome Variety Pack	2.9 oz.	280
Mini-Drums, Crispy	3.0 oz.	115
Mini-Drums, Herb 'n Spice	3.0 oz.	105
Nuggets	3.0 oz.	360
Party Pack, Batter Dipped	3.5 oz.	200
Party Pack, Dutch Frye	1.5 oz.	140
Rondelets, Cheese	3 oz.	660
Rondelets, Italian	3 oz.	570
Rondelets, Original	3 oz.	630
Thighs/Drums, Batter Dipped	3.5 oz.	240
Thighs/Drums, Dutch Frye	4 oz.	350

Processed

Eckrich

Breast of Chicken	1.3 oz. (2 slices)	420
Slender Sliced Chicken	2 oz.	780
Sliced Chicken Roll	2 slices	634

Klement's

Chicken Roll	2 oz.	290

Weaver

Chicken Bologna	2 oz.	520
Chicken Franks	1.6 oz. (1 link)	512
White Meat Chicken Roll	2 oz.	440

TURKEY
Canned

The Spreadables

Turkey Salad..............................2 oz...........................180

Swanson

Chunk Turkey ....2.5 oz. 380

Fresh
USDA

Breast, Roasted w/Skin ....4 oz. (⅙ breast) 60
Leg, Roasted w/Skin.................4 oz. (½ leg) 98

Frozen
Land O Lakes

Butter Basted Young Turkey ....3 oz.................................135
Buttermoist Turkey Roast,
  White (w/gravy) ....3 oz................................510
Buttermoist Turkey Roast,
  White/Dark (w/gravy).............3 oz...............................490
Diced Turkey, White/Dark
  Mixed ....3 oz................................590
Self-Basting (Broth) Young
  Turkey..................................3 oz................................145
Turkey Breast ....3 oz................................55
Turkey Breast Fillets
  w/Cheese..............................5 oz................................835
Turkey Drumsticks ....3 oz................................85
Turkey Ham................................3 oz................................845
Turkey Hindquarters Roast ......3 oz................................80
Turkey Patties............................2¼ oz. 330
Turkey Sticks ....2 oz. (2 sticks)..................295
Turkey Thighs ....3 oz................................75
Turkey Wings ....3 oz................................65
Young Turkey ....3 oz................................55

Swift Butterball

Dark Meat without Skin ....3.5 oz. 90

	PORTION	SODIUM (MG.)
Skin Only	3.5 oz.	100
White and Dark Meat with Skin	3.5 oz.	115
White Meat without Skin	3.5 oz.	130

Processed
Armour

	PORTION	SODIUM (MG.)
Armour Star Broth Baste Turkey, no sugar	4 oz.	185
Armour Star Broth Baste Turkey w/sugar	4 oz.	185
Armour Star Butter Baste Turkey	4 oz.	155
Armour Star Turkey Roast w/Gravy—White	3.7 oz.	655
Armour Star Turkey Roast w/Gravy—White & Dark	3.7 oz.	655
Avondale Cooked Turkey Roll —White	3.0 oz.	865
Avondale Cooked Turkey Roll —White & Dark	3.0 oz.	860
Cured Turkey Breast	4 oz.	1320
Gold Band Cooked Turkey Roll—White	3.0 oz.	605
Gold Band Cooked Turkey Roll—White & Dark	3.0 oz.	690
Magic Slice Cooked Turkey Roll—White	3.0 oz.	685
Magic Slice Cooked Turkey Roll—White & Dark	3.0 oz.	695
Turkey Bologna	4 oz.	1070
Turkey Cotto Salami	4 oz.	1080
Turkey Ham	4 oz.	1210
Turkey Hot Dogs	2 oz.	580
Turkey Meat Loaf	3.0 oz.	470
Turkey Pastrami	4 oz.	1305

Eckrich

	PORTION	SODIUM (MG.)
Slender Sliced Smoked Turkey	2 oz.	800

PORTION   SODIUM (MG.)

Sliced Turkey Roll	2 slices	580

Klement's

Turkey Breast	2 oz. (2 slices)	522

Land O' Lakes Oven Cooked Turkey Breast

Bronze Label	3 oz.	510
Gold Label Browned	3 oz.	635
Gold Label Skinless	3 oz.	715
Gold Label Skin On	3 oz.	640
Silver Label	3 oz.	565

Land O' Lakes Turkey Roll

Blue Label—Mixed	3 oz.	550
Blue Label—White	3 oz.	560
Red Label—Mixed	3 oz.	510
Red Label—White	3 oz.	530

Louis Rich

Chopped & Formed Turkey	2 oz.	540
Smoked Turkey	2 oz.	550
Smoked Turkey Wing Drummettes	2 oz.	606
Turkey, ground (after cooking)	2 oz.	40
Turkey Bologna	2 oz. (2 slices)	444
Turkey Breast, barbecued	2 oz.	320
Turkey Breast, oven roasted	2 oz. (2 slices)	380
Turkey Breast, smoked	2 slices	400
Turkey Cotto Salami	2 oz. (2 slices)	500
Turkey Drumsticks, smoked	2 oz.	750
Turkey Franks	1.5 oz. (1 link)	472
Turkey Franks (8 links/ 16 oz.)	2 oz. (1 link)	600
Turkey Ham (cured turkey thigh)	2 oz.	580
Turkey Luncheon Loaf	2 oz.	550
Turkey Pastrami	2 oz.	560

	PORTION	SODIUM (MG.)
Turkey Salami, cooked	2 oz. (2 slices)	502
Turkey Summer Sausage	2 oz.	666

Marval

Gourmet Supreme Turkey Breast	2 oz.	290
Marval Turkey Roll	2 oz.	250
Smoked Turkey Breast	2 oz.	445
Turkey Bologna	2 oz.	500
Turkey Franks	1.6 oz. (1 frank)	450
Turkey Ham	2 oz.	578
Turkey Pastrami	2 oz.	590

Oscar Mayer

Turkey Breast	1.5 oz. (2 slices)	590

Weaver

Turkey Ham	2 oz.	600
Turkey Roll	2 oz.	340

OTHER
Fresh
USDA

Duck, roasted, flesh & skin	4 oz.	67
Gizzard, poultry, simmered	4 oz.	68
Goose, roasted, flesh & skin	4 oz.	80
Heart, poultry, simmered	4 oz.	56
Liver, poultry, simmered	4 oz.	64

Seasonings

(see also Condiments, page 240 and Gravies and Sauces, page 328)

Dia-Mel

Salt-It	1 tsp.	1
Seasoned Salt-It	1 tsp.	1

PORTION   SODIUM (MG.)

Durkee

Beef Stew Mix	½ cup prepared	488
Chili Con Carne Mix	½ cup prepared	489
Chop Suey Seasoning Mix	½ cup prepared	182
Fried Rice Seasoning Mix	½ cup prepared	675
Ground Beef Mix	½ cup prepared	400
Ground Beef w/Onion Mix	½ cup prepared	550
Hamburger Seasoning Mix	½ cup prepared	506
Italian Meatball Mix	½ cup prepared	510
Pizza Flavor Sloppy Joe Mix	½ cup prepared	606
Sloppy Joe Mix	½ cup prepared	715
Taco Seasoning Mix	½ cup prepared	598
Texas Chili Seasoning Mix	½ cup prepared	470

Featherweight

Fish Seasoning	¼ tsp.	0
Garlic Salt Substitute	¼ tsp.	0
K Salt Substitute	¼ tsp.	0
Seasoned Salt Substitute	¼ tsp.	0

French's

Barbecue Seasoning	1 tsp.	70
Beef Flavor Stock Base	1 tsp.	500
Beef Stew Seasoning Mix	⅙ package	765
Celery Salt	1 tsp.	1505
Chicken Flavor Stock Base	1 tsp.	475
Chili-O Seasoning Mix	⅙ package	630
Dehydrated Vegetable Flakes	1 tbl.	20
Enchilada Seasoning Mix	¼ package	1130
Garlic Salt	1 tsp.	2050
Garlic Salt, Parslied	1 tsp.	1125
Ground Beef Seasoning w/Onions Mix	¼ package	440
Hamburger Seasoning Mix	¼ package	450
Hickory Smoke Salt	1 tsp.	1145
Instant Salad Onions	1 tbl.	2
Lemon & Pepper Seasoning	1 tsp.	805
Meat Marinade Mix	⅛ package	540
Meat Tenderizer	1 tsp.	1760

PORTION   SODIUM (MG.)

	PORTION	SODIUM (MG.)
Meat Tenderizer, Seasoned	1 tsp.	1550
Meatball Seasoning Mix	¼ package	825
Meatloaf Seasoning Mix	⅛ package	615
Onion Salt	1 tsp.	1590
Pepper, Seasoned	1 tsp.	5
Pizza Seasoning	1 tsp.	400
Salad Seasoning	1 tsp.	630
Salt, Imitation Butter Flavor	1 tsp.	1125
Seafood Seasoning	1 tsp.	1410
Seasoning Salt	1 tsp.	1280
Sloppy Joe Seasoning Mix	⅛ package	390
Taco Seasoning Mix	⅙ package	365

George Washington

	PORTION	SODIUM (MG.)
Brown Seasoning & Broth	0.1 oz. dry	1015
Brown Seasoning & Broth (Passover)	0.1 oz. dry	1110
Golden Seasoning & Broth	0.1 oz. dry	935
Golden Seasoning & Broth (Passover)	0.1 oz. dry	1015
Onion Seasoning & Broth	0.2 oz. dry	695
Vegetable Seasoning & Broth	0.2 oz. dry	715

Instead of Salt

	PORTION	SODIUM (MG.)
All Purpose	½ tsp.	6
Chicken	½ tsp.	3
Fish	½ tsp.	7
Steak & Hamburger	½ tsp.	4
Vegetables	½ tsp.	2

Shake 'n Bake

	PORTION	SODIUM (MG.)
Barbecue Chicken	¼ envelope	820
Original Chicken	¼ envelope	590

Snacks and Dips

CANDY	PORTION	SODIUM (MG.)

Amurol

Hard Candy	1 piece	0
Mints	1 piece	0

Campfire

Marshmallows	2 large or 24 mini	10

Chunky

Milk Chocolate	1 oz. (1 piece)	35
Original	1 oz. (1 piece)	23
Pecan	1 oz. (1 piece)	24

Crackerjack

Caramel Coated Popcorn Peanuts	1 oz.	85

Deran

Chocolate Coated Bridge Mix	1 oz.	25
Chocolate Malted Milk Balls	1 oz.	70
Milk Chocolate Covered Raisins	1 oz.	35
Peanut Clusters	1 oz.	15
Rum Wafers	1 oz.	15

Estee

Almond bar	2 sections	10
Bittersweet bar	2 sections	10
Chocolate Coated Raisins	6 candies	10
Coconut bar	2 sections	10
Crunch bar	2 sections	10
Estee-ets with Peanuts	5 candies	10
Fruit and Nut bar	2 sections	10
Gum Drops (licorice, fruit)	4 candies	5

	PORTION	SODIUM (MG.)
Hard Candy, Assorted	2 candies	5
Hard Candy, Peppermint	2 candies	5
Hard Candy, Tropi-Mix	2 candies	5
Lollipops	1 lollipop	5
Milk Chocolate bar	2 sections	10
Peanut Butter Cups	1 candy	20
Rolled Mints, all flavors	1 mint	0
Toasted Bran bar	2 sections	10
TV Mix	4 candies	10

Hershey

Golden Almond bar	1 oz. (1 bar)	20
Kisses	1 oz. (6 pieces)	25
Kit Kat bar	1 oz. (1 bar)	30
Krackel bar	1.2 oz. (1 bar)	50
Milk Chocolate bar	1 oz. (1 bar)	30
Milk Chocolate w/Almonds	1 oz. (1 bar)	25
Mr. Goodbar bar	1.3 oz. (1 bar)	20
Rolo	1 oz. (5 pieces)	65
Special Dark bar	1 oz. (1 bar)	1
Whatchamacallit	1 oz. (1 bar)	70

Kraft

Butter Mints	1 piece	5
Caramels	1 piece	20
Chocolate Fudgies	1 piece	20
Funmallows	1 piece	10
Marshmallows	1 piece	10
Miniature Funmallows	10 pieces	10
Miniature Marshmallows	10 pieces	10
Party Mints	1 piece	5
Peanut Brittle	1 oz.	145
Toffee	1 piece	20

Life Savers

Lollipop	1 piece	9
Mint	1 piece	0
Roll Candy, Butter Creme Mint	1 piece	6

PORTION   SODIUM (MG.)

	PORTION	SODIUM (MG.)
Roll Candy, Butter Rum	1 piece	10
Roll Candy, Butterscotch	1 piece	9
Roll Candy, Other Flavors	1 piece	2
Sours	1 piece	2
Sugar Free Breath Savers	1 piece	0

M&M/Mars

Mars bar	1.7 oz.	71
Milky Way bar	2.1 oz.	119
M&M's Peanut	1.7 oz.	29
M&M's Plain	1.7 oz.	41
Royals bar	1.5 oz.	36
Snickers bar	2 oz.	139
Starburst Fruit Chews	2 oz.	26
Summit bar	0.7 oz.	30
3 Musketeers bar	2.3 oz.	135
Twix Caramel bar	0.9 oz.	51
Twix Peanut Butter bar	0.9 oz.	74

Nabisco

Baby Ruth bar	1.8 oz. (1 bar)	100
Butterfinger bar	1.6 oz. (1 bar)	70
Reggie bar	2 oz. (1 bar)	220

Nestlé

Choco-Lite bar	1 oz. (1 bar)	25
Crunch bar	1 oz. (1 bar)	50
Milk Chocolate bar	1 oz. (1 bar)	25
Milk Chocolate w/Almonds bar	1 oz. (1 bar)	20
$100,000 bar	1.5 oz. (1 bar)	75

Pearson's

Caramel Nip	1.8 oz.	130
Changemaker Mints	0.3 oz.	15
Chocolate Parfait	1.8 oz.	20
Coffee Nip	1.8 oz.	130
Coffioca	1.8 oz.	20
Licorice Nip	1.8 oz.	130

	PORTION	SODIUM (MG.)
Mint Parfait	1.8 oz.	20
Nut Goodie	1.8 oz.	141
Round Mint Patties	1.4 oz.	62
Salted Nut Roll	4 oz.	451
Salted Nut Roll	1.8 oz.	204
Salted Nut Roll	1.3 oz.	142
Vend Mints	1.4 oz.	66

Reese's

Peanut Butter Cups	1.2 oz. (2 pieces)	110

CHEWING GUM

Amurol

Amurol Gum	1 stick	1

Beech-Nut

Beechies	1 piece	1
Gum	1 piece	1

Breath Savers

Sugar Free Gum	1 piece	0

Bubble Yum

Bubble Gum	1 piece	2
Sugarless Bubble Gum	1 piece	1

Care Free

Sugarless Gum	1 piece	1

Freedent

Peppermint	1 stick	0
Spearmint	1 stick	0

Hubba Bubba

Fruit Bubble Gum	1 stick	0
Grape Bubble Gum	1 stick	0

PORTION   SODIUM (MG.)

Original Bubble Gum	1 stick	0
Raspberry Bubble Gum	1 stick	0
Strawberry Bubble Gum	1 stick	0

Orbit

Peppermint	1 stick	0
Spearmint	1 stick	0

Replay

Gum	1 piece	0

Wrigley

Big Red	1 stick	0
Doublemint	1 stick	0
Juicy Fruit	1 stick	0
Spearmint	1 stick	0

CHIPS, PRETZELS, CHEESE BALLS, POPCORN

Buckeye

Ketchup and French Fry Flavor Potato Chips	1 oz.	230

Charles

Potato Chips, no salt added	1 oz.	10

Featherweight

Popcorn, unsalted	1 oz. unpopped	7
Pretzels, unsalted	3 pieces	5

Frito-Lay

Baken-ets Fried Pork Rinds	1 oz.	570
Cheese Flavored Popcorn	1 oz.	185
Cheese Peanut Butter Crackers	1.5 oz.	415
Cheetos, Crunchy	1 oz.	260
Cheetos, Nacho Flavored	1 oz.	275
Cheetos Puffed Balls	1 oz.	280

PORTION   SODIUM (MG.)

	PORTION	SODIUM (MG.)
Doritos Nacho Cheese Flavor Tortilla Chips	1 oz.	165
Doritos Sour Cream/Onion Tortilla Chips	1 oz.	165
Doritos Taco Flavored Tortilla Chips	1 oz.	185
Doritos Tortilla Chips	1 oz.	165
Fritos Bar-B-Q Flavor Corn Chips	1 oz.	245
Fritos Corn Chips	1 oz.	180
Fritos Corn Chips—Lights	1 oz.	180
Funyuns Onion Flavor Snack	1 oz.	260
Lay's Barbecue Flavor Potato Chips	1 oz.	320
Lay's Potato Chips	1 oz.	260
Lay's Sour Cream/Onion Potato Chips	1 oz.	335
Munchos Potato Crisps	1 oz.	235
Popcorn, Salted	1 oz.	220
Rold Gold Pretzel Rods	1 oz.	510
Rold Gold Pretzel Sticks	1 oz.	685
Rold Gold Pretzel Twists	1 oz.	410
Rold Gold Tiny Tim Pretzels	1 oz.	495
Ruffles Bacon/Sour Cream Potato Chips	1 oz.	325
Ruffles Barbecue Flavor Potato Chips	1 oz.	260
Ruffles Potato Chips	1 oz.	260
Toasted Corn Nuggets	1 oz.	192
Toasted Peanut Butter Crackers	1.5 oz.	385
Tostitos Nacho Flavor Round Tortilla Chips	1 oz.	205

General Mills

	PORTION	SODIUM (MG.)
Bugles	1 oz.	285
Nacho Cheese Bugles	1 oz.	285

Giant

	PORTION	SODIUM (MG.)
Potato Chips, No Salt Added	1 oz.	25
Pretzels, No Salt Added	1 oz.	10

PORTION   SODIUM (MG.)

Granny Goose

100% Natural Potato Chips, unsalted	1 oz.	10

Guy's

Baked Cheese Balls	1 oz.	320

Health Valley

Buenitos Tortilla Chips	1 oz.	100
Buenitos Tortilla Chips, no salt	1 oz.	6
Cheddar Lites	1 oz.	125
Cheddar Lites, no salt	1 oz.	35
Cheese Tortilla Strips	1 oz.	145
Corn Chips	1 oz.	90
Corn Chips, no salt	1 oz.	1
Corn Chips w/Cheese	1 oz.	115
Corn Chips w/Cheese, no salt	1 oz.	34
Country Chips	1 oz.	60
Country Chips, no salt	1 oz.	1
Country Ripples	1 oz.	60
Country Ripples, no salt	1 oz.	1
Dip Chips	1 oz.	85
Dip Chips, no salt	1 oz.	1
Golden Hawaiian, canned	1 oz.	80
Golden Hawaiian, no salt	1 oz.	10
Mini Whole Wheat Pretzels	1 oz.	200
Mini Whole Wheat Pretzels, no salt	1 oz.	10
Nacho Cheese & Chili Tortilla Chips	1 oz.	60
Natural Mini Pretzels	1 oz.	580
Potato Chips, Natural Flavor	1 oz.	60
Potato Chips, no salt	1 oz.	1
Small Corn Chips	1 oz.	100
Small Corn Chips, no salt	1 oz.	1
Small Potato Chips	1 oz.	60
Small Potato Chips, no salt	1 oz.	1

	PORTION	SODIUM (MG.)
Small Yogurt Chips	1 oz.	160
Vegetable and Herb Tortilla Strips	1 oz.	125
Yogurt and Green Onion Chips, canned	1 oz.	115
Yogurt and Green Onion Tortilla Strips	1 oz.	80
Sesame Whole Wheat Pretzels	1 oz.	220
Sesame Whole Wheat Pretzels, no salt	1 oz.	10

Jiffy Pop

	PORTION	SODIUM (MG.)
Koepper's Snappies	1 oz.	360
Popcorn, Butter Flavor	1.5 oz. unpopped	640
Popcorn, Regular	1.5 oz. unpopped	680
Popping Corn, White & Yellow	1.6 oz. unpopped	10

Kraft

	PORTION	SODIUM (MG.)
Handi-Snacks Cheez 'N Crackers	1 package	370
Handi-Snacks Peanut Butter N Cheez Crackers	1 package	245

Laura Scudder

	PORTION	SODIUM (MG.)
BBQ Potato Chips	1 oz.	232
Caramel Corn	1 oz.	102
Cheese Puffs	1 oz.	326
Corn Chips	1 oz.	235
Korn Kernals	1 oz.	167
Mini-Tacos	1 oz.	281
Pork Skins	1 oz.	568
Potato Chips	1 oz.	196
Scudderings	1 oz.	278
Tortilla Chips	1 oz.	204

Lite-line

	PORTION	SODIUM (MG.)
Nacho Cheese Flavored Tortilla Chips	1 oz.	165

PORTION   SODIUM (MG.)

Mister Salty

Dutch Pretzels	1 oz. (2 pieces)	580
Pretzel Sticks	1 oz. (91 pieces)	940
Pretzels	1 oz. (5½ pieces)	685

Nabisco

Buenos Nacho Cheese Flavor Tortilla Chips	1 oz. (13½ pieces)	260
Buenos Sour Cream & Onion Flavor	1 oz. (13½ pieces)	130
Cheese N Crunch	1 oz. (41 pieces)	130
Corn & Sesame Chips	1 oz. (16 pieces)	225
Corn Diggers	1 oz. (35½ pieces)	355
Doo Dads Mixed Snacks	1 oz. (57 pieces)	380
Flings Crispy Corn Curls	1 oz. (16 pieces)	380
Nacho Cheese Flavor Tortilla Chips	1 oz. (13 pieces)	265
Potato Chipsters	1 oz. (57 pieces)	585
Wheat Chips Whole Grain	1 oz. (35 pieces)	225

O&C

Potato Sticks	1 oz.	256
Snackin' Crisp	1 oz.	257

Old El Paso

Nachips	1 oz.	102

Old London

Pretzel Nuggets Bite-Size Pretzels	1 oz.	365

Orville Redenbacher

Gourmet Microwave Popcorn, Butter Flavor	4 cups, popped	210
Gourmet Microwave Popcorn, Natural Flavor	4 cups, popped	290
Gourmet Popcorn, Oil and Salt	0.9 oz. (4 cups)	700
Gourmet Popcorn, Plain	0.8 oz. (4 cups)	0

PORTION   SODIUM (MG.)

Ovaltine

	PORTION	SODIUM (MG.)
Fiddle Faddle	1.3 oz.	234
Poppycock, Original Flavor	1.5 oz.	158
Poppycock, Walnut Cashew Flavor	1.5 oz.	170
Screaming Yellow Zonkers	0.9 oz.	233

Pepperidge Farm

Snack Sticks, Cheese	1 oz. (8 sticks)	345
Snack Sticks, Rye	1 oz. (8 sticks)	385
Snack Sticks, Pumpernickel	1 oz.	380
Snack Sticks, Salted	1 oz. (8 sticks)	320
Snack Sticks, Sesame	1 oz.	345

Pepperidge Farm Pretzels

Butter Nuggets	1 oz. (20 pretzels)	785
Thin Sticks	1 oz. (24 pretzels)	775
Tiny Twists	1 oz. (15 pretzels)	815

Pillsbury

Butter Flavor Microwave Popcorn	4 cups popped	405
Microwave Popcorn	4 cups popped	405
No Salt Microwave Popcorn	4 cups popped	6

Planters

Cheez Balls & Cheez Curls	1 oz.	300
Corn Chips	1 oz.	220
Nacho Tortilla Chips	1 oz.	170
Pretzel Twists & Sticks	1 oz.	480
Stackable Potato Chips	1 oz.	210
Taco Tortilla Chips	1 oz.	170

Popped-Right

BBQ Flavored Corn Chips	1 oz.	166
Caramel Corn	1 oz.	210
Cheese Crunchy	1 oz.	276
Cheese Flavored Popcorn	1 oz.	201

PORTION   SODIUM (MG.)

Cheese Flavored Puffs	1 oz.	255
Corn Chips	1 oz.	197
Corn Pops	1 oz.	191
Nacho Flavored Tortilla Chips	1 oz.	180
Plain Tortilla Chips	1 oz.	104
Popcorn	1 oz.	172
Sugar Corn Topping	1 oz.	29
Taco Flavor Tortilla Chips	1 oz.	124

Pringle's

Cheez-ums	1 oz.	220
Light Potato Chips	1 oz.	145
Regular Potato Chips	1 oz.	215
Rippled Potato Chips	1 oz.	250

Seyfert's

Butter Pretzel Rods	1 oz.	530

Wise

Potato Chips, Natural Flavor	1 oz.	240
Potato Chips, no salt added	1 oz.	20
Pretzel Nuggets Bite-Size Pretzels	1 oz.	365

DIPS
Eastern

Country Garden	0.5 oz.	35

Frito-Lay

Enchilada Dip	3.1 oz.	625
Jalapeño Bean Dip	3.1 oz.	550

Health Valley

Avocado Dip	1 oz.	145
Herb Dip	1 oz.	135

Kraft

Avocado (Guacamole)	2 tbl.	215

	PORTION	SODIUM (MG.)
Bacon & Horseradish	2 tbl	200
Clam	2 tbl	230
French Onion	2 tbl	240
Garlic	2 tbl	160
Green Onion	2 tbl	170
Jalapeño Pepper	2 tbl	160

Kraft Ready-to-Serve

Bacon & Horseradish	1 oz.	255
Blue Cheese	1 oz.	175
Clam	1 oz.	185
Creamy Cucumber	1 oz.	130
French Onion	1 oz.	160
Jalapeño Pepper	1 oz.	205
Onion	1 oz.	160
Premium Buttermilk	1 oz.	280
Premium Buttermilk and Bacon	1 oz.	305
Premium Buttermilk and Onion Dip	1 oz.	230
Premium Creamy Onion	1 oz.	160

Land O Lakes

All Flavors	2 oz.	315

OTHER

Comet

Cones & Cups	0.2 oz. (1)	6
Sugar Cones	0.4 oz. (1)	35

Dromedary

Date Nut Roll	½ inch	160

Fruit Corners

Fruit Roll-Ups, Apple	½ oz. (1 roll)	5
Fruit Roll-Ups, Apricot	½ oz. (1 roll)	5
Fruit Roll-Ups, Cherry	½ oz. (1 roll)	5
Fruit Roll-Ups, Grape	½ oz. (1 roll)	5
Fruit Roll-Ups, Strawberry	½ oz. (1 roll)	5

General Mills

Breakfast Squares ....1.5 oz. (1 bar) 255

Nature Valley

Light & Crunchy Granola
  Snack, Cinnamon ....1 oz. (1 pouch) 170
Light & Crunchy Granola
  Snack, Honey Nut ....1 oz. (1 pouch) 160
Light & Crunchy Granola
  Snack, Oats'n Honey ....1 oz. (1 pouch) 170
Light & Crunchy Granola
  Snack, Peanut Butter ....1 oz. (1 pouch) 240
Light & Crunchy Granola
  Snack, Raspberry Glaze ......1 oz. (1 pouch) 220
Light & Crunchy Granola
  Snack, Vanilla Glaze..............1 oz. (1 pouch) 220

Pillsbury

Food Sticks, Chocolate..............1.4 oz. (4 sticks) 115

Soups

CANNED

Condensed

Campbell's

Bean w/Bacon ....11 oz. prepared.................1203
Beef ....11 oz. prepared.................1176
Beef Broth (bouillon).................10 oz. prepared.................1094
Beef Noodle ....10 oz. prepared.................1094
Beef Noodle, Homestyle ....10 oz. prepared.................1013
Beefy Mushroom ....10 oz. prepared.................1238
Black Bean ....11 oz. prepared.................1368
Cheddar Cheese ....11 oz. prepared.................1217
Chicken Alphabet......................10 oz. prepared.................1088
Chicken and Stars.....................10 oz. prepared.................1169
Chicken Broth............................10 oz. prepared.................1013

PORTION   SODIUM (MG.)

	PORTION	SODIUM (MG.)
Chicken Broth & Noodles	10 oz. prepared	1081
Chicken Broth & Rice	10 oz. prepared	1100
Chicken Gumbo	10 oz. prepared	1138
Chicken 'N Dumplings	10 oz. prepared	1244
Chicken Noodle	10 oz. prepared	1169
Chicken Noodle, Homestyle	10 oz. prepared	1138
Chicken Noodle O's	10 oz. prepared	1075
Chicken Vegetable	10 oz. prepared	1100
Chicken w/Rice	10 oz. prepared	1088
Chili Beef	11 oz. prepared	1238
Clam Chowder, Manhattan Style	10 oz. prepared	1075
Clam Chowder, New England	10 oz. prepared	1106
Clam Chowder, New England, made w/milk	10 oz. prepared	1175
Consomme (Beef)	10 oz. prepared	981
Cream of Asparagus	10 oz. prepared	1125
Cream of Celery	10 oz. prepared	1094
Cream of Chicken	10 oz. prepared	1075
Cream of Mushroom	10 oz. prepared	1175
Cream of Onion	10 oz. prepared	1044
Cream of Onion, made w/water and milk	10 oz. prepared	1081
Cream of Potato	10 oz. prepared	1169
Cream of Potato, made w/water and milk	10 oz. prepared	1200
Cream of Shrimp	10 oz. prepared	1131
Cream of Shrimp, made w/milk	10 oz. prepared	1206
Creamy Chicken Mushroom	10 oz. prepared	1175
Curly Noodle w/Chicken	10 oz. prepared	1200
French Onion	10 oz. prepared	1200
Golden Mushroom	10 oz. prepared	1138
Green Pea	11 oz. prepared	1217
Meatball Alphabet	10 oz. prepared	1213
Minestrone	10 oz. prepared	1163
Noodles and Ground Beef	10 oz. prepared	1056
Oyster Stew	10 oz. prepared	1056
Oyster Stew, made w/milk	10 oz. prepared	1134

	PORTION	SODIUM (MG.)
Pepper Pot	10 oz. prepared	1200
Scotch Broth	10 oz. prepared	1125
Spanish Style Vegetable (Gazpacho)	10 oz. prepared	744
Split Pea w/Ham & Bacon	11 oz. prepared	1114
Tomato	10 oz. prepared	938
Tomato, made w/milk	10 oz. prepared	1000
Tomato Bisque	11 oz. prepared	1155
Tomato Rice, Old Fashioned	11 oz. prepared	1073
Turkey Noodle	10 oz. prepared	1150
Turkey Vegetable	10 oz. prepared	1031
Vegetable	10 oz. prepared	963
Vegetable Beef	10 oz. prepared	1038
Vegetable, Old Fashioned	10 oz. prepared	1156
Vegetarian Vegetable	10 oz. prepared	944
Won Ton	10 oz. prepared	1094

Campbell's Soup for One

Burly Vegetable Beef	11 oz. prepared	1480
Clam Chowder, New England	11 oz. prepared	1365
Clam Chowder, New England, made w/Milk	11 oz. prepared	1420
Full Flavor Chicken Vegetable	11 oz. prepared	1500
Golden Chicken & Noodles	11 oz. prepared	1460
Old Fashioned Bean w/Ham	11 oz. prepared	1405
Old World Vegetable	11 oz. prepared	1495
Savory Cream of Mushroom	11 oz. prepared	1495
Tomato Royale	11 oz. prepared	1335

Dia-Mel

Chicken Noodle	8 oz.	20
Cream of Mushroom	8 oz.	30
Tomato	8 oz.	15
Vegetable Beef	8 oz.	20

Doxsee

Clam Chowder, Manhattan	6 oz.	445
Clam Chowder, New England	6 oz.	640

PORTION   SODIUM (MG.)

Snow's

Clam Chowder, Manhattan	7.5 oz.	635
Clam Chowder, New England, prepared w/milk	7.5 oz.	665
Corn Chowder, New England, prepared w/milk	7.5 oz.	640
Fish Chowder, New England, prepared w/milk	7.5 oz.	620
Seafood Chowder, New England, prepared w/milk	7.5 oz.	690

Ready to Serve

Campbell's Chunky Soups

Beef	9.5 oz.	1050
Beef (individual serving)	10.8 oz.	1190
Chicken Noodle	9.5 oz.	1070
Chicken Noodle (individual serving)	10.8 oz.	1210
Chicken Vegetable	9.5 oz.	1115
Chicken w/Rice	9.5 oz.	1060
Chili Beef	9.8 oz.	1020
Chili Beef (individual serving)	11 oz.	1155
Clam Chowder, Manhattan Style	9.5 oz.	1105
Clam Chowder, Manhattan Style (individual serving)	10.8 oz.	1255
Clam Chowder, New England Style	9.5 oz.	1020
Clam Chowder, New England Style (individual serving)	10.8 oz.	1155
Ham & Butter Bean (individual serving)	10.8 oz.	1190
Mediterranean Vegetable	9.5 oz.	1040
Minestrone	9.5 oz.	985
Old Fashioned Bean w/Ham	9.5 oz.	1030
Old Fashioned Bean w/Ham (individual serving)	11 oz.	1180

PORTION   SODIUM (MG.)

Old Fashioned Chicken	9.5 oz.	1205
Old Fashioned Chicken (individual serving)	10.8 oz.	1365
Old Fashioned Vegetable Beef	9.5 oz.	1090
Old Fashioned Vegetable Beef (individual serving)	10.8 oz.	1235
Sirloin Burger	9.5 oz.	1130
Sirloin Burger (individual serving)	10.8 oz.	1285
Split Pea w/Ham	9.5 oz.	1000
Split Pea w/Ham (individual serving)	10.8 oz.	1130
Steak & Potato	9.5 oz.	1115
Steak & Potato (individual serving)	10.8 oz.	1265
Stroganoff Style Beef (individual serving)	10.8 oz.	1315
Turkey Vegetable	9.4 oz.	1090
Vegetable	9.5 oz.	995
Vegetable (individual serving)	10.8 oz.	1125

Campbell's Low Sodium Soups

Chicken Broth	10.5 oz.	100
Chicken w/Noodles	10.8 oz.	90
Chunky Beef & Mushroom	10.8 oz.	75
Chunky Chicken Vegetable	10.8 oz.	100
Chunky Vegetable Beef	10.8 oz.	65
Cream of Mushroom	10.5 oz.	60
French Onion	10.5 oz.	60
Split Pea	10.8 oz.	25
Tomato w/Tomato Pieces	10.5 oz.	40

Crosse & Blackwell

Black Bean	7.5 oz.	880
Clam Chowder, Manhattan Style	7.5 oz.	1125
Clear Consomme	7.5 oz.	1060

PORTION   SODIUM (MG.)

	PORTION	SODIUM (MG.)
Crab Bisque	7.6 oz.	795
Cream of Mushroom Bisque	7.5 oz.	930
Cream of Vichyssoise	7.5 oz.	855
French Onion	7.5 oz.	925
Gazpacho	7.5 oz.	1105
Lentil	7.5 oz.	980
Lobster Bisque	7.5 oz.	780
Minestrone	7.5 oz.	875
New England Clam Chowder	7.5 oz.	1085
Red Consomme Madrilene	7.5 oz.	830
Shrimp Bisque	7.5 oz.	750
Split Pea	7.5 oz.	935

Featherweight

Chicken Noodle Soup	1 cup	40
Cream of Mushroom Soup	1 cup	15

Health Valley

Beef Broth	8 oz.	1170
Beef Broth, unsalted	8 oz.	110
Chicken Broth	8 oz.	1100
Chicken Broth, unsalted	8 oz.	110
Chunky Bean	8 oz.	620
Chunky Bean, unsalted	8 oz.	60
Chunky Clam Chowder	8 oz.	520
Chunky Clam Chowder, unsalted	8 oz.	210
Chunky Minestrone	8 oz.	640
Chunky Minestrone, unsalted	8 oz.	70
Chunky Split Pea	8 oz.	610
Chunky Split Pea, unsalted	8 oz.	100
Chunky Vegetable	8 oz.	650
Chunky Vegetable, unsalted	8 oz.	100
Clam Chowder	8 oz.	880
Clam Chowder, unsalted	8 oz.	300
Green Split Pea	8 oz.	1100
Green Split pea, unsalted	8 oz.	110
Lentil	8 oz.	440
Lentil, unsalted	8 oz.	80

	PORTION	SODIUM (MG.)
Minestrone	8 oz.	1240
Minestrone, unsalted	8 oz.	60
Mushroom	8 oz.	780
Mushroom, unsalted	8 oz.	110
Old Fashioned Bean	8 oz.	1140
Old Fashioned Bean, unsalted	8 oz.	70
Old Fashioned Potato	8 oz.	1080
Old Fashioned Potato, unsalted	8 oz.	110
Tomato	8 oz.	600
Tomato, unsalted	8 oz.	60
Vegetable	8 oz.	1110
Vegetable, unsalted	8 oz.	140

Heinz

Bean	7.5 oz.	249
Beef Noodle	7.3 oz.	280
Chicken Noodle	7.3 oz.	264
Chicken Rice	7.3 oz.	280
Cream of Mushroom	7.3 oz.	311
Cream of Tomato	7.5 oz.	290
Minestrone	7.5 oz.	311
Vegetable, made w/beef stock	7.5 oz.	248
Vegetable Beef	7.5 oz.	205

Luck's Country Soups

Bean w/Ham & Onions	8.5 oz.	775
Bean w/Sausage	8.5 oz.	800
Old Fashioned Bean w/Ham	8.5 oz.	1060
Pinto Bean Chowder	8.5 oz.	1170
Vegetable	8.5 oz.	1185

Progresso

Chickarina Soup	8 oz.	806
Escarole In Chicken Broth	8 oz.	565
Green Split Pea	8 oz.	839
Home Style Chicken	8 oz.	875
Lentil	8 oz.	1031

	PORTION	SODIUM (MG.)
Macaroni & Bean	8 oz.	1221
Minestrone	8 oz.	531
Tomato	8 oz.	929

Swanson

Beef Broth	7.3 oz.	840
Chicken Broth	7.3 oz.	960

DRY
Estee

Cream of Chicken	1 envelope	60
Cream of Mushroom	1 envelope	60
Cream of Tomato	1 envelope	50
Cream of Vegetable	1 envelope	60
Tomato Vegetable	1 envelope	50

Fearn Soya

Lentil Minestrone	1.9 oz. dry	515
Split Pea	1.8 oz. dry	537

Featherweight

Beef Bouillon Instant	1 tsp.	under 10
Beef Soup Base	0.2 oz.	7
Chicken Bouillon Instant	1 tsp.	5
Chicken Soup Base	0.2 oz.	1

Hain

Chicken Noodle Soup Mix, no salt	8 oz.	90
Cream of Mushroom Soup Mix, no salt	8 oz.	125
Cream of Vegetable Soup Mix, no salt	8 oz.	90
Minestrone Soup Mix, no salt	8 oz.	25
Tangy Tomato Soup Mix, no salt	8 oz.	35
Zesty Onion Soup Mix, no salt	8 oz.	70

PORTION   SODIUM (MG.)

Herb-Ox

Beef Broth & Seasoning, Low
   Sodium......................................0.1 oz. (1 packet) 5
Chicken Broth & Seasoning,
   Low Sodium ....0.1 oz. (1 packet) 10
Vegetable Broth & Seasoning,
   Low Sodium ....0.1 oz. (1 packet) 5

Lipton

Beef Flavor Mushroom ....8 oz. prepared......................995
Chicken Noodle ....8 oz. prepared................900
Country Vegetable ....8 oz. prepared......................995
Giggle Noodle............................8 oz. prepared......................925
Golden Mushroom w/Chicken
   Broth..8 fl. oz. 900
Noodle Soup w/Chicken
   Broth..8 oz. prepared......................925
Onion Mushroom ....8 oz. prepared......................995
Ring-O-Noodle ....8 oz. prepared......................855
Vegetable Beef Stock ....8 oz. prepared......................995
Vegetable Soup for Dip ....8 fl oz..................................995

Lipton Country Style Cup-A-Soup

Chicken Supreme ....6 fl. oz. 930
Harvest Vegetable ....6 fl. oz. 620
Hearty Chicken...........................6 fl. oz. 970
Virginia Pea................................6 fl. oz. 840

Lipton Cup a Soup

Beef Flavor Noodle ....6 oz. prepared......................780
Chicken (Cup-a-Broth) ....6 oz. prepared......................800
Chicken Noodle w/Meat ....6 oz. prepared......................770
Chicken Rice ....6 oz. prepared......................760
Chicken Vegetable.....................6 oz. prepared......................800
Cream of Chicken ....6 oz. prepared......................850
Cream of Mushroom ....6 oz. prepared......................810
Cream of Vegetable ....6 oz. prepared......................631
Green Pea...................................6 oz. prepared......................680
Onion ....6 oz. prepared......................860

PORTION   SODIUM (MG.)

Ring Noodle	6 oz. prepared	760
Spring Vegetable	6 oz. prepared	1070
Tomato	6 oz. prepared	650
Vegetable Beef	6 oz. prepared	930

Lipton Cup-A-Soup. Lots-a-Noodles

Beef Flavor	7 fl. oz.	730
Chicken Flavor	7 fl. oz.	810
Cream of Chicken	7 fl. oz.	750
Garden Vegetable	7 fl. oz.	730
Oriental Style	7 fl. oz.	860
Tomato Vegetable	7 fl. oz.	885

Lipton Trim Cup-A-Soup

Beef	6 fl. oz.	650
Beefy Tomato	6 fl. oz.	420
Chicken	6 fl. oz.	550
Herb Vegetable	6 fl. oz.	535

Lite-Line

Beef Flavor, Low Sodium Instant Bouillon	1 tsp.	10
Chicken Flavor, Low Sodium Instant Bouillon	1 tsp.	5

Swift Soup Starter

Beef Barley	12 oz. prepared	1279
Beef Noodle	12 oz. prepared	1379
Beef Vegetable	12 oz. prepared	1230
Chicken Noodle	12 oz. prepared	1451
Chicken Rice	12 oz. prepared	1257
Chicken Vegetable	12 oz. prepared	1338
Ham & Split Pea	12 oz. prepared	1241

Wyler's

Beef Flavor, Instant Bouillon	1 tsp.	930
Chicken Flavor, Bouillon Cubes	1 cube	850

FROZEN
La Choy

Won Ton Soup ....7.5 oz. (½ package)..........1050

Stouffer's

Cream of Spinach ....8 oz.................................885
New England Clam Chowder 8 oz.................................510
Split Pea w/Ham ....8¼ oz................................695

Sugars, Sweet Toppings, and Syrups

SUGAR AND SWEETENERS
Dia-Mel

Sweet'n-it......................................6 drops 1

Featherweight

Calorie Free Sweetening ....3 drops 1
Half Grain Saccharin.................1 tablet..................................4
Quarter Grain Saccharin ....1 tablet.................................2

French's

Cinnamon and Sugar ....1 tsp. 0

Pillsbury

Sprinkle Sweet ....1 tsp. 1
Sweet 10 ....⅛ tsp.....................................2

USDA

Honey..1 tbl......................................1
Molasses, Blackstrap ....1 tbl.....................................18
Molasses, Light ....1 tbl......................................3
Molasses, Medium....................1 tbl......................................7
Sugar, Granulated ....1 tbl......................................0
Sugar, Powdered ....1 tbl......................................0

PORTION   SODIUM (MG.)

Weight Watchers

Artificial Sweetener .... 1 packet 16

SWEET TOPPINGS

Dia-Mel

Chocolate Syrup .... 1 tbl. 3

Diet Delight

Chocolate Topping .... 1 tbl. 26

Hershey

Chocolate Flavored Syrup .... 1 tbl. 10
Chocolate Fudge Topping .... 1 tbl. 15

Kraft

Artificially Flavored
  Butterscotch Topping .... 1 tbl. 65
Caramel Topping .... 1 tbl. 45
Chocolate Flavored Caramel
  Topping .... 1 tbl. 40
Chocolate Flavored Topping .... 1 tbl. 25
Fudge Topping .... 1 tbl. 40
Marshmallow Creme .... 1 oz. 15
Pineapple Topping .... 1 tbl. 5
Strawberry Topping .... 1 tbl. 5
Walnut Topping .... 1 tbl. 5

Sleepy Hollow

Walnut Topping .... 1 tbl. 5

SYRUPS

Dia-Mel

Table Syrup .... 1 tbl. (1 packet) 10

Diet Delight

Pancake Topping .... 1 tbl. 27

Golden Griddle

Pancake Syrup............................1 tbl.........................20

Happy Jack

Pancake Syrup...........................0.7 oz. 15

Karo

Dark Corn Syrup ....1 tbl.........................40
Imitation Corn Syrup................1 tbl.........................20
Light Corn Syrup ....1 tbl.........................30
Pancake & Waffle Syrup ....1 tbl.........................35

Pack Train

Imitation Maple Syrup...............1 tbl. 3
"Old Fashion" Maple Syrup......1 tbl. 0

Safeway

Fruit Syrups (all varieties) ....1 tbl. 0

Vegetables, Beans, and Tofu

(see also Noodles, Potatoes, and Rice, page 355)

VEGETABLES
Canned

B in B

Mushrooms ....2 oz.........................190

Contadina

Stewed Tomatoes ....½ cup 232
Whole Tomatoes, Round and
  Pear ....½ cup 179

PORTION   SODIUM (MG.)

Del Monte

Asparagus, All Green, Spears and Tips	½ cup	355
Asparagus, Green Tipped and White	½ cup	355
Beets, Pickled, Crinkle Cut	½ cup	375
Beets, Pickled, No Salt Added	½ cup	150
Beets, Sliced, Crinkle Cut	½ cup	290
Beets, Sliced, No Salt Added	½ cup	100
Beets, Whole	½ cup	290
Carrots, Sliced, Diced Whole	½ cup	265
Corn, Golden, Cream Style	½ cup	355
Corn, Golden, Cream Style, No Salt Added	½ cup	under 10
Corn, Golden, Family Style, No Salt Added	½ cup	under 10
Corn, Golden, Sweet, Vacuum Packed	½ cup	355
Corn, Golden, Sweet, Vacuum Packed, No Salt Added	½ cup	under 10
Corn, Golden, Whole Kernel	½ cup	355
Corn, White, Cream Style	½ cup	355
Corn, White, Whole Kernel	½ cup	355
Green Beans, Cut	½ cup	355
Green Beans, French Cut, No Salt Added	½ cup	under 10
Green Beans, French Style	½ cup	355
Green Beans, No Salt Added	½ cup	under 10
Green Beans, Seasoned	½ cup	355
Green Beans, Whole	½ cup	355
Italian Beans	½ cup	355
Lima Beans	½ cup	355
Mixed Vegetables	½ cup	355
Peas and Carrots	½ cup	355
Peas, Seasoned	½ cup	355
Peas, Small Sweet	½ cup	355
Peas, Sweet	½ cup	355
Peas, Sweet, No Salt Added	½ cup	under 10

PORTION   SODIUM (MG.)

Sauerkraut	½ cup	775
Spinach, Chopped	½ cup	355
Spinach, Whole Leaf	½ cup	355
Tomato Wedges	½ cup	355
Tomatoes, Stewed	½ cup	355
Tomatoes, Stewed, No Salt Added	½ cup	45
Tomatoes, Whole Peeled	½ cup	220
Wax Beans, Cut	½ cup	355
Wax Beans, French Cut	½ cup	355
Zucchini	½ cup	485

Diet Delight

Asparagus	½ cup	5
Corn, Whole Kernel	½ cup	5
Green Beans	½ cup	2
Peas	½ cup	5
Peas & Carrots	½ cup	6
Tomatoes, Whole Peeled	½ cup	12

Edwards-Finast, No Salt Added

Beets, Sliced	½ cup	40
Carrots, Sliced	½ cup	30
Corn, Whole Kernel	½ cup	10
Green Beans, Cut	½ cup	10
Mixed Vegetables	½ cup	25
Peas, Sweet	½ cup	10

Featherweight

Asparagus, Cut Spears	½ cup	under 10
Beets, Sliced	½ cup	55
Carrots, Sliced	½ cup	30
Green Beans, Cut	½ cup	under 10
Green Beans, French Style	½ cup	under 10
Lima Beans, Green	½ cup	25
Mixed Vegetables	½ cup	25
Wax Beans, Cut	½ cup	under 10

Giant

Beets	½ cup	40

PORTION   SODIUM (MG.)

Corn, Golden Kernel	½ cup	10
Green Beans, Cut	½ cup	10
Green Beans, French Style	½ cup	10
Peas, Sweet	½ cup	15
Tomatoes, Baby, Sliced	½ cup	352
Tomatoes, California	½ cup	15

Green Giant

Asparagus Cuts	½ cup drained	315
Corn, Cream Style	½ cup, undrained	320
Corn, Golden Shoe Peg	½ cup drained	155
Corn, White, Vacuum Pak	½ cup drained	220
Corn, Whole Kernel	½ cup drained	230
Corn, Whole Kernel, Vacuum Pak	½ cup drained	180
Green Beans	½ cup drained	190
Green Beans, Kitchen Cut	½ cup drained	145
Green Beans, French Style Cut	½ cup drained	140
Mexicorn with Peppers	½ cup drained	285
Mushrooms	2 oz. drained	155
Mushrooms in Butter Sauce	2 oz. drained	110
Peas, Early June	½ cup drained	225
Peas, Early Sweet & Onions	½ cup drained	350
Peas, Mini, Sweet	½ cup drained	290
Peas, Sweet	½ cup drained	255
Peas, Sweet and Onions	½ cup drained	350
Three Bean Salad	½ cup undrained	540

Hunt's

Tomatoes, Italian	4 oz.	415
Tomatoes, Stewed	4 oz.	460
Tomatoes, Stewed, No Salt Added	4 oz.	20
Tomatoes, Whole Peeled	4 oz.	420
Tomatoes, Whole Peeled, No Salt Added	4 oz.	15

La Choy

Bamboo Shoots	¼ cup	under 7

PORTION SODIUM (MG.)

Bean Sprouts	⅔ cup	25
Chop Suey Vegetables	½ cup	315
Mixed Chinese Vegetables	½ cup	35
Water Chestnuts	¼ cup	under 15

Le Sueur

Asparagus Spears	½ cup drained	300
Corn, Whole Kernel	½ cup drained	230
Peas, Early June	½ cup drained	225
Peas, Mini Sweet	½ cup drained	290

Libby's

Corn, Sweet, Cream Style	½ cup undrained	265
Corn, Sweet, Whole Kernel	½ cup undrained	300
Green Beans, Blue Lake, Cut	½ cup undrained	345
Green Beans, Blue Lake, French Style	½ cup undrained	345
Peas, Green, Immature, Sweet	½ cup undrained	325
Pumpkin, solid pack	½ cup undrained	under 5
Sauerkraut	½ cup undrained	740
Sauerkraut, glass packed	½ cup	705

Libby's Natural Pack

Beets, Cut	½ cup	60
Beets, Diced	½ cup	60
Beets, Shoestring	½ cup	60
Beets, Sliced	½ cup	60
Beets, Small Whole	½ cup	60
Carrots, Diced	½ cup	30
Carrots, Sliced	½ cup	20
Corn, Whole Kernel	½ cup	under 10
Green Beans, Cut	½ cup	under 10
Green Beans, French Style	½ cup	under 10
Green Beans, Whole	½ cup	under 10
Lima Beans	½ cup	under 10
Mixed Vegetables	½ cup	under 10
Mushrooms	¼ cup	under 10

	PORTION	SODIUM (MG.)
Peas	½ cup	under 10
Peas and Carrots	½ cup	20
Spinach	½ cup	110
Succotash, Whole Kernel	½ cup	under 10
Wax Beans	½ cup	under 10

Luck's

	PORTION	SODIUM (MG.)
Collard Greens, Chopped, Seasoned w/Pork	3.8 oz.	160
Crowder Peas Seasoned w/Pork	3.8 oz.	420
Field Peas w/Snaps, Seasoned w/Pork	3.8 oz.	404
Greens, Cut, & Shelled Beans Seasoned w/Pork	4 oz.	300
Limas, Small Green, Seasoned w/Pork	3.8 oz.	436
Salad Greens Seasoned w/Pork	3.8 oz.	158
Turnip Greens w/Diced Turnip Seasoned w/Pork	3.8 oz.	168
White Acre Peas w/Snaps Seasoned w/Pork	½ cup	388

O&C

	PORTION	SODIUM (MG.)
Onions, Boiled	1 oz.	2

Stokely-Van Camp

	PORTION	SODIUM (MG.)
Asparagus, Cut	½ cup	345
Beets, Cut	½ cup	273
Beets, Diced	½ cup	262
Beets, Harvard	½ cup	125
Beets, Pickled, Sliced	½ cup	283
Beets, Pickled, Whole	½ cup	238
Beets, Sliced	½ cup	258
Beets, Whole	½ cup	248
Carrots, Diced	½ cup	310
Carrots, Sliced	½ cup	260

PORTION   SODIUM (MG.)

Corn, Cream Style Golden	½ cup	383
Corn, Golden, Liquid Pack	½ cup	310
Corn, Golden Vacuum Pack	½ cup	405
Corn, White	½ cup	283
Corn, White Cream Style	½ cup	365
Green Beans, Cut	½ cup	440
Green Beans, Sliced	½ cup	455
Green Beans, Whole	½ cup	455
Lima Beans, Green	½ cup	363
Mixed Vegetables	½ cup	123
Peas and Carrots	½ cup	278
Peas, Early	½ cup	370
Peas, Sweet	½ cup	300
Potatoes, Whole	½ cup	295
Pumpkin	½ cup	13
RSP Cherries	½ cup	20
Sauerkraut, Bavarian Style	½ cup	790
Sauerkraut, Chopped	½ cup	910
Sauerkraut, Shredded	½ cup	825
Shellie Beans	½ cup	423
Succotash	½ cup	270
Tomatoes, Stewed	½ cup	220
Tomatoes, Whole	½ cup	190
Turnip Greens, Chopped	½ cup	325
Wax Beans, Cut	½ cup	428
Wax Beans, Sliced	½ cup	423

Stop & Shop, No Salt Added

Beets, Sliced	½ cup	35
Carrots, Sliced	½ cup	35
Corn, Whole Kernel	½ cup	under 10
Green Beans, Cut	½ cup	under 10
Mixed Vegetables	½ cup	35
Peas, Medium	½ cup	under 10

Vlasic

Old Fashioned Sauerkraut	1 oz.	280

PORTION   SODIUM (MG.)

Fresh
USDA

	PORTION	SODIUM (MG.)
Artichokes	1 medium, cooked	36
Asparagus	½ cup, cooked	1
Avocado	1 whole	22
Bean Sprouts, Mung	½ cup	3
Beet Greens	½ cup, cooked	55
Beets	½ cup, cooked	37
Broccoli	1 stalk, raw	23
Broccoli	½ cup, cooked	8
Brussels Sprouts	½ cup, cooked	8
Cabbage, Green	½ cup, cooked	8
Cabbage, Green	½ cup, raw	4
Cabbage, Red	½ cup, raw	9
Carrots	1 raw	34
Cauliflower	½ cup, cooked	7
Cauliflower	½ cup, raw	9
Celery	1 stalk, raw	25
Chard	½ cup, cooked	72
Chicory	1 cup	6
Collards	½ cup	12
Corn	1 ear, cooked	1
Cucumber	7 slices	2
Dandelion Greens	½ cup, cooked	23
Eggplant	½ cup, cooked	1
Endive	1 cup, raw	7
Kale	½ cup, cooked	24
Kohlrabi	½ cup, cooked	5
Leeks	1 bulb, raw	1
Lettuce	1 cup	4
Lima Beans	½ cup, cooked	1
Mushrooms	½ cup, raw	4
Mustard Greens	½ cup, cooked	13
Mustard Greens	½ cup, raw	11
Okra	10 pods, cooked	2
Onions, Green	2 medium	2
Onions, Mature	½ cup, cooked	8
Parsley	1 tbl., raw	2
Parsnips	½ cup, cooked	10
Peas, Green	½ cup, cooked	1

PORTION   SODIUM (MG.)

Peppers, Hot......	1 pod, raw	7
Peppers, Sweet	1 pod, raw or cooked	9
Potatoes, Baked or Boiled	1 medium	5
Radish	4 small	2
Rutabaga	½ cup, cooked	4
Shallot	1 shallot	3
Snap Beans	½ cup, cooked	3
Snow Peas	½ cup	3
Spinach	½ cup, cooked	47
Spinach	1 cup, raw	49
Summer Squash	½ cup, cooked	3
Sweet Potatoes, Baked or Boiled	1 medium	20
Tomato	½ cup, cooked	5
Tomato	1 raw	14
Turnip Green	½ cup, cooked	9
Winter Squash	½ cup, baked, mashed	1
Yams, Baked or Boiled	1 medium	20

Frozen

Birds Eye

Artichoke Hearts	3.0 oz.	40
Asparagus, Cuts	3.3 oz.	5
Asparagus Spears	3.3 oz.	5
Beans and Spaetzl, Bavarian Style Recipe	3.3 oz.	420
Broccoli Cuts	3.3 oz.	25
Broccoli Florets	3.3 oz.	20
Broccoli Spears	3.3 oz.	20
Broccoli Spears, Baby	3.3 oz.	15
Broccoli w/Almonds	3.3 oz.	215
Broccoli, Baby Carrots and Water Chestnuts	3.3 oz.	25
Broccoli w/Cheese Sauce	5 oz.	505
Broccoli, Carrots & Pasta Twists	3.3 oz.	270
Broccoli, Cauliflower & Carrots	3.2 oz.	30
Broccoli, Cauliflower & Carrots w/Cheese	5 oz.	400

PORTION SODIUM (MG.)

	PORTION	SODIUM (MG.)
Broccoli, Cauliflower & Red Peppers	3.3 oz.	20
Broccoli, Chopped	3.3 oz.	20
Broccoli, Corn, Red Peppers	3.2 oz.	10
Broccoli, French Green Beans, Onions, Red Peppers	3.2 oz.	15
Broccoli and Water Chestnuts	3.3 oz.	215
Brussels Sprouts	3.3 oz.	15
Brussels Sprouts, Baby	3.3 oz.	10
Brussels Sprouts, Baby, w/Cheese Sauce	4.5 oz.	435
Brussels Sprouts, Cauliflower & Carrots	3.2 oz.	20
Carrots, Baby Sweet Peas, and Pearl Onions	3.3 oz.	60
Carrots, Whole Baby	3.3 oz.	45
Cauliflower	3.3 oz.	20
Cauliflower Florets	3.3 oz.	15
Cauliflower w/Almonds	3.3 oz.	270
Cauliflower w/Cheese Sauce	5 oz.	505
Cauliflower, Green Beans and Corn	3.2 oz.	10
Chinese Style Vegetables	3.3 oz.	360
Chinese Style Vegetables, Stir Fry	3.3 oz.	480
Corn on the Cob, Little Ears	2 ears	5
Corn on the Cob	1 ear	5
Corn, Big Ears on the Cob	1 ear	5
Corn, Green Beans w/Pasta twists	3.3 oz.	280
Corn, Sweet Whole Kernel	3.3 oz.	5
Corn, Tendertreat Sweet	3.3 oz.	5
Far Eastern Style Vegetables	3.3 oz.	390
Green Beans, Corn, Carrots and Pearl Onions	3.2 oz.	15
Green Beans, Cut	3.0 oz.	5
Green Beans, French, Cauliflower & Carrots	3.2 oz.	20
Green Beans, French Style Cut	3.0 oz.	5

PORTION   SODIUM (MG.)

	PORTION	SODIUM (MG.)
Green Beans, French w/Toasted Almonds	3.0 oz.	335
Green Beans, Whole	3 oz.	0
Green Peas & Pearl Onions	3.3 oz.	310
Green Peas & Potatoes w/Cream Sauce	2.6 oz.	480
Green Peas w/Cream Sauce	2.6 oz.	440
Italian Green Beans	3 oz.	5
Italian Style Vegetables	3.3 oz.	575
Japanese Style Vegetables	3.3 oz.	505
Japanese Style Vegetables, Stir Fry	3.3 oz.	570
Lima Beans, Baby	3.0 oz.	115
Lima Beans, Fordhook	3.3 oz.	100
Lima Beans, Tiny	3.3 oz.	145
Mexicana Style Vegetables	3.3 oz.	465
Mixed Vegetables	3.3 oz.	45
Mixed Vegetables w/Onion Sauce	2.6 oz.	350
New England Style Vegetables	3.3 oz.	410
Okra, Cut	3.3 oz.	5
Okra, Whole	3.3 oz.	5
Onions, Pearl	3.3 oz.	10
Onions, Small, Whole	4 oz.	10
Onions, Small w/Cream Sauce	3 oz.	335
Peas & Pearl Onions w/Cheese Sauce	5 oz.	460
Peas, Carrots and Pearl Onions	3.2 oz.	85
Peas, Sweet Green	3.3 oz.	130
Peas, Tender Tiny	3.3 oz.	120
Rice and Peas with Mushrooms	2.3 oz.	320
San Francisco Style Vegetables	3.3 oz.	395
Spinach and Water Chestnuts	3.3 oz.	275
Spinach, Chopped	3.3 oz.	80

PORTION   SODIUM (MG.)

	PORTION	SODIUM (MG.)
Spinach, Creamed	3.0 oz.	275
Spinach, Whole Leaf	3.3 oz.	90
Squash, Cooked Winter	4.0 oz.	0
Zucchini, Sliced Baby	3.3 oz.	5

Green Giant

	PORTION	SODIUM (MG.)
Asparagus Spears, Cut, in Butter Sauce	½ cup	725
Broccoli, Cauliflower, Carrots in Cheese Sauce	½ cup	465
Broccoli Cauliflower Medley	½ cup	470
Broccoli, Cut	½ cup	160
Broccoli Fanfare	½ cup	455
Broccoli in Cheese Sauce	½ cup	425
Broccoli in White Cheddar Cheese Sauce	½ cup	420
Broccoli Spears	½ cup	160
Broccoli Spears in Butter Sauce	½ cup	325
Brussels Sprouts in Butter Sauce	½ cup	275
Brussels Sprouts in Cheese Sauce	½ cup	475
Carrots, Crinkle Cut, in Butter Sauce	½ cup	315
Cauliflower Carrot Bonanza	½ cup	295
Cauliflower in Cheese Sauce	½ cup	450
Cauliflower in White Cheddar Cheese Sauce	½ cup	415
Chinese Style Vegetables	½ cup	280
Corn	½ cup	280
Corn, Cream Style	½ cup	315
Corn in Cream Sauce	½ cup	295
Corn, Niblets, in Butter Sauce	½ cup	280
Corn, Shoe Peg White, in Butter Sauce	½ cup	290
Frozen Like Fresh Broccoli Minispears	½ cup	100

	PORTION	SODIUM (MG.)
Green Beans	½ cup	175
Green Beans, French Style and Cut, in Butter Sauce	½ cup	355
Green Beans in Cream Sauce w/Mushrooms	½ cup	280
Japanese Style Vegetables	½ cup	155
Le Sueur Mini Peas, Onions & Carrots in Butter Sauce	½ cup	100
Lima Beans	½ cup	310
Lima Beans in Butter Sauce	½ cup	445
Mixed Vegetables	½ cup	220
Mushrooms in Butter Sauce	½ cup	240
Onions, Small, in Cheese Flavored Sauce	½ cup	400
Peas, Early and Sweet, in Butter Sauce	½ cup	490
Peas in Cream Sauce	½ cup	320
Peas, Mini, Pea Pods & Water Chestnuts in Butter Sauce	½ cup	410
Peas, Sweet	½ cup	280
Spinach	½ cup	350
Spinach in Cream Sauce	½ cup	395
Spinach, Leaf, Cut, in Butter Sauce	½ cup	465

Green Giant Polybag Vegetables

	PORTION	SODIUM (MG.)
Broccoli, Carrots Fanfare	½ cup	20
Broccoli, Cauliflower Supreme	½ cup	30
Broccoli Cuts	½ cup	10
Brussels Sprouts	½ cup	15
Cauliflower Cuts	½ cup	30
Cauliflower, Green Bean Festival	½ cup	30
Corn, Broccoli Bounty	½ cup	10
Early June Peas	½ cup	25
Green Beans	½ cup	5
Lima Beans	½ cup	30
Mixed Vegetables	½ cup	35

PORTION   SODIUM (MG.)

Nibblers Corn on the
  Cob...1 ear...........................under 10
Niblet Ears Corn on the Cob   1 ear 20
Niblets w/White Corn ....½ cup.....................................5
Sweet Pea Cauliflower
  Medley ....½ cup 35
Sweet Peas.................................½ cup 25

La Choy

Chinese Pea Pods.....................½ package (3 oz.)......under 10
Chinese Style Vegetables ....½ cup (3.3 oz.) 540

Mrs. Paul's

Corn Fritters ....4 oz. (2 fritters) 725
French Fried Onion Rings ....2¼ oz......................................275
Light Batter Zucchini Sticks......3 oz...630

Pepperidge Farm

Asparagus with Mornay
  Sauce.......................................7¼ oz. (1 pastry) 245
Broccoli with Cheese ....7¼ oz. (1 pastry) 455
Cauliflower and Cheese
  Sauce.......................................7¼ oz. (1 pastry) 465
Green Beans with Mushroom
  Sauce.......................................7¼ oz. (1 pastry) 300
Mexican Style Picante ....7¼ oz. (1 pastry) 395
Mushrooms Dijon.......................7¼ oz. (1 pastry) 415
Oriental Garden in Szechwan
  Spices ....7¼ oz. (1 pastry) 335
Ratatouille with Cheese ....7¼ oz. (1 pastry) 505
Spinach Almondine ....7¼ oz. (1 pastry) 325
Zucchini Provencal ....7¼ oz. (1 pastry) 290

Ore-Ida

Cob Corn....................................4.5 oz. .... under 10
Onions, Chopped.......................2 oz..30
Onion Ringers ....2 oz..200
Stew Vegetables ....3 oz..50

PORTION   SODIUM (MG.)

Stouffers

	PORTION	SODIUM (MG.)
Broccoli in Cheddar Cheese Sauce	4½ oz.	970
Creamed Spinach	4½ oz.	855

Winter Garden

	PORTION	SODIUM (MG.)
Asparagus Spears	3.2 oz.	6
Bavarian Style	3.2 oz.	13
Blackeye Peas	3.2 oz.	5
Broccoli & Cauliflower	3.2 oz.	19
Broccoli and New Potatoes	3.2 oz.	14
Broccoli Cuts	3.2 oz.	0
Brussels Sprouts	3.2 oz.	13
California Style	3.2 oz.	20
Carrots, Crinkle Sliced	3.2 oz.	42
Cauliflower Florets	3.2 oz.	18
Collards, Chopped	3.2 oz.	44
Corn-on-Cob	1 ear	3
Corn, Cut	3.2 oz.	4
Creme Peas	3.2 oz.	6
Crookneck Squash, Sliced	3.2 oz.	2
Crowder Peas	3.2 oz.	5
English Peas and New Potatoes	3.2 oz.	37
Espanol Style	3.2 oz.	22
Field Peas with Snaps	3.2 oz.	6
Green Beans and New Potatoes	3.2 oz.	5
Green Beans, Cut	3.2 oz.	5
Green Beans, French Cut	3.2 oz.	3
Green Peas	3.2 oz.	93
Italian Style	3.2 oz.	6
Kale, Chopped	3.2 oz.	14
Leaf Spinach, Cut	3.2 oz.	74
Lima Beans, Baby	3.2 oz.	160
Mixed Vegetables	3.2 oz.	51
Mustard Greens, Chopped	3.2 oz.	27
New England Style	3.2 oz.	17
New Potatoes, Whole	3.2 oz.	6

	PORTION	SODIUM (MG.)
Northwest Style	3.2 oz.	22
Okra and Tomatoes	3.2 oz.	6
Okra, Whole or Cut	3.2 oz.	3
Oriental Style	3.2 oz.	19
Peas & Carrots	3.2 oz.	58
Peas & Cauliflower	3.2 oz.	44
Purple Hull Peas	3.2 oz.	6
Speckled Butter Beans	3.2 oz.	19
Squash, Cooked	3.2 oz.	3
Turnip Greens, Chopped	3.2 oz.	11
Turnips, Chopped	3.2 oz.	14
Western Style	3.2 oz.	3
White Acre Peas	3.2 oz.	6
Yams, Sliced	3.2 oz.	6
Yams, Whole	3.2 oz.	3
Zucchini Squash, Sliced	3.2 oz.	6

BEANS
Canned

B&M

Red Kidney Baked Beans	9 oz. (1 cup)	873
Small Pea Baked Beans	9 oz. (1 cup)	954
Yellow Eye Baked Beans	9 oz. (1 cup)	1089

Campbell's

Barbecue Beans	8 oz.	1110
Home Style Beans	8 oz.	1150
Old Fashioned Beans	8 oz.	1065
Pork & Beans in Tomato Sauce	8 oz.	945

Del Monte

Burrito Filling Mix	1 cup	1800
Refried Beans	1 cup	1060
Spicy Refried Beans	1 cup	960

Dennison's

Lima Beans w/Ham	7.5 oz.	935

PORTION SODIUM (MG.)

Friend's

Red Kidney Baked Beans ....9 oz. (1 cup) 1320
Small Pea Baked Beans ....9 oz. (1 cup) 1270
Yellow Eye Baked Beans ....9 oz. (1 cup) 1470

Health Valley

Boston Baked Beans ....8 oz.................................550
Boston Baked Beans, no
  salt ....8 oz..................................50
Vegetarian Beans w/Miso..........8 oz.................................850

Heinz

Pork 'N' Beans..........................8 oz.................................745
Vegetarian Beans.....................8 oz.................................980

Hunt's

Chili Beans ....8 oz.................................920
Pork and Beans........................8 oz.................................800
Red Kidney Beans ....8 oz.................................800

Libby's

Black-eyed Peas ....1 cup 560
Deep Brown Pork and Beans
  in Molasses Sauce ....1 cup 515
Deep Brown Pork and Beans
  in Tomato Sauce ....1 cup 850
Deep Brown Vegetarian Beans
  in Tomato Sauce ....1 cup 810

Luck's

Blackeye Peas & Corn,
  Seasoned w/Pork....................7.5 oz. 882
Blackeye Peas, Seasoned
  w/Pork (7-oz. can)..................7 oz. (1 can)......................788
Blackeye Peas, Seasoned
  w/Pork (15-oz. can)...............7.5 oz. 819
Blackeye Peas, Seasoned
  w/Pork (29-oz. can)...............7.5 oz. 1060

PORTION   SODIUM (MG.)

Great Northern Beans,
Seasoned w/Pork
(7-oz. can)..............................7 oz. (1 can) 723
Great Northern Beans,
Seasoned w/Pork (15-oz.
can)...7.5 oz. 734
Great Northern Beans,
Seasoned w/Pork (29-oz.
can)...7.5 oz. 667
Hot Chili Beans ....7.5 oz. 680
Lima Beans, Giant, Seasoned
w/Pork (7 oz. can)................7 oz. (1 can) 840
Lima Beans, Giant, Seasoned
w/Pork (15-oz. can)..............7.5 oz. 819
Mixed Beans, Seasoned
w/Pork (29-oz. can)..............7.5 oz. 850
Mixed Beans, Pinto and
Great Northern, Seasoned
w/Pork ....7.5 oz. 755
Navy Beans, Seasoned
w/Pork ....7.5 oz. 755
October Beans, Seasoned
w/Pork (15-oz. can)..............7.5 oz. 819
October Beans, Seasoned
w/Pork (29 oz. can)..............7.5 oz. 569
Pinto Beans, Seasoned
w/Pork (7-oz. can) ....7 oz. (1 can) 755
Pinto Beans, Seasoned
w/Pork (15-oz. can)..............7.5 oz. 787
Pinto Beans, Seasoned
w/Pork (29-oz. can)..............7.5 oz. 563
Pinto Beans with Onions,
Seasoned w/Pork...................7.5 oz. 691
Red Kidney Beans, Seasoned
w/Pork ....7.5 oz. 914
Red Kidney Beans, Special
Cook.......................................7.5 oz. 574
Speckled Butter Beans,
Seasoned w/Pork...................7.5 oz. 797
Yelloweye Beans, Seasoned
w/Pork ....7.5 oz. 1041

PORTION   SODIUM (MG.)

Old El Paso

Refried Beans	9 oz.	929
Refried Beans w/Sausage	9 oz.	796

Van Camp's

Baked Beans	8 oz.	1020
Beanee Weenee	8 oz.	925
Brown Sugar Beans	8 oz.	640
Butter Beans	8 oz.	710
Kidney Beans, Dark Red	8 oz.	830
Kidney Beans, Light Red	8 oz.	675
Mexican Style Chili Beans	8 oz.	725
New Orleans Style Kidney Beans	8 oz.	935
Pork and Beans	8 oz.	995
Red Beans	8 oz.	925
Vegetarian Style Beans	8 oz.	945
Western Style Beans	8 oz.	900

Dry
USDA

Blackeye Peas	1 cup, cooked	12
Chick-peas	1 cup, cooked	13
Great Northern Beans	1 cup, cooked	5
Kidney Beans	1 cup, cooked	4
Lentils	1 cup, cooked	4
Lima Beans	1 cup, cooked	4
Navy Beans	1 cup, cooked	3
Pinto Beans	1 cup, cooked	4
Soybeans	1 cup, cooked	4
Split Peas	1 cup, cooked	5

TOFU
USDA

Tofu (2½ × 2¾ × 1 in.)	4.2 oz. (1 piece)	9

Yogurt

Colombo

Plain	8 oz. (1 cup)	159
Strawberry	8 oz. (1 cup)	147

Dannon

Flavored (Coffee, Lemon, Vanilla)	8 oz. (1 cup)	70-90
Fruit	8 oz. (1 cup)	70-125
Plain	8 oz. (1 cup)	115

Land O Lakes

Lite 'n Creamy Yogurt	8 oz. (1 cup)	153

Light 'n Lively

Black Raspberry	8 oz. (1 cup)	119
Blueberry	8 oz. (1 cup)	143
Blueberry-Vanilla	8 oz. (1 cup)	140
Lemon-Lime	8 oz. (1 cup)	188
Plain	8 oz. (1 cup)	159
Strawberry Fruitcup	8 oz. (1 cup)	112

Lite-line

Cherry Vanilla Flavor, Natural, 1% milk fat	8 oz.	160
Lemon Flavor, Natural, 1% milk fat	8 oz.	115
Pineapple Flavor, Natural, 1% milk fat	8 oz.	115
Plain, 1½% milk fat	8 oz.	145
Strawberry Flavor, Natural, 1% milk fat	8 oz.	145

Yami

Fruit Lowfat	8 oz. (1 cup)	136
Plain Lowfat	8 oz. (1 cup)	161

Yoplait

Apple Cinnamon, Breakfast	6 oz. (¾ cup)	95
Banana, Custard Style	6 oz. (¾ cup)	95
Berries, Breakfast	6 oz. (¾ cup)	95
Blueberry, Custard Style	6 oz. (¾ cup)	95
Citrus Fruits, Breakfast	6 oz. (¾ cup)	95
Coffee, Custard Style	6 oz. (¾ cup)	110
Fruit Flavors, Original Style	6 oz. (¾ cup)	105
Lemon, Custard Style	6 oz. (¾ cup)	95
Orchard Fruits, Breakfast	6 oz. (¾ cup)	95
Plain	6 oz. (¾ cup)	135
Plain with Honey, Custard Style	6 oz. (¾ cup)	110
Raspberry, Custard Style	6 oz. (¾ cup)	95
Strawberry, Custard Style	6 oz. (¾ cup)	95
Tropical Fruits, Breakfast	6 oz. (¾ cup)	95
Vanilla, Custard Style	6 oz. (¾ cup)	110

References

CHAPTER 1

Hypertension and Heart Attack, Stroke, and Heart Failure

Kannel, W. B. "Importance of Hypertension as a Major Risk Factor in Cardiovascular Disease." In *Hypertension: Physiopathology and Treatment*, edited by J. Genest, E. Koiw and O. Kuchel. New York: McGraw-Hill, 1977.

Incidence of Hypertension

High Blood Pressure Coordinating Committee. *New Hypertension Prevalence Data and Recommended Public Statements*. National Heart, Lung and Blood Institute, February 1978.

Jacobson, M., and B. F. Liebman. "Dietary Sodium and the Risk of Hypertension." *New England Journal of Medicine* 303, no. 14 (1980): 817.

"Proposal on High Blood Pressure Widens 'Greater Risk' Group." *The Washington Post*, 24 October 1982.

Sodium Need

Committee on Dietary Allowances. *Recommended Dietary Allowances* Washington, D.C.: National Academy Press, 1980, p. 178.

Dahl, L. "Salt Intake and Salt Need." *New England Journal of Medicine* 258 (1958): 1152.

Kempner, W., "Treatment of Hypertensive Vascular Disease with Rice Diet." *American Journal of Medicine* 4 (1948): 545.

CHAPTER 2

Decline in Heart Attack and Stroke Deaths

Levy, R. I., and J. Moskowitz. "Cardiovascular Research: Decades of Progress, a Decade of Promise." *Science* 217 (1982): 121.

Effects of High Blood Pressure

Freis, E., and G. B. Kolata. *The High Blood Pressure Book: A Guide for Patients and their Families.* Sausalito, Calif.: Painter Hopkins, 1979.

Galton, L. *The Silent Disease: Hypertension.* New York: Crown, 1973.

Kannel, W. B. "Importance of Hypertension as a Major Risk Factor in Cardiovascular Disease." In *Hypertension: Physiopathology and Treatment,* edited by J. Genest, E. Koiw, and O. Kuchel. New York: McGraw-Hill, 1977.

Working Group of Arteriosclerosis of the National Heart, Lung, and Blood Institute. *Arteriosclerosis, 1981.* Vol. 1 (NIH Publication No. 81-2034). Washington, D.C.: U.S. Department of Health and Human Services, 1981.

Hypertension Detection and Follow-up Study

Hypertension Detection and Follow-up Program. "The Effect of Treatment on Mortality in 'Mild' Hypertension." *New England Journal of Medicine* 307 (1982): 976.

Kolata, G. B. "Treatment Reduces Deaths from Hypertension." *Science* 206 (1979): 1386.

MRFIT Study

Kolata, G. "Heart Study Produces a Surprise Result." *Science* 218 (1982): 31.

Multiple Risk Factor Intervention Trial Research Group. "Multiple Risk Factor Intervention Trial: Risk Factor Changes and Mortality Results." *Journal of the American Medical Association* 248 (1982): 1465.

Reserpine

Food and Drug Administration. "Professional Labeling for Reserpine Drugs; Revised Labeling." *Federal Register* Vol. 48, No. 64 (1 April 1983).

Selacryn

Food and Drug Administration. "Ticrynafen Recalled." *FDA Drug Bulletin* Vol. 10, No. 1 (February 1980).

Sodium During Pregnancy

Lindheimer, M. D., and A. I. Katz. "Sodium and Diuretics in Pregnancy." *New England Journal of Medicine* 288 (1973): 891.

Pike, R. L., and H. A. Smickilas. "A Reappraisal of Sodium Restriction during Pregnancy." *International Journal of Gynaecology and Obstetrics* 10 (1972): 1.

VA Study

Veterans Administration Cooperative Study Group on Antihypertensive Agents. "Effects of Treatment on Morbidity in Hypertension: Results in Patients with Diastolic Blood Pressures Averaging 115 through 129 mm Hg. *Journal of the American Medical Association* 202 (1967): 116.

Veterans Administration Cooperative Study Group on Antihypertensive Agents. "Effects of Treatment on Morbidity in Hypertension: Results in Patients with Diastolic Blood Pressures Averaging 90 through 114 mm Hg. *Journal of the American Medical Association* 213 (1970): 1143.

CHAPTER 3

Alcohol

Friedman, G. D., et al. "Alcohol, Tobacco, and Hypertension." *Hypertension* 4 (Supplement 3) (1982): III-143.

Animal Studies

Cherchovich, G. M., et al. "High Salt Intake and Blood Pressure in Lower Primates (Papio hamadryas). *Journal of Applied Physiology* 40 (1976): 601.

Corbett, W., et al. "Utilization of Swine to Study the Risk Factor of an Elevated Salt Diet on Blood Pressure." *American Journal of Clinical Nutrition* 32 (1979): 2068.

Dahl, L., et al. "Effects of Chronic Excess Salt Ingestion: Evidence that Genetic Factors Play an Important Role in Susceptibility to Experimental Hypertension." *Journal of Experimental Medicine* 115 (1962): 1173.

Calcium

McCarron, D. A. "Calcium, Magnesium, and Phosphorus Balance in Human and Experimental Hypertension." *Hypertension* 4 (Supplement 3) (1982): III-27.

Epidemiological Studies

Blackburn, H., and R. Prineas. "Diet and Hypertension: Anthropology, Epidemiology, and Public Health Implications." *Progress in Biochemical Pharmacology* 19 (1983): 31.

Page, L. B. "Epidemiologic Evidence on the Etiology of Human Hypertension and Its Possible Prevention." *American Heart Journal* 91 (1976): 527.

Page, L. B., et al. "Antecedents of Cardiovascular Disease in Six Solomon Islands Societies." *Circulation* 49 (1974): 1132.

Sasaki, N. "The Relationship of Salt Intake to Hypertension in the Japanese." *Geriatrics* 19 (1964): 735.

Exercise

Leon, A., and H. Blackburn. "Physical Activity and Hypertension." In *Hypertension*. Vol. 1 of *Cardiology,* edited by P. Sleight and E. Freis. London: Butterworth, 1982.

Human Studies: Salt Loading

Luft, F., et al. "Sodium Sensitivity and Resistance in Normotensive Humans." *American Journal of Medicine* 72 (1982): 726.

Human Studies: Moderate Sodium Restriction

Daugherty, S. A., J. Z. Miller, M. H. Weinberger, and C. E. Grim. "Blood Pressure Response to Dietary Sodium Changes in Young Identical Twins and Their Families." Abstract presented at American Heart Association 55th Scientific Session, 15-18 November, 1982.

MacGregor, G. A., et al. "Double-blind Randomised Crossover Trial of Moderate Sodium Restriction in Essential Hypertension." *Lancet* 1 (1982): 351.

Miller, J. M., et al. "Blood Pressure Response to Dietary Sodium Restriction in Normotensive Adults." *Hypertension* 5 (September/October 1983).

Morgan, T., et al. "Hypertension Treated by Salt Restriction." *Lancet* 1 (1978): 227.

Parjis, J., et al. "Moderate Sodium Restriction and Diuretics in the Treatment of Hypertension." *American Heart Journal* 85 (1973): 22.

Obesity

"Hypertension and Obesity." *New England Journal of Medicine* 298 (1978): 46.

Reisen, E., et al. "Effect of Weight Loss Without Salt Restriction on the Reduction of Blood Pressure in Overweight

Hypertensive Patients." *New England Journal of Medicine* 298 (1978): 1.

Tuck, M., et al. "The Effect of Weight Reduction on Blood Pressure, Plasma Renin Activity, and Plasma Aldosterone Levels in Obese Patients." *New England Journal of Medicine* 304 (1981): 930.

Sasaki, N. "High Blood Pressure and the Salt Intake of the Japanese." *Japanese Heart Journal* 3 (1962): 313.

Polyunsaturated Fats

Iacono, J. M., et al. "Reduction of Blood Pressure Associated with Dietary Polyunsaturated Fat." *Hypertension* 4 (Supplement 3) (1982): III-34.

Potassium

MacGregor, G., et al. "Moderate Potassium Supplementation in Essential Hypertension." *Lancet* 2 (1982): 567.

Meneely, G. R., and H. Battarbee. "High Sodium-Low Potassium Environment and Hypertension." *American Journal of Cardiology* 38 (1976): 768.

Watson, R. L., et al. "Urinary Electrolytes, Body Weight, and Blood Pressure: Pooled Cross-sectional Results Among Four Groups of Adolescent Females." *Hypertension* 2 (part 2) (1980): I-93.

Sodium and Hypertension: Review Articles

Dahl, L. K. "Salt Intake and Hypertension." *Hypertension: Physiopathology and Treatment,* edited by J. Genest, E. Koiw, and O. Kuchel. New York: McGraw-Hill, 1977.

Freis, E. "Salt, Volume, and the Prevention of Hypertension." *Circulation* 53 (1976): 589.

Tobian, L. "The Relationship of Salt to Hypertension." *American Journal of Clinical Nutrition* 32 (Supplement) (1979): 2739.

CHAPTER 4

Percent Salt from the Shaker

Altschul, A., and J. Grommet. "Food Choices for Lowering Sodium Intake." *Hypertension* 4 (Supplement 3) (1982): III-116.

Salt in Drinking Water

Calabrese, E. J., and R. W. Tuthill. "Elevated Blood Pressure and High Sodium Levels in the Public Drinking Water." *Archives of Environmental Health* (September/October 1977): 200.

Sodium Losses With and Without Exercise

Committee on Dietary Allowances. *Recommended Dietary Allowances*. Washington, D.C.: National Academy Press, 1980, p. 170.

Costill, D. L., et al. "Water and Electrolyte Replacement During Repeated Days of Work in the Heat." *Aviation, Space and Environmental Medicine* 46 (1975): 795.

Taylor, H. L., et al. "The Effect of the Sodium Chloride Intake on the Work Performance of Man During Exposure to Dry Heat and Experimental Heat Exhaustion." *American Journal of Physiology* 140 (1944): 439.

CHAPTER 5

Congressional Hearings on Salt

U.S. House of Representatives: Subcommittee on Health and the Environment. Hearing: *Sodium and Potassium Content Labeling*, 25 September 1981.

U.S. House of Representatives. Subcommittee on Investigations and Oversight. Hearing: *Sodium in Food and High Blood Pressure*, 13-14 April 1981.

FDA's Labeling Proposal

Food and Drug Administration. "Declaration of Sodium Content of Foods and Label Claims for Foods on the Basis of Sodium Content and GRAS Safety Review of Sodium Chloride." *Federal Register* 47, no. 118 (18 June 1982): 26580.

Industry Lobbying

"On Sodium, Low-Key Lobbying Wins." *The Washington Post*, 26 March 1982.

Official Recommendations to Eat Less Salt

Committee on Dietary Allowances. *Recommended Dietary Allowances*. Washington, D.C.: National Academy Press, 1980, p. 178.

Select Committee on GRAS Substances. *Evaluation of the Health Aspects of Sodium Chloride and Potassium Chloride as Food Ingredients*. Contract No. FDA 223-75-2004 (1979).

U.S. Department of Agriculture and U.S. Department of Health,

Education and Welfare. *Nutrition and Your Health: Dietary Guidelines for Americans.* February 1980.

U.S. Department of Health, Education and Welfare. *Healthy People: The Surgeon General's Report on Health Promotion and Disease Prevention.* DHEW (PHS) Publication No. 79-55071 (1979).

CHAPTER 6

Avoid Excess Alcohol

Eckardt, M. J. et al. "Health Hazards Associated with Alcohol Consumption." *Journal of the American Medical Association* 246 (1981): 648.

Eat Foods Rich in Calcium

Heaney, R. P., et al. "Calcium Nutrition and Bone Health in the Elderly." *American Journal of Clinical Nutrition* 36 (1982): 986.

Eat Foods Rich in Vitamins A and C

Committee on Diet, Nutrition and Cancer. *Diet, Nutrition, and Cancer.* Washington, D.C.: National Academy Press, 1982.

Eat Less Fat

Committee on Diet, Nutrition and Cancer. *Diet, Nutrition, and Cancer.* Washington, D.C.: National Academy Press, 1982.

Hausman, P. *Jack Sprat's Legacy: The Science and Politics of Fat and Cholesterol.* New York: Richard Marek, 1980.

Nutrition Committee of the American Heart Association. "Rationale of the Diet-Heart Statement of the American Heart Association." *Circulation* 65 (1982): 339-A.

Eat Less Sugar

Liebman, B., and G. Moyer. "The Case Against Sugar," *Nutrition Action,* December 1980.

Eat More Fiber

Anderson, J. W. *Diabetes: A Practical New Guide to Healthy Living.* New York: Arco, 1981.

Burkitt, D. *Eat Right—To Stay Healthy and Enjoy Life More.* New York: Arco, 1979.

General Nutrition

U.S. Department of Agriculture and U.S. Department of Health,

Education and Welfare. *Nutrition and Your Health: Dietary Guidelines for Americans*. February 1980.

U.S. Department of Health, Education and Welfare. *Healthy People: The Surgeon General's Report on Health Promotion and Disease Prevention*. DHEW (PHS) Publication No. 79-55071 (1979).

Taste for Salt

Bertino, M., G. K. Beauchamp, and K. Engleman. "Long Term Reduction in Dietary Sodium Alters the Taste of Salt." *American Journal of Clinical Nutrition* 36 (1982): 1134.

Low-Sodium Cookbooks

Craig Claiborne's Gourmet Diet, by Craig Claiborne. New York: Times Books, 1980.

Cooking Without a Grain of Salt, by Elma W. Bagg. New York: Bantam Books, 1972.

Cooking Without Your Salt Shaker, American Heart Association, Dallas: 1978.

How to Live 365 Days a Year the Salt Free Way, by P. Brunswich, D. Love, A. Weinberg. New York: Bantam Books, 1978.

The Live Longer Now Cookbook, by Jon N. Leonard and Elaine A. Taylor. Foreword by Nathan Pritikin. New York: Grosset & Dunlap, 1977.

Living with High Blood Pressure, by Joyce Daly Margie, M. S. and James C. Hunt, M.D. HLS Press, 1978.

Living without Salt, by Karin Baltzell and Terry Parsley. Elgin, Illinois: The Brethren Press, 1982.

The Low Salt/Cholesterol Cookbook, by Myra Waldo. New York: Putnam's, 1972.

The Pritikin Program of Diet and Exercise, by Nathan Pritikin and Patrick McGrady. New York: Grosset & Dunlap, 1979.

Secrets of Salt-Free Cooking, by Jeanne Jones. San Francisco: 101 Productions, 1979.

Tasting Good: The International Salt-Free Diet Cookbook, by Merle Schell. New York: Bobbs-Merrill, 1981.

This index does not include every brand name food listed in this book. To find the sodium content of a particular food, start by checking the Table of Contents, pages vii and viii. Only foods that may be difficult to find using that table are listed here.

INDEX